POLYMER ELECTRONICS

OXFORD MASTER SERIES IN PHYSICS

The Oxford Master Series is designed for final year undergraduate and beginning graduate students in physics and related disciplines. It has been driven by a perceived gap in the literature today. While basic undergraduate physics texts often show little or no connection with the huge explosion of research over the last two decades, more advanced and specialized texts tend to be rather daunting for students. In this series, all topics and their consequences are treated at a simple level, while pointers to recent developments are provided at various stages. The emphasis is on clear physical principles like symmetry, quantum mechanics, and electromagnetism which underlie the whole of physics. At the same time, the subjects are related to real measurements and to the experimental techniques and devices currently used by physicists in academe and industry. Books in this series are written as course books, and include ample tutorial material, examples, illustrations, revision points, and problem sets. They can likewise be used as preparation for students starting a doctorate in physics and related fields, or for recent graduates starting research in one of these fields in industry.

CONDENSED MATTER PHYSICS

1. M.T. Dove: *Structure and dynamics: an atomic view of materials*
2. J. Singleton: *Band theory and electronic properties of solids*
3. A.M. Fox: *Optical properties of solids, second edition*
4. S.J. Blundell: *Magnetism in condensed matter*
5. J.F. Annett: *Superconductivity, superfluids, and condensates*
6. R.A.L. Jones: *Soft condensed matter*
17. S. Tautz: *Surfaces of condensed matter*
18. H. Bruus: *Theoretical microfluidics*
19. C.L. Dennis, J.F. Gregg: *The art of spintronics: an introduction*
21. T.T. Heikkilä: *The physics of nanoelectronics: transport and fluctuation phenomena at low temperatures*
22. M. Geoghegan, G. Hadziioannou: *Polymer electronics*

ATOMIC, OPTICAL, AND LASER PHYSICS

7. C.J. Foot: *Atomic physics*
8. G.A. Brooker: *Modern classical optics*
9. S.M. Hooker, C.E. Webb: *Laser physics*
15. A.M. Fox: *Quantum optics: an introduction*
16. S.M. Barnett: *Quantum information*

PARTICLE PHYSICS, ASTROPHYSICS, AND COSMOLOGY

10. D.H. Perkins: *Particle astrophysics, second edition*
11. Ta-Pei Cheng: *Relativity, gravitation and cosmology, second edition*

STATISTICAL, COMPUTATIONAL, AND THEORETICAL PHYSICS

12. M. Maggiore: *A modern introduction to quantum field theory*
13. W. Krauth: *Statistical mechanics: algorithms and computations*
14. J.P. Sethna: *Statistical mechanics: entropy, order parameters, and complexity*
20. S.N. Dorogovtsev: *Lectures on complex networks*

Polymer Electronics

MARK GEOGHEGAN

University of Sheffield

GEORGES HADZIIOANNOU

University of Bordeaux 1

OXFORD
UNIVERSITY PRESS

Great Clarendon Street, Oxford, OX2 6DP,
United Kingdom

Oxford University Press is a department of the University of Oxford.
It furthers the University's objective of excellence in research, scholarship,
and education by publishing worldwide. Oxford is a registered trade mark of
Oxford University Press in the UK and in certain other countries

First Edition published in 2013
Impression: 1

British Library Cataloguing in Publication Data
Data available

ISBN 978–0–19–953382–4 (hbk.)
      978–0–19–953383–1 (pbk.)

Printed and bound by
CPI Group (UK) Ltd, Croydon, CR0 4YY

Preface

This book originates from a course entitled 'Optoelectronic properties of polymers' presented to Masters level students at the École Européenne de Chimie, Polymères et Matériaux de Strasbourg, and previously at the Chemistry Department at the University of Groningen by one of the authors (Hadziioannou). The paucity of such courses is due mainly to the absence of a suitable text. In this spirit we decided to write a book which we hope will encourage the teaching of polymer electronics. Although the field is new, with a Nobel Prize in 2000 providing ample justification of its scientific importance, it has attracted much interest among scientists involved in providing cheap and (literally) flexible display technology, as well as solar cells and integrated circuits.

The target audience for this book is senior undergraduates and Masters level students of physics, electronic engineering, chemistry, and materials science. The book will probably be of more interest to the first two of these, but the interdisciplinary nature of the subject means that we have tackled it with a view to providing the broadest readership possible. Whether or not we have succeeded is not for us to judge, but if the book encourages new courses then we shall be happy with what we have achieved.

The book introduces the subject with examples of where polymer electronics might be important and a discussion of why some polymers can conduct. This theme is extended in the second and third chapters before we discuss optical properties of conjugated polymers and charge transport in the following two chapters. Chapter 6 contains an introduction to the relevant polymer chemistry, largely based on cross-coupling reactions, which are the workhorse of synthesis in polymer electronics. This chapter is perhaps surprising in a series for physics students, but experience suggests that physicists and chemists need to work together more and more in these areas, so we hope this chapter will be helpful. In any case, it can be skipped if it is not felt necessary. Very little elsewhere in the book relies on it. The physics of polymers as applied to these semiconducting polymers is the focus of the next two chapters, and a whole chapter is devoted to thin films because most devices are made in a thin film form. With the underlying physics (and chemistry) covered, we conclude with two chapters on devices. Thin-film transistors are covered first, followed by optoelectronic devices, which includes photovoltaic cells. The ordering reflects our belief as to how the subject might be taught, but we have tried to make each chapter as self-contained as possible so that the interested reader may be able to learn about specific topics without reading the whole book. Each chapter (except Chapter 1) concludes with some (mostly numerical) exercises and references to further reading.

All responsibility for mistakes and unclear explanations lies with us, and we are very grateful to our colleagues who supplied feedback on different chapters before submission. Many errors were spotted and improvements made as a result of the careful thoughts, criticisms, and comments from William Barford, Nigel Clarke, Eric Cloutet, Frans De Scryver, Martin Grell, Ahmed Iraqi, David Lidzey, George Malliaras, and Daniel Taton.

We should like to thank Sönke Adlung and colleagues at Oxford University Press for helping to ensure that this book was finished, for their professional approach to publishing it, and for their tolerance of the seemingly enormous delays in its preparation.

Mark Geoghegan and Georges Hadziioannou, July 2012

Timeo Danaos et dona ferentis. (Virgil, Aeneid II.)

I was asked by Georges to help turn his course on polymer electronics into a book in San Francisco in 2005. I declined. I visited Strasbourg the following year for a meeting on confined polymer films, and Georges, who was working there at the time, asked me again. Georges is very hospitable, and for anyone on the receiving end of his hospitality he is very hard to refuse. So this has been something of a long journey, but I think it has been worth it.

Georges hosted me for several visits to Strasbourg and Bordeaux, so I am particularly grateful to his wife Evelyne for her hospitality and kindness during these stays.

Mark Geoghegan, Sheffield (July 2012)

This book is the result of teaching, for almost twenty years, to Masters students at the University of Groningen, Université Louis Pasteur (Strasbourg, France), and most recently at the Université Bordeaux. It benefited also from special lectures I gave while invited, regularly or otherwise, during the last two decades at the Katholieke Universiteit Leuven (Belgium), the University of Patras (Greece), Universität Bayreuth (Germany), and the Ecole Normale Supèrieure (Paris, France).

The teaching content of the lectures, from early on, was always a moving target following the latest results of research developments in the field of organic and polymer electronics. This enrichment of the content of the lectures over the past twenty years would not have been possible without the help and constructive criticism of my doctoral students: in particular, Georges Malliaras, the late Bert de Boer, Henk Bolink, Sjoerd Veenstra, Guillaume Bonnet, and Guillaume Fleury. I am very grateful to them all.

I should like to thank Dr Paul van Hutten for helping to ensure that this course could see the light of day when it was first developed in Groningen, and to my daughter Céline for transcribing the videotaped lectures.

The process of developing the course over the past twenty years was rich and enjoyable, but the journey during the last six years, for turning the course into a book with Mark as captain, was even richer and delightful. Thank you Mark for your monumental contributions at all levels.

Georges Hadziioannou, Bordeaux (July 2012)

Contents

Polymer electronics

1

Polymer or plastic electronics has the feel of an oxymoron. Metals conduct, and copper, for example, is a fine conductor. Some materials are semiconductors, and the semiconductor industry is based around silicon. This technology is many years old, is mature, and satisfies all our needs, so why should we replace it?

It is probably true to say that silicon technology can satisfy our immediate needs. It is certainly not true to suggest that it needs replacing. Polymer electronics is important in many areas because polymers cost much less to work with than silicon. Conventional silicon technology requires very expensive ultrahigh vacuum equipment to be able to perform the necessary deposition of nearly atomically thin layers. Once the equipment is in place it is very difficult to change the processes that are used. To upgrade a manufacturing process from one chip to another with a different kind of architecture is a massive capital expenditure, and therefore a significant risk for any company. Polymers, on the other hand, are easy to process. After synthesis they can be stored on a shelf, often in no special environment, and often for long periods without degradation. They can be dissolved in solvent, and films can be easily made by straightforward processes known as spin-casting, dipcoating, or doctor-blading, all of which are discussed in Section 8.3. These films do not need to be made using expensive equipment, and have other advantages in that they are flexible. Imagine a newspaper updated in real time by wireless electronics. In Steven Spielberg's film *Minority Report*, this vision was presented as reality in the year 2054. Silicon electronics cannot do this because silicon is not flexible, and neither are related technologies. A route to achieving flexible electronic newspapers with polymers is very plausible. (In fact it is probably more plausible than the newspaper having pages; if pages can be updated in real time, why should we need them?) In fact the first flexible displays have been demonstrated, but it is not currently possible to use them without a rigid case so the advantages of flexibility are not apparent to the user. Nevertheless, it is not expected to be too long before flexible electronic display devices will appear on the market.

Polymeric materials can be used for all the different areas of electronics. If we are to have flexible displays, we need light emitting diodes (LEDs); if these displays are to provide circuitry we shall need transistors; if power is required we should need solar cells or other photovoltaic devices. Polymers can supply all these needs. The ability to connect them all together is a wholly different challenge, and one which is a

(a)

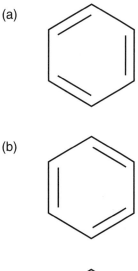

(b)

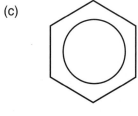

(c)

Fig. 1.1 (a) The chemical structure of benzene can be shown as having alternating single and double covalent bonds linking the carbon atoms. (b) The same structure but with the double covalent (π) bonds where single covalent bonds were shown in (a). Both of these have the same energy. (c) The π bonds can move around the molecule with no impediment, so these are usually indicated by a ring.

significant drag on progress up to now. These difficulties are challenges that must be beaten, and do not mean that the technology is doomed. Incredibly, we can rise to this challenge with the humble ink-jet printer! Indeed, commercial polymer devices are presently with us. Some cameras and mobile telephones have polymer displays. Flexible 'paper' displays are being produced, but these are only prototypes, which is a qualification that should not be needed should a second edition of this book be published.

Nevertheless, the perception that polymers are insulators is a prevalent one, but with a little knowledge of chemistry it is clear that this need not be a surprise. Benzene is a perfect example of an organic conductor. Not a very interesting conductor, but one whose electronic structure reveals much about how organic materials conduct. Benzene can be drawn with covalent bonds, but this can be done two equivalent ways as shown in Fig. 1.1. We have the carbon atoms bonded to both of their neighbours, and also to a hydrogen atom. This necessitates alternating double and single bonds, which is known as *conjugation*. These double and single bonds are known as π and σ bonds respectively, and are a prerequisite for semiconducting behaviour; in fact semiconducting polymers are often referred to as conjugated polymers, although not all semiconducting polymers are conjugated, e.g. poly(vinyl carbazole), which we meet in Section 5.2. The difficulty in assigning fixed double bonds to the ring is clear; why should the electrons be associated with one bond and not the adjacent one? Actually, the electron is said to be delocalized and these six electrons can move freely, without resistance, along the ring. A benzene molecule can in this way be said to be a conductor, albeit a quite useless one. Nonetheless, we can see how organic chemistry shows us that electrons can move along a molecule, and this is a message of basic importance to the behaviour of conducting and semiconducting organic materials. We shall return to this chemistry soon, but first we shall provide some context by discussing the history of polymer electronics.

1.1 A history of polymer electronics

Science is largely incremental—a fact implicitly recognized by Newton when he referred to his standing on the shoulders of giants. A starting point for a history of polymer electronics could begin with X-ray experiments measuring bond lengths of benzene to reveal that the electrons are in fact delocalized. However, a more instructive breakthrough would be the discovery, reported in 1977, that doped polyacetylene conducted electricity.

Imagine we could take a pair of scissors to a benzene molecule and straighten it. Imagine then that we add more of these straightened benzene molecules to create a very long molecule. This is polyacetylene. The delocalized π electrons should therefore be able to travel along the chain, from one end to the other, giving rise to metallic behaviour. In fact, this does not happen, due to what is known as the Peierls instability, which we shall treat later (Section 2.4.4). However, even though polyacetylene does not exhibit metallic behaviour, it does have a band gap and therefore behaves like a semiconductor, although technically it is not a true semiconductor. In fact, for these reasons, (undoped) polyacetylene was known to have an elevated conductivity due to its bonding, many years before the breakthrough by Shirakawa.

Whilst developing a novel synthesis of polyacetylene, Hideki Shirakawa in Tokyo accidentally used too much catalyst, and the polyacetylene synthesized, betraying a silvery colour. At a different reaction temperature, the polymer had another metallic colour, this time rather like copper. History tells us that serendipity has much to play in the role of scientific discovery. In fact, Shirakawa had synthesized two (pure) different forms of polyacetylene, *cis* and *trans*, and these are shown in Fig. 1.2. That it has new metal-like colours indicates that the polymer interacts with light in a different way to polymers previously synthesized. Most polymers have no colours. A Styrofoam cup might be white, but pure polystyrene is transparent; it does not interact with optical wavelengths. A polymer that looks metallic should have very different optical properties, which interested Alan MacDiarmad and Alan Heeger in Pennsylvania. By doping the *trans* polymer with iodine, it was observed that the conductivity of polyacetylene increased massively. In fact a value of 3000 S m^{-1} was measured. For comparison, the conductivity of copper at room temperature is about 60 MS m^{-1}, which might seem a large difference but the *trans*-polyacetylene is still metallic. A rule of thumb is that metallic behaviour occurs with conductivities greater than 1000 S m^{-1}. Very heavy doping of polyacetylene can reveal conductivities similar to that of copper.

The synthesis of polyacetylene in these forms and the discovery of metallic conduction in the doped polymer was one of the *great* triumphs of polymer science and resulted in the award of the Nobel Prize for chemistry to Shirakawa, MacDiarmad, and Heeger in the year 2000. However, polymerization of polyacetylene remained quite difficult, and a breakthrough was made in Durham by the group led by James (Jim) Feast, with a synthetic route known as ring-opening metathesis polymerization (ROMP). The polymerization used by Shirakawa was in itself not too difficult, but the polymer could not readily be processed for use in electronic devices because of its shape. Polyacetylene is a very rigid polymer, and does not dissolve easily, which makes it rather difficult to work with. In its simplest form the Durham-route polyacetylene synthesis (Fig. 1.3) is the chemical scissors to benzene that we proposed earlier, although benzene itself was not used in the synthesis. This precursor polymer can be easily dissolved, for example in isopropanol, and films can be created by using a process known as *spin-coating*, which is a readily available and inexpensive method of creating films from a polymer solution (Section 8.3.3). Heating the precursor polymer at elevated temperatures causes it to be converted into polyacetylene.

Although the Durham-route polyacetylene does not have the conduc-

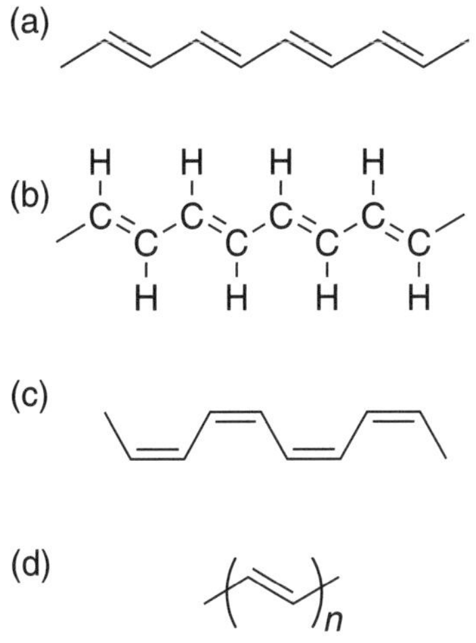

Fig. 1.2 (a) *trans*-polyacetylene. (b) Chemical structure showing explicitly the location of carbon and hydrogen atoms. In most chemical representations, these atoms are omitted. We shall generally follow standard chemical notation in this book. (c) *cis*-polyacetylene. (d) The monomer is shown here in brackets. n is the polymerization index, and the larger n is, the bigger the polymer. The molecular weight of the polymer is given by n times 26 g mol^{-1} and the mass of whatever groups terminate either end of the polymer.

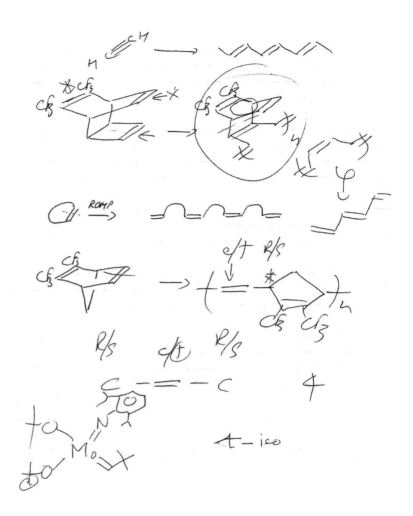

Fig. 1.3 The Durham route to poly-acetylene as described by Jim Feast. In 1991 he gave a plenary lecture entitled *Romping in Durham* at a polymer conference in Lancaster. The synthesis of polyacetylene was summarized for the benefit of a young Mark Geoghegan (attending his first conference as a Ph.D. student) in his lab book, from which this figure is taken.

tivity of the Shirakawa material, the ability to make good-quality films makes it useful in devices. The earliest devices were field-effect transistors (FETs), which were created by making a polyacetylene film on an oxide layer grown on a silicon wafer. The oxide layer makes what is known as the gate (switch), and lithographic techniques were used to lay gold source and drain contacts onto the oxide layer. The polyacetylene precursor was then spin coated. We shall consider the behaviour of FETs in Chapter 9. These early transistors were made in the Cavendish Laboratory in Cambridge by the group of Richard Friend and it was the same group a couple of years later that was to provide the next breakthrough in polymer electronics with the discovery of optoelectronic behaviour in poly(*p*-phenylene vinylene), often abbreviated to PPV. In fact serendipity played its role here too. The authors were trying to make improved FETs, and in performing the tests, one of the team, Jeremy Burroughes, spotted a green–yellow glow from between the electrodes.

These particular PPV devices were actually rather poor emitters, and many people might have struggled to see the light coming out. Light emission from a polymer is not surprising insofar as semiconducting materials often display electroluminescence. However, this was the first such discovery, and due to the fact that the efficiency of this PPV is very poor, a huge amount of research effort was invested in the field, with large increases in performance occurring over a very short period of time. In fact, the importance of the PPV discovery was not lost on the team, and Burroughes, Friend, and Donal Bradley were awarded a patent on the basis of this discovery. A company, Cambridge Display Technology (CDT), was formed in 1992, two years after the publication of the original article. In 2007 it was sold to Sumitomo Chemical for $285 million, which clearly demonstrates the importance of recognizing the commercial value of that little green glow.

The seventeen years or so that had arisen since the description of the first polymer LED and the sale of CDT demonstrates the importance of the original patent, as well as the expertise within the company. It also demonstrates the slow pace of transformative technology, because products based upon polymer technologies are still not available. Small-molecule organic light-emitting diodes (OLEDs) have become the doyens of the market due to their image quality and relatively cheap production. Active matrix OLEDs (AMOLEDs) are touted as a selling point in many new smart phones. This perhaps is a recognition that small molecules might have a different method of transporting charge, but production techniques involved in their preparation still have much in common with inorganic semiconductors. Small-molecule organic materials cannot be processed with the ease of polymers, which, although a long-term advantage for those who wish to deploy polymers in their devices, is a medium-term disadvantage, because the processing of the materials needs to be developed, rather than merely altered.

The discovery of the high conductivity of doped polyacetylene provoked a great deal of effort in searching for other polymers with similar properties. One such polymer was poly(3,4-ethylene dioxythiophene) (PEDOT) which was being developed by Bayer in the late 1980s, which not only had a conductivities in excess of 300 S m^{-1}, but was also very stable. Originally synthesized by electropolymerization (discussed for another polymer in Section 6.9), it later became possible to synthesize PEDOT in a water-based complex with polystyrene sulfonic acid. This formulation was much more tractable, and applicable to standard film-forming techniques such as doctor-blading and spin-casting (Section 8.3). Conductivities in excess of 1000 S m^{-1} have also been measured. This complex was marketed as Baytron, and later as Clevios™. The complex, usually denoted PEDOT:PSS, is also manufactured by Agfa, who have a great interest in using it as an antistatic agent to cover photographic film; it is marketed under the name Orgacon™. PEDOT also has good transparency properties, which make it useful as an electrode (anode) in optoelectronic devices. As such, it is the most successful semiconducting or conducting polymer available to date.

1.2   Future applications of polymer electronics

In considering future applications of polymer-based devices, one should not underestimate the power of incumbent technologies; manufacturing facilities exist, and it is surely desirable to only have to modify these for new products than to start again. Expertise is widespread in established areas, so a chain of knowledge exists from the purest research to final production. Innovation in current technologies should not be underestimated either; the microcomputer appeared as a commercially viable product in the mid-to-late 1970s with chips such as the MOS 6502 (1975) and the Intel 8086 (1976). The latter powered early market leading computers such as the Commodore PET and the Apple II, whereas the 8086 was still being used in the late 1980s inside some IBM PS/2 machines. The 8086 had a transistor channel length (Section 9.3) of 3 μm. Latest technology is based around 22 nm—a decrease in size of a factor of more than 100 in just over 30 years. The commensurate increase in performance is huge, but the key point is that microprocessor circuits were designed with optical lithography techniques in mind; these established techniques have not changed much, but currently are capable of being used with ultraviolet light.

Polymer electronics is not going to provide a significant contribution to high-performance computing, but applications in logic circuits and optoelectronics must still compete with incumbent technologies. Nevertheless, polymer electronics is in a state to capitalize on its potential. Products have moved beyond prototype stage, and manufacturing facilities for display devices are now available. As an aside, it is also possible to make polymer transistors with channel lengths shorter than 50 nm. Their performance is not comparable to inorganic thin film transistors because charge transport in organic media is not as impressive as transport in their inorganic counterparts. High hopes rest in areas where the barriers to incumbent technologies are already high, such as light tiles. Here, polymers can be coated on surfaces quite easily, and they can be used for large-area lighting. Major companies such as Osram and Philips are investing heavily to exploit this area.

As we have seen, polymer electronics and optoelectronics is not going to replace inorganic electronics, certainly not for high-performance products. However, polymers have advantages in the area of processability and flexibility that are unmatched by inorganic materials. Polymers are ideally suited for high-volume production of commodity products.

Some areas of polymer electronics that could be significant in the future include superconducting polymers and polymers for spintronics applications. These are covered in this chapter (Sections 1.2.4 and 1.2.5 below) but not elsewhere in the book, because at the time of writing their importance in the field is perhaps to the periphery of the subject. Of course, predicting the future is a fool's game, so we look forward to seeing these areas dominate the subject in future years.

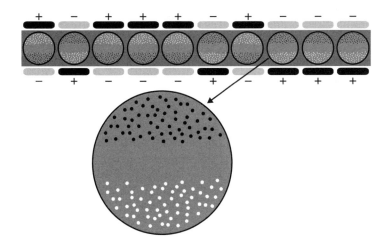

Fig. 1.4 Cross-sectional diagram of the major components of an electrophoretic display. An electrophoretic display contains two parallel arrays of electrodes, which define the pixels. Between these pixels is a fluid (either a polymer or an oil) containing microcapsules, which contain charged (coloured) particles. These particles are attracted to the electrode of the opposite charge when it is switched on, resulting in a rearrangement of the particles within the microcapsules, as shown in the lower image. The location of the particles within the microcapsules defines the on/off state of the pixel.

1.2.1 Lighting and display technology

Polymer light-emitting diodes are likely to be reaching the market in high numbers some time soon. Active matrix display devices (i.e. those whereby a transistor stores the display state of each pixel) are available in different forms, and very few issues need to be addressed. Colour reproduction is good, device lifetime is excellent, although reproducibility is variable; ultimately, the technology will gain traction when the price comes down. One would expect early displays to be on mobile telephones and cameras simply because these are smaller than other displays. Nevertheless, computer screens and televisions are perfectly possible; their performance should in principle be at least as good as the best displays currently on the market.

An area that would give polymers the edge over existing technologies is electrophoretic displays, which offer a route to a flexible monochromatic screen that has sometimes been referred to as 'electronic paper'. They consist of microcapsules containing charged particles dispersed in an oil, and these are attracted to or repelled from the electrodes. Whether or not the particle is attracted to a pixel determines whether that particular pixel is on or off. The microcapsules are dispersed in a fluid (usually an oil or a polymeric fluid) between two parallel electrodes (Fig. 1.4), one of which must be transparent. Reference to polymer electrophoretic displays does not generally mean the encapsulation of microcapsules, but rather refers to flexible substrates; in any case, polymer electrophoretic displays do not take advantage of the optoelectronic properties of polymers. A useful property of such displays is that removing the power does not send the particles away from the electrodes, so they can be read without power. They are *passive* displays, because there is no source of light within the device and are thus read like a book. These devices have extremely good contrast, and being passive, cause less eye strain than LED-based devices, and require very little power. Electrophoretic displays are on the market in many forms as e-readers, the most famous being perhaps Amazon's Kindle™. Flexible electrophoretic displays have made it to the prototype stage; the Readius, which originated from Philips, contained a small (13 cm) rollable display. The spin-out company responsible for the Readius (Polymer Vision) continues to develop rollable technologies, despite a brief period of bankruptcy. Another electronic reader, from Plastic Logic also contained a flexible display, although this was encased in a rigid enclosure. Plastic Logic has since demonstrated a flexible device, but the advantage of this newer technologies lies not so much with flexibility, but rather with the form factor; gadgets that are very thin and lightweight are particularly appealing.

Of course, a device containing the flexible substrate of the new electrophoretic displays, with polymer light-emitting diodes is in principle possible. Such devices will include colour, and can be read in low-light conditions. The winning technology is very much hard to predict; electrophoretic displays are a promising area, but traditional technologies will not go away without a fight. Flexible glass is now an established technology, which could well house AMOLED displays.

1.2.2 Transistors

The arrival of all-polymer transistors is the cornerstone behind polymer electrophoretic displays and flexible electronics. Polymer transistors are mainly expected to see application in low-cost high-volume areas of technology such as radio frequency identification tags, as well as display technology. These transistors would not need expensive circuitry for these applications, and could indeed be printed (standard ink-jet printing technology is sufficient) onto the substrate. Transistors are an interesting technical challenge because they are the basic components of electronic circuits, and they must be connected to make logic circuits. Ink jet printing is popular because one can use materials such as PEDOT:PSS for the electrodes and use a standard printer to coat electrodes onto surfaces for different devices. One is not restricted to prearranged structures, but can choose the most appropriate circuit design for the device to be created. It is also possible to use the lithographic methods currently in place in the inorganic semiconductor industry.

Fig. 1.5 Printing processes can be used as a means for the mass production of polymer circuitry, such as those in radio frequency identification tags. Image courtesy and copyright of PolyIC GmbH & Co.

1.2.3 Radio frequency identification tags

Radio frequency identification tags (known generally as RFID) are essentially a replacement for the ubiquitous barcodes. (RFID can be written to, does not require a line-of-sight interrogation, and can store substantial information; none of this is the case for bar codes.) Their application to polymers is very real, with prototypes currently available. RFID is the classic example of printed electronics (Fig. 1.5), because mass production will require a printing process.

Currently, RFID tags based on silicon technology are quite popular, in areas such as animal tagging and 'electronic' passports, but the cost is too great to replace the barcodes currently used on commodity products in stores; a production cost of maybe $0.01 per tag is something of a goal. It is hoped that organic RFID tags may eventually achieve such a target price. RFID that can work over large distances could present a very useful means of inventory management in large warehouses, which is known in the industry as the 'smart shelf'. For this to occur the RFID tag would need to respond to a probing signal over distances of metres. RFID over long distances does not necessarily require external power sources if multiple antennae (readers) are used. RFID technology therefore requires an antenna, which sends out information to the tag, and also, using a transceiver, receives the information sent by the tag. The radio waves sent out by the antenna are a source of energy for the tag, which means that it works only when interrogated, and usually does not need an external power supply.

The main part of the tag is a transponder, and it is here that polymer electronics is expected to be important because the production technology should be some form of printing. The transponder is generally an electronic circuit sandwiched between two polymer substrate layers. The circuit requires a series of transistors, and typically a form of polythiophene (Section 3.3) would be the active layer.

1.2.4 Superconductivity

As impressive as the semiconducting properties of many conjugated polymers are, it would be an enormous achievement to demonstrate superconductivity in a polymeric system. Superconductivity in organic systems is a very current area of research, with hybrid systems involving C_{60} and other unsaturated (containing carbon–carbon double or triple bonds) organic molecules with alkali metals being quite popular. The alkali metals are important because they donate electrons to the organic component. However, in normal superconductors conduction is in pairs of electrons, the so-called *Cooper pairs*, where electrons flow without resistance aided by lattice vibrations (phonons). This is contained within the Bardeen–Cooper–Schrieffer (BCS) theory of superconductivity. In organic materials, it has been suggested that the interactions between electrons and bound pairs of electrons and holes (excitons) can mediate superconductivity. This is much more energetic than the BCS mechanism, and presents the possibility of an increased barrier to the thermal disturbance of this mechanism. No organic superconductors have been observed at high temperatures, and no mechanism of organic superconductivity has been identified, although it is believed that the BCS mechanism may well not be responsible in organic media.

Superconductivity in polymers was reported many years ago amidst some fanfare. A field-effect transistor based on a semiconducting polymer, poly(3-hexylthiophene), was shown to superconduct at around 2 K. However, the results were fabricated and had to be retracted in one of the biggest cases of scientific fraud reported. This does not mean that the ideas underlying the faked experiments are invalid; it is just that polymer superconductivity has not yet been demonstrated. Certainly polymers present a particularly difficult challenge because of their disorder. In metals a superconducting transition happens as the temperature is decreased and the conductivity increases. At a transition temperature, superconductivity occurs. The origin of the increase in conductivity with decreasing temperature in metals is loosely because the time between collisions of electrons that would disturb the motion of free electrons increases with decreasing temperature, which is simply due to the electrons moving more slowly, giving rise to an increased conductivity (eqn 2.6). In polymers this is tricky because there are too many defects present.

1.2.5   Spintronics

Electrons, being fermions, have more than one state of spin. The two possible spin states, $S = \pm 1/2$ are perfect for applications in computing and memory storage. Spintronics, which is a contraction of spin transport electronics, is a technology requiring a means of creating polarized electrons (it suffices to simply have a measurable and meaningful excess of one spin over the other) that may be input and detected or read out. The benefit of such a technology is that it is not necessary for electrons to traverse a circuit, but rather flip their spins, with a commensurate saving in both time and energy. Rather than using a the presence or absence of current to represent a bit 1 or 0, this is obtained from a spin of $+1/2$ or $-1/2$. In fact spintronics might also have applications in light-emitting diodes. Polymer LEDs have efficiencies limited by the spins associated with the excited states that emit light (Section 4.12). Spintronics may be one means of controlling those spin states to improve efficiencies.

1.2.6   Biological hybrid systems

The integration of electronics with biological systems is an area where organic electronics will play a serious role simply because biology is based on organic systems. A key developing area is that of biosensors, whereby *in situ* diagnostic systems report the behaviour of various functions. Simple biosensing is well advanced with current technology, so blood pressure and pulse can easily be measured using standard technologies, but more subtle and sensitive investigations will require flexible technologies. It is even possible to incorporate RFID tags which act as sensors inside the body. Polymer substrates would therefore be required to match the skin as it moves, in order for devices to work reliably; weak muscle movements could thus be detected, as well as providing greater resolution to cardiovascular behaviour, or activity within the brain. The great advantage of such technologies is that they could be worn throughout the day, and could send signals wirelessly that may be recorded. Such devices will require polymers, because flexibility and comfort without them is likely to be cost-prohibitive. However, it is not fully necessary to use polymer electronic materials within the device, although there would be clear advantages in doing so, largely from material processing perspectives.

The development of polymer-based biosensors in pathogen and disease sensing is also an area of development. Antibodies can be bound to polymers and can be used to recognize specific pathogens. (This binding is known as conjugation, and should not be confused with the alternating double and single covalent bonds that concern us in this book.) Again, biochips such as these might not require polymer electronic properties, but the electronic requirements of this kind of device would not be prohibitive to organic electronics, and so the mass production of disposable devices in the future may well be based completely on polymeric materials. Nevertheless, hybrid devices made of polymers and inorganic semiconductors are probably going to dominate the early stages of development of biosensing technologies.

1.3 Challenges

The future of polymer electronics is likely to involve the development of the ideas developed above, and numerous others besides. Of course, there are challenges to be faced, because otherwise these technologies would be present. Polymeric materials can have lifetime issues, and their performance could be improved with better electron transport, but most of all, purification is a problem.

Polymers are generally robust when it comes to charge transport dominated by holes, which travel deep down in the energy levels of a device, below the band gap. To interfere with hole transport is therefore much more difficult than to interfere with electron transport. That is not to say it cannot be done. Put plenty of impurities and defects in a device and it is likely to be of a very poor quality. Nevertheless, there is a discrepancy between the quality of hole-transporting polymers and the quality of electron-transporting polymers. This is because electrons move in the lowest unoccupied molecular orbital, which is just above the band gap. This means that the electron is quite reactive; it does not cost a huge amount of energy to form new bonds. Electron transporting materials are therefore prone to oxidation. In most cases this problem is treatable. Electron transporting polymers are generally situated next to the cathode, since this is where electron injection occurs. Given that the cathode is usually a layer of a metal such as calcium which is evaporated onto the polymer a certain degree of protection is provided. This layer is then very often further protected by the deposition of an aluminium layer onto the calcium. Under these circumstances, the device is said to be *encapsulated*. Nevertheless, some gas ingress may occur, which limits the protection provided by encapsulation. A badly designed device will remain susceptible to damage. For example, excessive heat generation in devices, usually through a process known as *Joule heating*, which is simply the dissipation of heat through a resistance, can supply enough energy to damage materials through oxidation, amongst other routes.

Electron-transporting materials might well be encapsulated in devices, but they are more limited in their performance characteristics. Electron and hole transport should ideally be balanced in optoelectronic devices such as photovoltaic cells and LEDs, and logic circuits cannot be effective without good characteristics for both electrons and holes. An example of a simple logic circuit that will illustrate this point is given in Section 9.6. The problems of electron stability and mobility are linked, and developments to address one problem may well help the other. Recent progress has been made in providing good-quality electron-transporting polymers through the use of polymers with both donor and acceptor groups. Donor groups are electron-rich and thus have useful hole-transporting properties (Section 3.7); acceptors are electron deficient and can improve electron mobility. The combination of the two creates a molecular environment in which injected electrons, for example, at the cathode, are not impeded by interactions with other electrons and can have high mobility through the device.

Small-molecule organic electronics provides high-quality charge transport for both electrons and holes. Mobilities are also higher than those for polymers in the best materials. These are not necessarily problems for polymer electronics. because not all applications require these qualities. However, the improvement in small molecule organic electronics is essentially insurmountable insofar as it concerns deposition routes. It is possible to deposit highly ordered structures of carbon-based molecules through epitaxial routes. The best mobilities are achieved in molecules such as pentacene using high-vacuum deposition techniques, and some comparative examples are included in Section 2.1. High-vacuum methods generally guarantee purity, which is a prerequisite for the best performance. Polymers are notoriously difficult to purify, and it will only be when this challenge is overcome that we can expect to see polymers in everyday devices such as mobile phones. Even so, technology does not stand still, and solution-processable small molecule organic devices are actively being developed, and these are expected to compete with polymers in the future.

1.4   Further reading

A history of the broader field of molecular electronics is included in the text by Petty (2007). The book may also be of interest as a complementary text to this book. Printing processes (roll-to-roll printing)are covered in an exhaustive review by Søndergaard et al. (2013).

Electronic structure and band theory

In this chapter we review the theory that gives rise to semiconducting behaviour and show how this applies to polymers. It might seem a little incongruous in a book on polymer electronics to start with charge transport in metals and inorganic semiconductors. A basic treatment of charge transport in metals via the free electron model is useful because it introduces the concept of the *Fermi energy*. It goes without saying that the physics of the origin of the band gap that gives rise to semiconducting behaviour is important, especially as it allows us to highlight the differences between organic and inorganic semiconductors.

2.1   Conductivity

Electrical conductivity, σ is a measure of the ability of a material to allow charges to flow or to conduct electricity under an applied electric field E. It is formally defined by

$$\sigma = j/E \qquad (2.1)$$

where j is the flux of charge carriers (usually holes or electrons). The flux can be defined by

$$j = nqv \qquad (2.2)$$

where n is the number density of the charges, v is their drift velocity, and q is the charge on each of the carriers, usually $\pm e$, depending on the charge of the carriers, and where e is the basic electronic charge. The units of conductivity can be derived from the formulae above; since n, q, v, and E having units m^{-3}, C, m s^{-1}, and V m^{-1}, σ have respective units Ω^{-1} m^{-1}. We generally write the units of conductivity in their more common form of S m^{-1}, where S is the siemens, the unit of conductance and $1\,S = 1\,\Omega^{-1}$. Another parameter that we shall need very often is the carrier mobility, μ, which is given by

$$\mu = v/E \qquad (2.3)$$

although this is not a formal definition. The conductivity is often assumed to be an intrinsic property of the material in question, but it is actually dependent on temperature, and this dependence should be

Table 2.1 Electrical conductivities and carrier mobilities of some materials at ~ 293 K.

Material	Conductivity (S m^{-1})	Mobility (m^2 V^{-1} s^{-1})	Dominant charge car
Copper	6.0×10^7	0.0030	electron
Gold	4.5×10^7	0.0030	electron
Aluminium	3.8×10^7	1.3×10^{-4}	hole
Doped polyacetylene	1.5×10^7	0.02	hole
Doped pentacene	1.1×10^4	1.6×10^{-5}	hole
Pentacene	2×10^{-6}	0.0035	hole
Silica (SiO$_2$)	$< 10^{-18}$	—	—

stated. If a material has different forms, then these are often stated too, and we need look no further than polyacetylene, for which the *cis* and *trans* forms have different properties to exemplify this. We also note that impurities can dramatically affect mobility. Most impurities are unwanted and will cause the conductivity to decrease. The addition of certain materials, called *dopants*, can cause the conductivity to dramatically increase. Virtually all semiconducting polymers require doping in order to conduct significantly. Those that do not are known as *synthetic metals*—a phrase that is not restricted to polymers. We shall describe how doping improves conduction in Section 2.4.5.

Many materials have mobilities for both holes and electrons, in which case the flux (eqn 2.2) requires summation over the different carriers. The mobility given by eqn (2.3) is a property of the individual carriers and not the material itself. Obviously, in a calculation of the flux, the movement of both electrons and holes will offset each other; however, under an applied electrical field a hole travelling in one direction is equivalent to an electron travelling in the opposite direction, so it is important that the *charge* flux be calculated rather than a simple summation of the two carriers.

The conductivities and mobilities for some materials are listed in Table 2.1. For comparison, pentacene, an organic semiconductor, and silica, a common insulator, are included in the table. Note the improvement in conductivity by some 10 orders of magnitude when pentacene is doped in iodine vapour. The carrier mobility for pentacene *decreases* when the pentacene is doped; this is because it spoils crystallinity, which aids the transport properties of the material.

If a charge is placed in an applied field, it will experience a force

$$m_q \frac{\mathrm{d}^2 x}{\mathrm{d}t^2} = qE, \tag{2.4}$$

where m_q is the mass of the charge carrier and $\mathrm{d}^2 x/\mathrm{d}t^2$ is its acceleration, which we can integrate to obtain the drift velocity

$$v = \frac{qE\tau}{m_q}, \tag{2.5}$$

where τ is the mean time between collisions. The assumption here is that the charges are travelling with thermal motion with a velocity due to the applied electric field superimposed on this Brownian motion. It is the Brownian motion that is giving rise to the collisions, and these are assumed to be with the atoms that make up the conductor. We thus obtain for the conductivity

$$\sigma = \frac{nq^2\tau}{m_q}. \tag{2.6}$$

Equation (2.6) is the main conclusion from the *Drude model* of *electron* transport in metals. It is instructive but is nevertheless incorrect. Electrons do not interact with atoms because *wave–particle duality* allows them to behave as waves. (An important exception is when the atoms are in a lattice that gives rise to *diffraction*, but that need not be considered here.)

2.2   The free electron model

The understanding of semiconductor behaviour is possibly best considered in terms of the failure of free electron model to predict conductivity in many different materials. The free electron model is the basis of our understanding of conductivity in metals, and although it has its successes, it has limited applicability. One result of this is the development of *band theory*, which we consider in Section 2.3.

The Drude model fails because of the assumption that electrons interact strongly with the lattice. If we treat electrons as waves, we can understand their behaviour. As well as the absence of an interaction with the lattice, we assume that electrons do not interact with each other. In other words the electrons move in a uniform potential, which we can set at zero. The energy E of the electron (with wave function ψ) can be obtained by solution of the time-independent Schrödinger equation,

$$\frac{-\hbar^2}{2m_e}\nabla^2\psi = E\psi, \tag{2.7}$$

where m_e is the mass of the electron and $\hbar$ is the Planck constant h divided by 2π. In Cartesian coordinates, it can be readily shown that the wave function has a simple standing wave solution,

$$\psi(x, y, z) = A\sin(k_x x)\sin(k_y y)\sin(k_z z), \tag{2.8}$$

where A is a constant, and our boundary conditions dictate that the wave function is zero at the sides of the specimen. Here k_x, k_y, and k_z are the wave numbers of the electron in the relevant direction. We remind the reader that the wave function has important physical meaning because $\psi(r)\psi(r)^*$ is the probability that the electron finds itself at a displacement r from the origin. The magnitude of the wave vector k is found from the solution to the eqn (2.7):

$$E = \frac{\hbar^2}{2m_e} \left(k_x^2 + k_y^2 + k_z^2\right) = \frac{\hbar^2 k^2}{2m_e}. \tag{2.9}$$

If we assume the specimen to be a cube of side L, then it is clear that the lowest-energy solution corresponds to an electron wavelength, $\lambda = 2L$. The next highest energy is with wavelength L, followed by $3L/2$, $2L$, and so on. The boundary conditions here are those where x, y, and $z = 0$ or L, and $\psi = 0$. This is a *standing wave* solution. We are not really interested in standing waves in this book, but rather *travelling waves*. To allow for both standing waves and travelling waves, our length L becomes a ring, with the conditions at $x = 0$ being the same as $x = L$, and similarly for the other axes. Travelling waves can move around the ring and there will be no problem with boundary conditions; in fact these are referred to as *periodic boundary conditions*. If we are to work with a ring, our standing wave boundary conditions change because an electron wavelength of $\lambda = 2L$ is no longer a valid solution because the function is no longer smooth and continuous at the limits of x. Neglecting $\lambda = \infty$, which has no useful analogue in a finite material, the lowest energy solution is for a wavelength $\lambda = L$, with higher energy states with wavelengths given by L/n, where n is an integer. Such periodic boundary conditions are illustrated in Fig. 2.1. The use of periodic boundary conditions changes nothing about the physics of the free electron model, but the reader should be aware of the possible confusion in the wavelength of the lowest energy state.

Returning to our cube of side L, it is possible for two electrons to have the longest wavelength of $2L$. Because electrons are *fermions*, the Pauli exclusion principle prohibits more than two electrons occupying the same state. Two electrons are allowed because they can have different spins, and thus are not identical. The next energy state is not of two electrons having wavelength $\lambda = L$ but rather having $\lambda = L$ in one direction, and $2L$ in the other two. Because there are three possible directions for the wavelength λ, there are three different states associated with this second-lowest energy state. The possibility of both spin up and spin down means that six electrons can have this state. We write the energy as

$$E = \frac{\hbar^2 \pi^2}{2m_e L^2} \left(n_x^2 + n_y^2 + n_z^2\right). \tag{2.10}$$

where n_x, n_y, and n_z are integers. The lowest electron energy state therefore has energy $E = 3\hbar^2\pi^2/2m_e L^2$ and the next state has $E = 3\hbar^2\pi^2/m_e L^2$, corresponding to $n_x = 2$, with the other two values of n equal to unity, or any equivalent combination.

At absolute zero of temperature, the electron energy states are all filled up; the occupancy for each state is one. An important concept is that of the *density of states*, which describes the number of energy states with an energy lying between E and $E + \mathrm{d}E$. This is a rather unhelpful concept when one considers the lower-lying states. A look at Fig. 2.2 shows how the number of electrons of a given state behave with

Fig. 2.1 Travelling waves can be treated in a similar fashion to standing waves if they are plotted in a circular geometry. One difference is that we can consider standing waves of half-wavelengths, but travelling waves require whole wavelengths along a ring of length L.

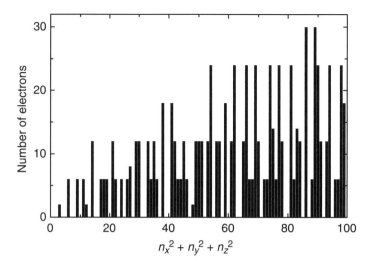

Fig. 2.2 Only a discrete number of electrons may occupy any one energy level. These are given by eqn (2.10). Here, the number of electrons is plotted as a function of $n_x^2 + n_y^2 + n_z^2$. Although the energy levels may appear random for the first few values of $n_x^2 + n_y^2 + n_z^2$, by the time $n_x^2 + n_y^2 + n_z^2 = 100$, we begin to see the appearance of the density of states as discussed in the text, and presented in Fig. 2.3.

energy. In fact, one might initially think that it is not possible for more than twelve electrons to occupy any one energy state, since there are six different permutations of three different numbers for n_x, n_y, and n_z, and the two different spin states mean that there are twelve possibilities for any one energy. However, at higher energies there are more combinations of $n_x + n_y + n_z$ that give an integer, and fewer numbers that are excluded. For example, there are twelve arrangements that give $n_x^2 + n_y^2 + n_z^2 = 54$: three containing (6, 3, 3); three containing (5, 5, 2); and six containing (7, 1, 2), which, with the two spin states, allows for twenty-four electrons with energy $27\hbar^2\pi^2/m_e L^2$. Clearly then, the density of states increases with energy as we go to higher energies. Given that in 50 g of potassium there are 7.7×10^{23} free (valence) electrons, it does not make sense to consider these energy states as a discrete function as shown in the bar chart in Fig. 2.2. To this end we need to treat the number of states as a function of energy as continuous, which we do in reciprocal space. Imagine a Cartesian coordinate system in k-space, with axes, k_x, k_y, and k_z. If we draw a spherical shell around a portion of k-space, it will enclose

$$g\left(k\right) \mathrm{d}k = 2 \times \frac{1}{8} 4\pi k^2 \mathrm{d}k \times V/\pi^3 = \frac{Vk^2\mathrm{d}k}{\pi^2}, \qquad (2.11)$$

where V is the volume of the sample. The factor V/π^3 represents the axes having energy states at $k = \pi/L$ on each of the three axes, so the volume per state, π^3/V is used to normalize the density of states. The factor $1/8$ ensures that we only consider positive k, otherwise our results would not be meaningful; a wave with negative k is identical to the corresponding wave with positive k. We can convert from k to E by using

$$\mathrm{d}E = \frac{\hbar^2 k}{m_e} \mathrm{d}k, \qquad (2.12)$$

which can be obtained directly from eqn (2.9). The density of states can then be rewritten as

$$g\left(E\right)\mathrm{d}E = \sqrt{2m_e^3 E}\,\frac{V}{\pi^2\hbar^3}\mathrm{d}E. \qquad (2.13)$$

We now see the mathematics of calculating the number of electrons in a small band of energy, so it is a relatively straightforward task to consider all of the electrons. If the metal has N free electrons, at absolute zero temperature the density of states is related to N by

$$\int_0^{\epsilon_F} g\left(E\right)\mathrm{d}E = N. \qquad (2.14)$$

This equation essentially describes the energy levels of the electrons in a metal as a continuum. We have seen in Fig. 2.2 how the number of electrons changes with energy for a very small number of electrons; here we show the same thing (see Fig. 2.3), but where the number of electrons is large, and cannot be treated discretely. This equation (2.14) also introduces a very important parameter, the Fermi energy, ϵ_F. Because we do not consider eqn (2.14) for systems with only a few hundred electrons, we do not need to worry about it contradicting eqn (2.10).

The free electron model tells us that valence electrons in a metal have energy states based on their wavelength. The treatment begins by filling lower energy states first. When all of the electrons have had an energy assigned to them, the density of states is complete. We stated above that eqn (2.14) applies at absolute zero. This should be clear, because it would be impossible for any electron to lower its energy by finding an unoccupied state, since all states are occupied. However, thermal energy can be used to excite electrons with energies close to the Fermi energy into higher states. The probability that this happens at (absolute) temperature T is given by the *Fermi–Dirac* probability distribution (Fig. 2.4),

$$f_{\mathrm{FD}}\left(E\right) = \frac{1}{1 + \exp\left(\dfrac{E - \epsilon_F}{k_B T}\right)}. \qquad (2.15)$$

where k_B is Boltzmann's constant.

At absolute zero $f_{\mathrm{FD}}(E) = 1$ for electron energies less than the Fermi energy. For electron energies greater than the Fermi energy $f_{\mathrm{FD}}(E) = 0$, which means that there is zero chance of an electron occupying a state with an energy greater than the Fermi energy at absolute zero. The Fermi–Dirac distribution is often referred to as the electron *occupancy*. The total number of electrons with an energy E at temperature T is then

$$N = \int_0^{\infty} f_{\mathrm{FD}}\left(E\right) g\left(E\right)\mathrm{d}E, \qquad (2.16)$$

which, at absolute zero, is identical to eqn (2.14). The number of electrons with an energy less than or equal to ϵ may be given by

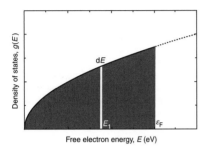

Fig. 2.3 The density of states of electrons with energy E increases with $E^{1/2}$. The number of electrons with an energy between E_1 and $E_1 + \mathrm{d}E$ is given by $g(E_1)\mathrm{d}E$. At absolute zero of temperature the maximum energy of a free electron is given by the Fermi energy, ϵ_F, above which energy no further states can be occupied.

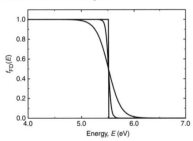

Fig. 2.4 The Fermi–Dirac distribution for gold at absolute zero, 300 K, and 1337 K, its melting point. The distribution becomes broader with increasing temperature. At $E = \epsilon_F$, $f_{\mathrm{FD}}(\epsilon_F) = 0.5$

$$N\left(E < \epsilon\right) = \int_0^\epsilon f_{\text{FD}} g\left(E\right) \mathrm{d}E \qquad (2.17)$$

and the number of electrons with an energy between E and $E + \mathrm{d}E$ is

$$N\left(E\right) = f_{\text{FD}}\left(E\right) g\left(E\right) \mathrm{d}E. \qquad (2.18)$$

If we take the density of states to be defined only at absolute zero, $g(E) = 0$ for $E > \epsilon_F$. This would mean that there is no contribution to eqns (2.16) and (2.18) from the Fermi–Dirac distribution at energies above the Fermi energy. In order that these equations be meaningful, the function for the density of states, $g(E)$ is extended beyond ϵ_F. The astute student will note that the area under this curve will change as a function of temperature, and so the total number of electrons, N would consequently not be fixed using this methodology. This is actually a consequence of the simplification of assuming ϵ_F to be a chemical potential in the Fermi–Dirac distribution, and is discussed in more detail in other (advanced) texts.

There is a certain amount of mathematics here, and it is important that we understand the physics. In an applied electric field, free electrons attain more energy simply due to the electrostatic interaction between the field and the charge. In order to conduct, these electrons need to be able to occupy higher energy states. In the free electron model, these states are readily available, and the model is best illustrated by eqn (2.11), where the density of states in k-space is proportional to the square of the electron wave vector k.

2.3 Band theory

The band gap is responsible for semiconductor behaviour. In metals, the range of energy states into which an electron can be excited in order to conduct. In a semiconductor, there is an energy barrier, the band gap, preventing conduction is limited only by the possibility of ionization. This energy barrier is not particularly large (typically 0.7–3 eV) and can be controlled relatively easily, for example by doping, the deliberate addition of impurities that donate (n-type) or accept (p-type) electrons. (For most semiconductors, thermal excitation is not enough given that $k_{\text{B}} T = 0.025 \, \text{eV}$ at room temperature and semiconductor band gaps can be as large as 3 eV.) Pure semiconductors therefore are insulators at absolute zero, with a conductivity that increases with temperature. Metals, on the other hand, experimentally display a conductivity that decreases with temperature. The difference between insulators and semiconductors is that insulators require an unfeasibly large energy to raise the charge carrier energy to one at which it might conduct.

2.3.1 Lattice periodicity and Bloch theory

The starting point to a theory of band structure is again the Schrödinger equation, but different treatments develop the mathematics in different directions. The *nearly free electron model*, for example, treats the positive charges on the ions in the crystal lattice as a perturbation on the uniform (zero) field in which the electrons can move. The *Kronig-Penney model*, which we shall discuss here, is a more recent development, and includes the explicit addition of a position dependent potential $U(x, y, z)$, which represents the positive charges of the crystal lattice. The Schrödinger equation is then

$$\frac{-\hbar^2}{2m_e}\nabla^2\psi + U\left(r\right)\psi = E\psi. \tag{2.19}$$

Solutions to this equation are straightforward providing that $U(r)$ is periodic, which, given a crystal lattice, must be the case. We then have $U(\mathbf{r}) = U(\mathbf{r}+\mathbf{T})$, where $\mathbf{T}$ is a translation vector on the lattice made up of an integer number of primitive lattice vectors. Solutions to equations of this kind are generally referred to as Bloch functions, after Felix Bloch, who contributed much to our understanding of conductivity during his doctoral work in Leipzig. Bloch demonstrated that eqn (2.19) might have as its solution equations of the form

$$\psi_k\left(\mathbf{r}\right) = u_k\left(\mathbf{r}\right)\exp\left(i\mathbf{k}\cdot\mathbf{r}\right) \tag{2.20}$$

where $\mathbf{k}$ is the wave vector of the electrons, $\mathbf{r}$ is a position on the lattice, $\mathbf{r} = x\mathbf{i} + y\mathbf{j} + z\mathbf{k}$, and $u_k(\mathbf{r})$ is any function with the periodicity of the lattice. The Bloch theorem also states that electron wave functions in crystals must satisfy

$$\psi_k\left(\mathbf{r} + \mathbf{T}\right) = u_k\left(\mathbf{r}\right)\exp\left(i\mathbf{k}\cdot\mathbf{T}\right), \tag{2.21}$$

which, when substituted into eqn (2.20) leads to

$$u_k\left(\mathbf{r} + \mathbf{T}\right) = u_k\left(\mathbf{r}\right) \tag{2.22}$$

and so mimics the periodicity of $U(r)$ in eqn (2.19). It is easy to confuse the function $u_k\left(\mathbf{r}\right)$ with $\psi_k\left(\mathbf{r}\right)$ and assume that $\psi_k\left(\mathbf{r} + \mathbf{T}\right) = \psi_k\left(\mathbf{r}\right)$. The wave function $\psi_k\left(\mathbf{r}\right)$ does not necessarily match the periodicity of the lattice, and is chosen so as to satisfy eqn (2.20), where $u_k\left(\mathbf{r}\right)$ matches the periodicity of the lattice.

To prove the Bloch theorem we need to convert the potential due to the crystal lattice into its Fourier components. This is not difficult; a crystal lattice has periodicity so it can be expanded simply as

$$U\left(\mathbf{r}\right) = \sum_{G>0} U_G\left(\exp\left(i\mathbf{G}\cdot\mathbf{r}\right) - \exp\left(-i\mathbf{G}\cdot\mathbf{r}\right)\right) = 2\sum_{G>0} U_G\cos\left(\mathbf{G}\cdot\mathbf{r}\right). \tag{2.23}$$

where $\mathbf{G}$ is a reciprocal lattice vector. The summation over reciprocal lattice vectors depends significantly only on the smallest values of $\mathbf{G}$. On the reciprocal lattice, the nearest, next nearest, and next next nearest neighbours will usually be more than adequate.[1] It would be equally valid for the cosine to be replaced with a sine, the difference simply

<hr>

[1] The reader should recall that all crystal lattices have their own reciprocal lattices in k-space. Since $\mathbf{T}$ is a lattice vector, the Laue condition, $\exp\left(i\mathbf{T}\cdot\mathbf{G}\right) = 1$, applies.

relates to whether or not an ion is located at the origin. We have chosen to subtract one exponential from the other in order to obtain the sine, which means that $U(0) = 0$. (We note that the conventional practice of $U(\infty) = 0$ would not be helpful here.) Substituting eqn (2.23) into the Schrödinger eqn (2.19) yields

$$\left(\frac{-\hbar^2}{2m_e}\nabla^2 + \sum_{G>0} U_G\left(\exp\left(i\mathbf{G}\cdot\mathbf{r}\right) - \exp\left(-i\mathbf{G}\cdot\mathbf{r}\right)\right)\right)\psi(\mathbf{r}) = E\psi(\mathbf{r}).$$

$$(2.24)$$

The form of $\psi(\mathbf{r})$ may be taken to be

$$\psi(\mathbf{r}) = \sum_k \Psi_k \exp\left(i\mathbf{k}\cdot\mathbf{r}\right). \tag{2.25}$$

Substituting eqn (2.24) into eqn (2.25), we obtain after cleaning up,

$$\frac{-\hbar^2 k^2}{2m_e}\Psi_k + \sum_{G>0} U_G\left(\Psi_{k-G} - \Psi_{k+G}\right) = E\Psi_k. \tag{2.26}$$

The summation in k from eqn (2.25) has disappeared because eqn (2.26) must hold for each value of k. Equation (2.23) presents the potential as a sine wave, giving it the important qualities of being real and periodic. To demonstrate the sine wave, we restricted ourselves to positive G. However, if we remove the restriction that $G > 0$ and allow all G, in the same way that we allowed all k in eqn (2.25), we can eliminate ψ_{k+G} and simplify eqn (2.26) to give

$$\left(\frac{-\hbar^2 k^2}{2m_e} - E\right)\Psi_k + \sum_G U_G\Psi_{k-G} = 0. \tag{2.27}$$

The importance of eqn (2.27) is that any solution for $\psi(\mathbf{r})$ must relate Ψ_{k-G} and Ψ_k. This means that if we take one Fourier component of $\psi(\mathbf{r})$, which, in keeping with our notation, we denote $\psi_k(\mathbf{r})$, then

$$\psi(\mathbf{r}) = \sum_G \Psi_{k-G} \exp\left(i\left(\mathbf{k} - \mathbf{G}\right)\cdot\mathbf{r}\right). \tag{2.28}$$

We can now take u_k from eqn (2.20) and write

$$u(\mathbf{r}) = \sum_G \Psi_{k-G} \exp\left(-i\mathbf{G}\cdot\mathbf{r}\right), \tag{2.29}$$

enabling us to rewrite eqn (2.28) as

$$\psi_k(\mathbf{r}) = u_k(\mathbf{r})\exp\left(i\mathbf{k}\cdot\mathbf{r}\right), \tag{2.30}$$

which is identical to eqn (2.20). We can easily demonstrate eqn (2.29) to satisfy the periodicity requirement of eqn (2.22) by substituting in $u_k(\mathbf{r} + \mathbf{T})$ and using the Laue condition. By obtaining eqn (2.29) we have been able to prove the Bloch theorem. We can now formally state Bloch's theorem:

If $\psi(\mathbf{r})$ is a single electron solution to the Schrödinger eqn (2.19), then it must have the form given by eqn (2.20) for a reciprocal lattice wave vector k, where the function $u_k(\mathbf{r})$ is given by eqn (2.22).

Although Bloch theory is presented as building on the free electron model, there are a couple of important differences that should be noted. In the free electron theory, the wave vector $\mathbf{k}$ may be treated as having a complete range of positive and negative values, but in the Bloch theory, we do not need to consider wave vectors with values $|k| > 2\pi/a$. Formally, this means that Bloch theory is restricted to the *first Brillouin zone*. Another important difference between Bloch theory and the free electron model concerns the electron momentum. Since the free electron model treats electrons as non-interacting particles it is reasonable to attribute a momentum $\hbar\mathbf{k}$. In the Bloch theory, $\hbar\mathbf{k}$ is known as the *crystal momentum* (or the *quasi-momentum*) and should not be considered as a physical momentum of an electron. In fact $\mathbf{k}$ is mathematically used as a means of referring to the electron location within the first Brillouin zone.

2.3.2   The Kronig–Penney model

The Bloch model provides us with rules for the behaviour for electrons in a periodic potential. We need to consider electronic behaviour in different parts of the lattice. What are the electrons doing in the neighbourhood of the ion cores? What about the region between these cores? To achieve such an analysis, we need a model for the potential, and the Kronig–Penney model gives us a rather simple model that we can use to explain the origin of the semiconductor band gap by solving the Schrödinger equation, whilst simultaneously satisfying the requirements of the Bloch theorem. The starting point is to assume that the ion cores are described by periodic square wells, which can be approximated to delta functions if each ion is assumed to take negligible space on the lattice.

The square well that we use is illustrated in Fig. 2.5. For simplicity, we restrict ourselves to one dimension (x) here. The potential $U(x) = 0$ exists where $0 < x < a$, and $U(x) = U(x + a + b)$ must also apply to reflect the periodicity of the potential. Where $a < x < b$, we have $U(x) = U_0$. Clearly $U(x) = 0$ corresponds to regions where the potential is lower and therefore to the ion centres, which have an electrostatic attraction for the electrons. In the region where $U(x) = 0$, we can have solutions to the Schrödinger equation given by

$$\psi_0(x) = A \exp(ik_0 x) + B \exp(-ik_0 x), \qquad (2.31)$$

and similarly, when $U(x) = U_0$,

$$\psi_1(x) = C \exp(ik_1 x) + D \exp(-ik_1 x), \qquad (2.32)$$

where the subscripts 0 and 1 indicate the potential being zero or U_0. The solutions to eqns (2.31) and (2.32) are those for travelling waves,

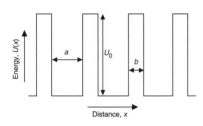

Fig. 2.5 Electrons cannot be considered free because they have to navigate a periodic potential caused by the lattice ions. In semiconductors this is important because it gives rise to energy bands. A simple periodic potential that gives rise to energy bands is a square well potential. The fixed locations of the ions raise the electronic energy by U_0 over a distance b, which largely confines the electrons to regions between the ions of width a. Clearly, the potential barriers presented by lattice ions do not have a square well form, but the simplicity of this model does allow an understanding of the origin of semiconductor behaviour.

which would have the periodic boundary conditions of the ring shown in Fig. 2.1. There are two sets of boundary conditions to the waves: they must be both continuous and smooth at the junction of the two potentials. This means that both the wave functions and their gradients must be equal at the boundary. At $x = 0$, these can be stated as

$$A + B = C + D \tag{2.33}$$

and

$$k_0 \left(A - B \right) = k_1 \left(C - D \right). \tag{2.34}$$

We can also apply the boundary conditions at $x = a$ to give

$$A \exp \left(i k_0 a \right) + B \exp \left(-i k_0 a \right) = \\ \left(C \exp \left(-k_1 b \right) + D \exp \left(k_1 b \right) \right) \exp \left(-k \left(a + b \right) \right). \tag{2.35}$$

Notice that in eqn (2.35) we have used the Bloch requirement that $\psi \left(x \right) = \psi \left(x + a + b \right) \exp \left(-ik \left(a + b \right) \right)$, so, although we have evaluated the left hand side of the equation at $x = a$, the right-hand side has been evaluated at $x = -b$. Such a procedure is necessary to impose the periodicity on the boundary conditions. We can also apply the same logic to the gradient in the wave function at $x = a$:

$$k_0 A \exp(i k_0 a) - B \exp(-i k_0 a) = \\ k_1 (C \exp(-k_1 b) - D \exp(k_1 b)) \exp(-k(a + b)). \tag{2.36}$$

These equations can be solved in a matrix formulation;

$$\begin{pmatrix} 1 & 1 & 1 & 1 \\ k_0 & -k_0 & k_1 & -k_1 \\ \exp(i k_0 a) & \exp(-i k_0 a) & \exp(-i(k_1 b + k(a + b))) & \exp(i(k_1 b - k(a + b))) \\ k_0 \exp(i k_0 a) & -k_0 \exp(-i k_0 a) & k_1 \exp(-i(k_1 b + k(a + b))) & -k_1 \exp(i(k_1 b - k(a + b))) \end{pmatrix} \begin{pmatrix} A \\ B \\ C \\ D \end{pmatrix} = 0 \tag{2.37}$$

which can be rewritten by setting the determinant of the 4×4 matrix equal to zero. The solution to this equation requires simple algebra but is rather time-consuming to do with pen and paper. The result is

$$4 k_0 k_1 \cos \left(k \left(a + b \right) \right) - \\ \left(k_0 + k_1 \right)^2 \cos \left(k_0 a + k_1 b \right) + \left(k_0 - k_1 \right)^2 \cos \left(k_0 a - k_1 b \right) = 0. \tag{2.38}$$

Next we solve the Schrödinger eqn (2.19) with $U(x) = 0$ to obtain

$$\frac{\hbar^2 k_0^2}{2 m_e} = E \tag{2.39}$$

and with $U(x) = U_0$ to obtain

$$\frac{\hbar^2 k_1^2}{2 m_e} = E - U_0. \tag{2.40}$$

By subtracting eqn (2.40) from eqn (2.39) we have a relationship between k_0, k_1, and U_0 given by

$$\frac{\hbar^2 \left(k_0^2 - k_1^2 \right)}{2 m_e} = U_0, \tag{2.41}$$

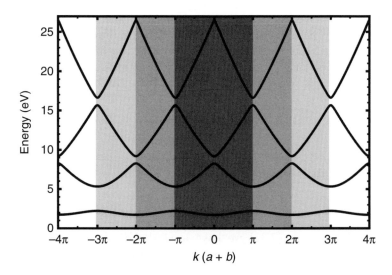

Fig. 2.6 The dispersion relation for the Kronig–Penney model, revealing the first three band gaps. The simulation here is for a periodic potential of 0.5 nm, half of which is at $E = U_0 = 5.0$ eV and the other half $E = 0$. The first Brillouin zone at $-\pi < k(a + b) < \pi$ is shown in the darkest shading, which decreases in darkness towards for further Brillouin zones. The fourth Brillouin zone is not shaded.

which now enables us to plot the dispersion relation linking k and the electron energy, which is shown in Fig. 2.6. To achieve this one needs to solve eqn (2.38) using eqns (2.39) and (2.41). It is sometimes written that the free electron model intersects the Kronig–Penney model, with the lowest energy dispersion relation meeting the free electron model at the edge of the first Brillouin zone $(k(a + b) = \pi)$, and the next second dispersion relation crossing the free electron model at the edge of the second Brillouin zone $(k(a + b) = 2\pi)$, and so on. In fact this is only true when $b \to 0$ and $U_0 \to \infty$.

The link between the free electron model and band theory can be considered by plotting together the dispersion relations obtained for the two models. If we set $b = 0$ in the Kronig-Penney model, only at the limits of each Brillouin zone do we see departures from $E \propto k^2$. Metallic behaviour is accommodated by how much these bands are filled. If one considers the first Brillouin zone, the electrons (or holes) would have a lattice defined by a primitive cell (the minimum volume that when repeated in three dimensions defines the structure of the crystal) of size a^3 or, as necessary, $(a + b)^3$. The allowed wave vectors of the charge carriers is then given in each of the three dimensions by $k = \pm 2\pi/L$, $\pm 4\pi/L$, and so on. Here, L is the size of the crystal and the limits on k are such that periodic boundary conditions are satisfied (Section 2.2). The maximum value of k in the first Brillouin zone is $\pm \pi/a = \pm N\pi/L$ for N charge carriers, where $N = La$. Taking the positive and negative solutions means that there are $2N$ possible wave vectors in the first Brillouin zone. If we have a monovalent metal (e.g. sodium) then the the density of states allows conduction because electrons can be excited into higher energies because there are twice as many states as there are charge carriers (Fig. 2.7). A divalent metal, such as calcium, will have this Brillouin zone completely full because each of the $2N$ carriers in their lowest energy state will correspond to the available wave vectors. However, such metals have negative band gaps, which means that the minimum energy associated with second Brillouin zone is lower than the Fermi level for that metal (Fig. 2.7). When the minimum energy of the second Brillouin zone exceeds the (full) level we have semiconducting (or insulating) behaviour. The reader should be careful in the use of Fermi levels to describe semiconducting behaviour. In a semiconductor, the Fermi level is generally (but not exclusively) located at the midpoint of the band gap. The filled region in an inorganic semiconductor, which does not contribute to conduction, is known as the valence band, or the highest occupied molecular orbital (HOMO) in organic semiconductors. The lowest energy of the other side of the band gap is known, in an inorganic semiconductor, as the valence band, or the lowest unoccupied molecular orbital (LUMO) in organic semiconductors. The Fermi energy is often defined as the highest occupied energy state at absolute zero, but with energy bands this is inappropriate. One may consider the Fermi energy to be the energy at which the occupancy (i.e Fermi–Dirac distribution) is 0.5. This changes for metals at absolute zero, since the occupancy falls precipitously from unity to zero at the Fermi energy and the probability of occupancy of the Fermi level is 0.5 for all $T > 0$. For a semiconductor, the Fermi energy must fall in the middle of the band (Fig. 2.8). If the semiconductor is doped, so that there is an excess of electrons, the Fermi energy moves towards the conduction band. If the semiconductor is doped p-type, then the Fermi energy is shifted towards the valence band.

The electronic structure that we can calculate from the Kronig-Penney model (Fig. 2.6 shows how a *direct band gap semiconductor* might look.) In fact, there are semiconductors with indirect band gaps. In an indirect band gap, the peak of a valence band does not correspond to the minimum of the nearest conduction band, but it is shifted by a certain wave vector. This complicates semiconductor physics. Fortunately for us, conjugated polymers are generally direct band gap materials, so we shall not consider this subject in any more detail.

2.4 Energy bands in polymers

The band gap in polymers is more complicated than in inorganic semiconductors, and in other ways it is easier to understand because, for example, we need not concern ourselves with indirect band gaps and related phenomena. Ultimately, however, polymers are rarely pure, which complicates much of the physics described in this section and mean that a full understanding of the semiconducting properties of polymers is lacking, at least in comparison to inorganic semiconductors. The description provided below implies that alternating π and σ bonds, the conjugation, is all that matters, but in fact other mechanisms of charge transport are also important (Section 5.2). Nevertheless, we need to understand the origin of the band gap in these conjugated systems, and so we need to consider the behaviour of molecular orbitals. The chemistry of bonding in polymer electronics is based around the principles of what is known

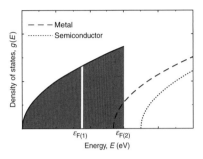

Fig. 2.7 The density of states within the lowest energy band is full for a divalent metal and a semiconductor. In the case of a divalent metal the Fermi level $\epsilon_{F(2)}$ exceeds the lowest energy of the second band and so no additional energy is required to excite an electron into this second band. For a semiconductor the minimum energy of the second band exceeds $\epsilon_{F(2)}$ by a small amount. If this difference in density of states (i.e. the band gap) is greater than 3 eV one considers the material to be an insulator. For monovalent metals, conduction can take place entirely within the first energy band, because the density of states is only half-filled, i.e. as far as $\epsilon_{F(1)}$.

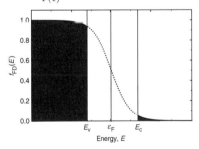

Fig. 2.8 The Fermi–Dirac distribution is continuous in energy, but in semiconductors there is a band gap delimited by E_v and E_c. At $T > 0$, electrons may be excited above the band gap, if the temperature is large enough. Here, the shaded region shows the occupancy as a function of energy. The Fermi energy ϵ_F is situated in the middle of the band.

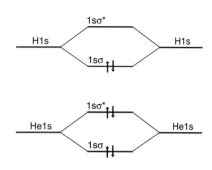

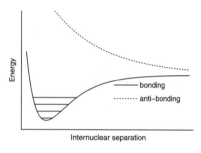

Fig. 2.9 Energy levels of the $1s\sigma$ states of H_2 (top) and He_2 (bottom) with respect to the ground (1s) states of the individual atoms. The $1s\sigma_z{}^*$ state of He_2 is an anti-bonding orbital, and is less stable than the two individual atoms.

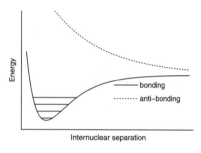

Fig. 2.10 Bonding orbitals give rise to a minimum in the energy of the system at the equilibrium distance between two nuclei in a diatomic molecule such as H_2. No such minimum exists in the case of the anti-bonding orbital. For such an orbital the energy is continually lowered until the separation of the two atoms is infinite. In the bonding orbital the atoms vibrate around the mean internuclear separation (four vibrational energy levels are shown) but in the anti-bonding orbital the two atoms would separate without any electronic transition involved.

as the *linear combination of atomic orbitals*, with which we start this discussion.

2.4.1 Linear combination of atomic orbitals

The reason two atoms bond to form a molecule is to lower their overall energy. If they cannot lower their energy by bonding, they will not bond. Sometimes *activation energies* need to be overcome in order for bonding to occur. We first consider H_2, because of its simplicity. In Fig. 2.9, we show the energy-level diagrams for two isolated hydrogen atoms and for the combined molecule. There are two levels shown in the combined molecule. The lower energy state is for a stable molecule. The state of greater energy is a valid solution to the Schrödinger equation but leads to the molecule dissociating. These are known as electron bonding and anti-bonding orbitals respectively. In Fig. 2.10 we show potential energy as a function of distance for the two situations. Both curves include a repulsive term due to internuclear repulsion. The stable curve includes an attractive term due to the electrons being shared by the molecule, whilst the unstable curve has an additional repulsive term due to the electrons being located outside the two nuclei. (In larger diatomic molecules the internuclear repulsion is supplemented by repulsion between core electrons.) The stable state consists of the two electrons between the nuclei, and the unstable state consists of the two electrons outside the two nuclei. Specifically, the most probable location of the two electrons is between or outside the two nuclei. The linear combination of atomic orbitals model suggests that each electron belongs to both nuclei.

The linear combination is the addition of the orbitals (wave functions) for each atom, so we have the electronic wave function given by $\psi = \phi_A + \phi_B$ for the stable molecule and $\psi = \phi_A - \phi_B$ for the unstable molecule. Since the electrons are shared between the two nuclei, the stable situation corresponds to the most probable position of the electrons being between the two nuclei. In order to satisfy the exclusion principle, the electrons have opposing spins. This may seem counterintuitive: why do the electrons not repel each other? In fact, the system is stable with the two electrons between the nuclei because the electrons are shared between the two nuclei. Because each electron belongs to *both* nuclei, the molecule will not dissociate. If the two electrons are most likely on opposite sides of the nuclei, electron A has negligible interaction with nucleus B, and *vice versa*. Such a situation would leave to the dissociation of the molecule into two hydrogen atoms because the two protons can repel each other. It is necessary that the most probably location of the electrons be situated on a line passing through the centre of both nuclei in order that molecule have no dipole moment. (For symmetry reasons all homonuclear diatomic molecules have no dipole moment.) Similarly, the mean position of the electrons must be centrally located between the two nuclei.

We can extend the model to the diatomic helium molecule, He_2 (Fig. 2.9).

The first two 1s electrons of helium are allocated the lowest energy state, which is the bonding orbital. The remaining two 1s electrons cannot be placed in this orbital on account of the exclusion principle. They are therefore placed in the anti-bonding orbital, which explains why He_2 is not a stable molecule. (Generally, anti-bonding orbitals have a larger contribution towards instability than bonding orbitals do towards stability, so the first two 1s electrons in the bonding orbital cannot compensate for the two 1s electrons in the anti-bonding orbital.)

We now consider how carbon forms bonds. Carbon atoms have their six electrons with both 1s and 2s orbitals filled. That leaves two electrons in the p orbitals, which are denoted by $2p_x$ and $2p_y$. These are in principle both available for bonding, whereas the 1s and 2s states are filled. We could put two electrons in, say $2p_x$, leaving $2p_y$ empty, but this would not be able to form covalent bonds. However, even taking this into account, such a structure only leaves two 2p electrons available for bonding, but carbon is well known to be contribute four electrons for the formation of bonds.

There are three alternatives for bonding in carbon depending on the molecule formed. These involve the *hybridization* of the 2p and 2s orbitals, the form of which is different for the alkanes (sp^3 hybridization), alkenes (sp^2 hybridization), and alkynes (sp hybridization). We consider that for the alkenes and the sp^2 hybrid orbitals first. One of the 2s electrons is 'promoted' to a p_z state in order to participate in the hybridization. The other 2p electrons and the remaining 2s electron hybridize from the linear combination of 2p and 2s wave functions forming orbitals with an energy in-between those of the 2p and 2s orbitals. There are three sp^2 orbitals and one p_z orbital. This p_z electron did not participate in the hybridization, and its nomenclature as p_z is merely to distinguish it from the p_x and p_y electrons that were used to form the sp^2 hybrid orbitals. The three sp^2 hybrid orbitals are planar, and separated by an angle of 120°. By contrast, in methane, there are four sp^3 hybrid orbitals in a tetrahedral structure. Here, a 2s electron is promoted to a p_z orbital, and the remaining 2s electron and the three 2p electrons hybridize to form four sp^3 hybrid orbitals.

If we combine two sp^2-hybridized carbon atoms we get a bonding situation appropriate for ethylene (Fig. 2.11a). One hybrid orbital of each atom is used to form a bond with the other carbon atom (Fig. 2.12), each orbital contributing one electron; the two other hybrid orbitals are used to bond the hydrogen atoms. This leaves the 2p electron which is used to form the rest of the covalent C=C bond. The three sp^2 hybridized orbitals of each carbon atom are used to form σ bonds, whereas the $2p_z$ orbitals overlap to form π bonds. The σ bonds are in the plane of the molecule, whereas the π bonds are perpendicular to this plane.

The other bond in ethylene, which is of a higher energy than the σ bonds, is the π bonds. The p_z electron associated with each carbon atom sits perpendicular to the sp^2 orbitals, above the plane of the molecule. In ethylene the two p_z electrons are close enough to form their own bond, known as the π bond. This bond is situated above and below the planar

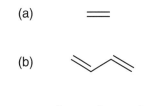

Fig. 2.11 Chemical structures for alkenes of increasing complexity: (a) ethylene (ethene), (b) butadiene, and (c) hexatriene.

Fig. 2.12 The carbon atoms in ethylene each contribute three sp^2 hybrid orbitals. Two of these are used to bond with the hydrogen atoms, and the remaining one is used to form one of the covalent bonds between the two carbon atoms. All three resulting σ bonds are coplanar.

ethylene molecule and is due to the overlap of the p_z electronic wave functions. The bond is weak, at least compared to the σ bonds, and it is this weakness that makes ethylene and other unsaturated organic materials relatively reactive. A π bond formed using two p_y orbitals also exists at higher energies, and is the highest-energy stable configuration of ethylene. Above these energies is an energy gap before the anti-bonds formed from pairs of either p_z or p_y electrons. The difference between these *frontier orbitals* is the organic equivalent of the semiconductor band gap; the energy gap separates the lowest unoccupied (π bonds) and highest occupied (π^* anti-bonds) molecular orbitals, which are known as the LUMO and HOMO respectively (Fig. 2.13). Note that there are no p_x orbitals forming π bonds in ethylene. These cannot exist because a π bond cannot be parallel to a σ bond, which for the C–C bond, exists along the x-axis.

The difference in energy between the two frontier orbitals in ethylene is greater than 4 eV (in fact it is 7.6 eV), which is so large that ethylene cannot be considered a semiconductor. We now ask ourselves, what happens if we increase the complexity of the molecule by extending the chain? The next molecule is butadiene, which has four carbon atoms. The chemical structure for butadiene is included in Fig. 2.11. Butadiene, like ethylene, has all of its σ bonds filled, but also has four π bonds, including two π_z bonds and two π_z^* anti-bonds, and again, each of these π_z and π_z^* bond-anti-bond pairs corresponds to a linear combination of two p_z electrons. Hexatriene, unsurprisingly has another two π bonds, with six in all. These correspond to three π_z bonds and three π_z^* anti-bonds, reflecting a general rule that for $(CH)_{2N}$, there are N π_z bonds and N π_z^* anti-bonds. Each of these bonds means an energy level in the HOMO and each of the anti-bonds means an energy level in the LUMO. By adding $(CH)_2$ units to the growing chain, we add an extra energy level in both the HOMO and LUMO. These extra bonds in the LUMO are both higher and lower in energy than the smaller alkenes. The same is also the case for the HOMO. The energy levels gradually become closer together—a phenomenon which continues until the infinite polymer chain band gap is approached (Fig. 2.13). In this way, increasing complexity reduces the band gap.

If we consider acetylene, sp hybridization is relevant. Each carbon atom has six electrons in total, and two of these fill the 1s state. The next two electrons are allocated to the 2s state, which in a linear combination for a diatomic molecule is similar to the 1s electrons; the two electrons are either located between the two nuclei, for a bonding orbital, or are located on opposite sides of the nuclei, for an anti-bonding orbital. In the sp hybridized system, the 2s and 2p orbitals hybridize to form two 2p orbitals ($2p_y$ and $2p_z$), and two sp orbitals; again the two 1s electrons remain unaffected. The remaining two electrons are the p electrons, and these do not all sit on a line joining the nuclei, as is the case for the s electrons (p_x electrons would sit on a line between the two nuclei, but they cannot be used to form π bonds). The 1s, 2s, and $2p_x$ electron energy states, when combined to form molecular orbitals, make up σ

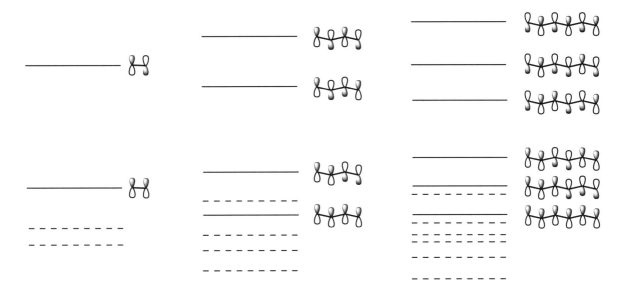

Fig. 2.13 Ethylene (left) has one π bond and one π^* anti-bond (both represented by solid lines) separated by a band gap of close to 7.6 eV rendering it an insulator. The extension to butadiene adds another π bond and anti-bond. The two π bonds and π^* anti-bonds are at lower and higher energies than their equivalents in ethylene, with a consequently reduced band gap to 5.7 eV. Another bond and anti-bond appears for the hexatriene molecule, causing a yet further decrease in the band gap to 4.9 eV. The dashed lines represent the σ bonds, and the solid lines the π bonds.

bonds. Anti-bonding orbitals of this type are possible and are referred to as σ^*-(anti)bonds.

For *trans*-polyacetylene, sp^2 hybridization is important. In Fig. 2.14 we show the location of the 2p$_z$ electrons used to form the π bonds, which are out of the plane of the chain; the probability of finding this electron on the main chain is zero; a characteristic of these orbitals. If one-dimensional metallic behaviour were possible, these orbitals would strongly overlap and the 2p$_z$ electrons would be delocalized.

In the case of polyacetylene, the three hybridized electrons and the remaining 2p$_z$ electron are relevant for its bonding. If we consider a chain of 2N carbon atoms (i.e. N monomers), we have 8N orbitals, which fortunately do not cause insurmountable problems in an analytical treatment because we can take them to be lying in a periodic potential by metaphorically tying together the ends of the chain, to create a ring, as in Fig. 2.1. We collect in eqn (2.42) the atomic orbitals together for each carbon atom to form the molecular orbital,

$$\phi(\mathbf{r}) = \sum_i u_i(\mathbf{r})\phi_i(\mathbf{r}). \qquad (2.42)$$

The coefficients $u_i(\mathbf{r})$ reflect the symmetry of the molecule. The molecular orbital can then be summed around the chain to give the wave function

$$\psi(k) = \sum_{n=1}^{N} \phi_n \exp(ikna). \qquad (2.43)$$

Fig. 2.14 As for the smaller alkenes shown in Fig. 2.13, the 2p$_z$ electrons of polyacetylene are located out of the plane of the molecule. Positive wave functions are shown shaded, and negative ones unshaded. The lowest energy state of these has the electron wave functions alternating so that the neighbours of a negative wave function are positive and *vice versa*. This is shown in the lower image and corresponds to a π bond. In the upper image, positive and negative wave functions align adjacent to each other. This is a π^* configuration, and is a higher energy state.

Here we do not need to specify vectors for the summation, because the chain is one-dimensional. The periodic potential created by the series of $2N$ carbon atoms has periodic boundary conditions that require there to be no discontinuity in the wave function circumnavigating the ring. This means that we can define a potential that is, at $n = 0$ identical to that at $n = N$, and consequently, at $n = 1$ identical to that at $n = N+1$. (The $2N$ carbon atoms fit onto a potential of periodicity N because the potential repeats over every two carbon atoms; single and double bond conjugation means that the periodicity cannot be repeated over every carbon atom.) As a result

$$\exp(ikNa) = 1, \qquad (2.44)$$

has the effect of quantizing k, because eqn (2.44) is only satisfied when

$$kNa = 2\pi j, \qquad (2.45)$$

where j is an integer. Here, a is not a carbon–carbon bond but a repeat unit (monomer) length and so is the distance between every second carbon atom on the chain backbone. The physics presented here is very rich, but a little effort dedicated to its understanding yields great rewards. We are considering a single polymer of N monomers and the electronic wave functions must be extraordinarily complex, but we have in fact reduced it to a much more digestible function (eqn 2.43), which has N allowed values of k. We discard the additional solutions provided by adding $2\pi/a$ to k because these have no physical meaning; there need not be an infinite number of solutions for each atomic orbital in eqn (2.43). The values of k that we consider are restricted to the first Brillouin zone, just as we noted in our discussion of Bloch theory in Section 2.3.1. In fact, the treatment of the wave functions in polyacetylene and other conjugated polymers is another form of Bloch theory. We do not have a three-dimensional periodic crystal lattice, but our polymer chain is periodic in one-dimension. Similarly, the de Broglie relation applies and $\hbar k$ is the crystal momentum of an electron, which is also equal to h/λ. We can then define the first Brillouin zone to lie in the k-range $-\pi/a < k < \pi/a$.

We recall that the probability of an electron in a given location is proportional to $\psi\psi^*$ and also that in a simple diatomic molecule, such as H_2, bonding depends on the location of the electrons with respect to the two nuclei. Putting these two points together, we see that wave functions given by eqn (2.43) tell us about bonding in the polymer. As a first step in this direction we consider the effect of *nodes* in the bonding of polymers. A node is a point on a standing wave whose amplitude does not vary. When $k = 0$ there are no nodes, because the wavelength is infinite. Usually (but not always) $k = 0$ corresponds to a strongly bonding wave function, and $k = \pi/Na$ a strongly anti-bonding wave function. In general, the number of solutions to the Bloch functions is the same as the number of electrons available to contribute to conduction; this is the same as in benzene whereby the number of solutions to the Bloch functions is simply the number of delocalized electrons, i.e. six.

An important point concerns the value of N used in the above discussion. It is tempting to use N as the number of monomer units in an entire polymer chain, but in fact a chain of N monomers ($2N$ carbon atoms) refers only to that part of the polymer chain that is not deformed in any way. Here, N corresponds to the number of monomers in one *conjugation length*, which is the length over which band transport (i.e. the transport of electrons along the chain) is applicable. Better chemistry generally results in longer conjugation lengths and better hole or electron mobilities along the chain.

2.4.2   Energy bands

The total energy of the electron, unperturbed by any periodic potential, follows from the de Broglie relation,

$$E = \hbar^2 k^2 / 2m_e. \tag{2.46}$$

Here the electrons can be considered free, with a parabolic potential as a function of k, and are treated the same way as free electrons in metals. The periodic potential for a conjugated polymer system can be represented as in Fig. 2.15, and is given by the Su–Schrieffer–Heeger model after those responsible for its development. Electron energies within the band gap (E_g) are forbidden, but electrons whose energies lie in bands above the gaps are free to conduct. The principle here is the same as that for the Kronig–Penney model described in Section 2.3.2, but there are some differences. For example, in the band theory of inorganic semiconductors, the Brillouin zones can be related to the reciprocal lattice. Conjugated polymers, whose conduction can be treated as one-dimensional, cannot be described in such a manner. The Su–Schrieffer–Heeger model is simple compared to other models, and predicts an energy for the electron given by,

$$E = \epsilon_F \pm \sqrt{E_0^2 \cos^2\left(ka\right) + \left(E_g/2\right)^2 \sin^2\left(ka\right)}, \tag{2.47}$$

where $2E_0$ is the band width, and the Fermi level (ϵ_F) is located at the centre of the band gap.

In fact, that there is a band gap in a polymer such as polyacetylene is relatively easy to understand. The π electrons we can imagine as being delocalized and so free to move around the backbone of the polyacetylene. However, in polyacetylene the bonding is of alternating single and double bonds. It turns out that this has a lower energy than delocalizing the π electrons. Alternating single and double bonds means that there are separate bond lengths involved, for C–C and for C=C bonds (This explains why the polyacetylene monomer has the chemical formula C_2H_2 rather than CH.) The energy required to switch the C–C and C=C bonding around is the band gap for polyacetylene, which is approximately 1.5 eV.

Polyacetylene is an example of an unsaturated polymer, which refers to a polymer containing carbon–carbon (C=C) double bonds; conjugated polymers (i.e. those containing alternating single and double bonds

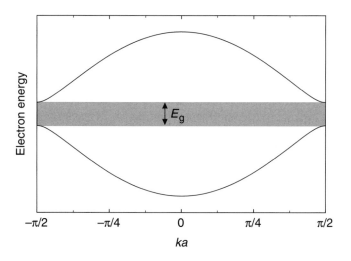

Fig. 2.15 The Su–Schrieffer–Hegger theory (tight binding approximation) predicts a periodic potential for polyacetylene for polyacetylene with the form given by eqn (2.47). The band gap, $E_g = \epsilon_L - \epsilon_H$, where ϵ_H and ϵ_L are the (energy) limits of the HOMO and LUMO respectively. The Fermi level (ϵ_F) lies in the middle of the band gap.

–C=C–C=...) are always unsaturated, but not all unsaturated polymers (e.g. polybutadiene) are conjugated. These double bonds, when located along the chain backbone, usually convey a greater degree of stiffness to such polymers that is not found in polymers containing only carbon–carbon single bonds such as polystyrene or polyethylene, which are examples of saturated polymers. (Saturation does not refer solely to polymers; smaller organic molecules may also be saturated.)

By solving the Schrödinger equation for a periodic function, we have seen in Section 2.3.2 how bands can originate in semiconductors. We have seen from the free electron model how mobile electrons can exist and be available for conduction. These ideas will be compared and contrasted with the behaviour in polymers. For metals, it was sufficient, at least in simple theory, to consider the valence electron as free to move in a periodic potential. In metals, conduction is due to the d electrons, and these form part of the σ bonds. In a polymer the σ bonds remain untouched, even when doped. The polymer backbone remains unaffected by the electronic behaviour, which could not be the case if the σ bonds were responsible for electronic transitions. Ultimately, the extended π system along chain backbones (conjugation) is responsible for electronic transitions in polymers, providing the crucial difference between the electronic behaviour of organic and inorganic semiconductors.

2.4.3   Su–Schrieffer–Heeger theory

The origin of the band gap in polyacetylene has been elegantly explained in a comparatively simple theory, which considers a *hopping potential* describing the energetic costs of moving an electron from one monomer to another. The full energy also includes terms due to the potential energy of stretching or compressing the bonds (i.e. neighbouring monomers are treated as a spring) and the corresponding kinetic energy. The question as to how a band gap arises becomes much more tractable when the chain is considered to be perfectly dimerized. This means that we consider two CH units so that we can work with two different bond lengths, and we assume these to be the same around the rest of the chain. The undimerized chain, by contrast, would contain one equilibrium bond length and there would be no bond alternation.

Su–Schrieffer–Heeger theory requires that the electrons interact with the lattice through *electron–phonon* coupling. This means that the electrons cannot move independently of the chain backbone. An important assumption by Su, Schrieffer, and Heeger is that electrons do not interact with each other.

We start by writing the energy summed over each of the N CH groups making the chain,

$$H = -\sum_n \left(t_0 + (-1)^n 2\alpha u\right) \left(c_{n+1}^\dagger c_n + c_n^\dagger c_{n+1}\right) + 2NKu^2, \quad (2.48)$$

where we have used the nomenclature from the original paper: u is the perturbation to the bond spacing in the unperturbed system, i.e. the difference in bond length between the (longer) single and (shorter) double bond is $2u$; K is an effective spring constant representing the behaviour of σ bonds ($2NKu^2$ is an elastic energy due to bond distortion); t_0 is an energy associated with moving electrons from one monomer to its neighbour; α is an electron–phonon coupling parameter, and accounts for the perturbation due to different bond lengths; and c_n and $c_{n+1}^\dagger$ are fermion (i.e. here π or $2p_z$ electrons) annihilation and creation operators, which allow for electron movement from monomer to monomer. As such $t_0 \left(c_{n+1}^\dagger c_n + c_n^\dagger c_{n+1}\right)$ represents the transfer of an electron between one π orbital to another, which consequently means the transfer of an electron between different sites, because different π orbitals are on different carbon atoms. Strictly H is a Hamiltonian, because it is an operator (or the sum of operators). These can overcomplicate this discussion, but the reader need not fear the concept of these operators in this discussion; the creation and annihilation operators, which recognize that when one electron arrives at one CH unit, it is lost (annihilated) from its neighbour. These operators allow the Schrödinger equation to be written in a simpler form, and in this respect eqn (2.48) is simply the time-independent Schrödinger equation for π electrons in dimerized polyacetylene.

Moving electrons from one site to another is a manifestation of their kinetic energy, but a Hamiltonian requires a potential energy term. As well as the elastic energy due to bond distortion, there is also a term due to the alternating double and single bonds having different bond lengths. Because the π bonds are not here located on an aromatic ring with similar bond lengths either side of the carbon atom, the energy of the system is changed, and this perturbation due to nearest-neighbour hybridization has an associated energy given by

$$\Delta_n = -\alpha \left(u_{n+1} - u_n\right) = -2\left(-1\right)^n u = (-1)^n \Delta, \quad (2.49)$$

where $\Delta = -2u$, and $2u = u_{n+1} - u_n$; u_n is the deviation from the bond length in the linear (undimerized) chain, and may take positive or negative values depending on whether the bond to which refers is a single ($u_n > 0$) or a double ($u_n < 0$).

The chain is held together by σ bonds which are always present, and so they are not associated with creation and annihilation operators. Nevertheless, they are affected by bond distortions and so also contribute a potential energy of $K\left(u_{n+1} - u_n\right)^2/2$, which is $2NKu^2$ over the whole chain.

Polyacetylene density of states

The density of states, ρ is the number of electronic states at a given energy, or $\mathrm{d}N/\mathrm{d}E$. However, we can use the periodic boundary conditions described in Fig. 2.1 and treat the polymer chain as a circle of radius r. In this case, $N = 2\pi r/a$, where $2a$ is the size (length) of the dimer. Using this logic,

$$\rho = \frac{\mathrm{d}N}{\mathrm{d}E} = r\frac{\mathrm{d}k}{\mathrm{d}E} = \frac{Na}{2\pi}\frac{\mathrm{d}k}{\mathrm{d}E} = \frac{Na}{2\pi\mathrm{d}E/\mathrm{d}k}. \tag{2.50}$$

We then simply take the derivative of eqn (2.47), defining $E_k = E - \epsilon_\mathrm{F}$ to simplify the working,

$$\frac{\mathrm{d}E}{\mathrm{d}k} = \frac{\mathrm{d}E_k}{\mathrm{d}k} = \frac{1}{E_k}\left(\epsilon_k\frac{\mathrm{d}\epsilon_k}{\mathrm{d}k} + \Delta_k\frac{\mathrm{d}\Delta_k}{\mathrm{d}k}\right), \tag{2.51}$$

where we define $\epsilon_k = E_0\cos\left(ka\right)$ and $\Delta_k = \left(E_\mathrm{g}/2\right)\cos\left(ka\right)$. We then have

$$
\begin{aligned}
\frac{\mathrm{d}E}{\mathrm{d}k} &= \frac{\sin\left(ka\right)\cos\left(ka\right)}{E_k a}\left(\left(\frac{E_\mathrm{g}}{2}\right)^2 - E_0^2\right) \\
&= \frac{1}{E_k a}\left(\Delta_k^2\cot\left(ka\right) - \epsilon_k^2\tan\left(ka\right)\right) \\
&= \frac{a}{E_k}\sqrt{\left(\Delta_k^2\cot\left(ka\right) - \epsilon_k^2\tan\left(ka\right)\right)\left(\Delta_k^2\cot\left(ka\right) - \epsilon_k^2\tan\left(ka\right)\right)} \\
&= \frac{a}{E_k}\sqrt{\left(\Delta_k^2\cot^2\left(ka\right) - \epsilon_k^2\right)\left(\Delta_k^2 - \epsilon_k^2\tan^2\left(ka\right)\right)} \\
&= \frac{a}{E_k}\sqrt{\left(\Delta_k^2\operatorname{cosec}^2\left(ka\right) - E_k^2\right)\left(E_k^2 - \epsilon_k^2\sec^2\left(ka\right)\right)} \\
&= \frac{a}{E_k}\sqrt{\left(\left(\frac{E_g}{2}\right)^2 - E_k^2\right)\left(E_k^2 - E_0^2\right)}.
\end{aligned}
\tag{2.52}
$$

The reader will note that this result is written in a way that eliminates the explicit use of $\cos\left(ka\right)$ and $\sin\left(ka\right)$, but that does not mean these functions do not contribute to $\mathrm{d}E/\mathrm{d}k$, because they are found in E_k. The density of states thus becomes

$$\rho = \frac{NE_k}{2\pi}\left(\left(\left(\frac{E_g}{2}\right)^2 - E_k^2\right)\left(E_k^2 - E_0^2\right)\right)^{-1/2}, \tag{2.53}$$

which is plotted in Fig. 2.16. Note that only states satisfying $|E_0| > |E_k| > |E_g/2|$ can contain an electron. This corresponds to the electron energies shown in Fig. 2.15. This density of states is a one-electron density of states, which is in contrast to those shown in Figures 2.3 and 2.7, which contain all the free electrons, and where the energy levels are filled from the lowest upwards. Here we are considering only one electron, because we do not have free or nearly free electrons; an electron is associated with a monomer.

For completeness, we note that it can be shown that the values used here can be related to those in eqn (2.48) by $8\alpha u = E_g$ and $2t_0 = E_0$.

2.4.4   Solitons

We are now aware of the different bond lengths in polyacetylene, caused by the alternation of double and single bonds, each of which has a different length. We also know that polyacetylene is not like benzene. The Peierls instability prevents the equivalence of the two structures causing delocalization, as shown in Fig. 1.1. In the absence of delocalization, polyacetylene is degenerate, because there is an equivalence between the two structures shown in Fig. 2.17. The degeneracy is such that it is not easily possible for the polymer to change between each of the two states via the conjugated state because a significant energy is required to switch between the two degenerate states. To switch between the two states on one monomer would require an energy which, as a minimum, was commensurate with the band gap. If this energy were considerably lower and comparable to thermal energy, then (pure) polyacetylene would display metallic or near-metallic behaviour.

Nevertheless, the polymer is not perfect and may well have kinks or other imperfections that allow charge carriers to cross the band gap. In terms of the chemistry of the chain, this is where two single bonds are joined, breaking the conjugation. When this is the case, a bi-radical is created. Here, a double bond is broken leaving two unpaired electrons. (The unpaired electrons mean that a CH group is connected to both of its neighbours by single bonds, and so simple bond-counting arguments show this to be a radical.) These radicals can move by exchanging single and double bonds, and are called solitons. The exchange of single and double bonds does not require that 1.5 eV (the band gap) be supplied for each monomer as the radical moves along the chain. The two radicals created are a 'soliton–anti-soliton' pair, and can move apart along the chain, as shown in Fig. 2.18. There is no energy barrier for this process, because we are considering delocalized electrons. There is nothing particularly different about the soliton and anti-soliton, but it is helpful to consider them as such, because they can annihilate each other. Because a (neutral) soliton is a radical with one unpaired electron, it has spin 1/2, and an anti-soliton spin -1/2, to conserve total spin. (The reader with a strong solid-state physics background might like to make the analogy for the movement of solitons along the chain with the movement of dislocations in a crystal. Dislocations, like solitons, move with no (or

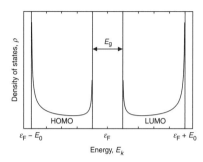

Fig. 2.16 The one-electron density of states for a dimerized chain as predicted by the Su–Schrieffer–Heeger theory.

(a)
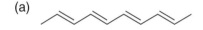

(b)
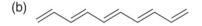

(c)

Fig. 2.17 The single and double bonds of polyacetylene have different lengths, so both of the structures (a) and (b) are degenerate states of the same molecule. The conjugated structure (c) is not a viable structure in the absence of doping, because of the normally prohibitive energy required to jump from (a) to (b) and *vice versa*.

(a)

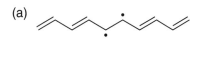

(b)

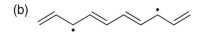

Fig. 2.18 A soliton and anti-soliton created together (e.g. thermally) may move from their point of creation in (a) in separate directions along the chain as in (b).

negligible) energy barriers (described by Sir Neville Mott as rather like a ruck in a carpet), and dislocations, like solitons, can annihilate each other when those of equal and opposite Burgers vector, rather than spin, come into contact.)

Solitons are also created in chains by defects. A kink in the chain, for example, will lower the energy required to create a soliton–anti-soliton pair, although thermal energy is still required to generate the pair. The energy and bond length distortion associated with a soliton means that solitons in neighbouring chains may align, so soliton domain walls may be created; polyacetylene, for example, is known to exhibit soliton domain walls. Because solitons are free to move, they do not need to be associated with an anti-soliton.

The spin on the soliton is because of its association with an electron. However, the electron is countered by the charges on the chain so the soliton does not have charge. As we have discussed, solitons can move along the chain backbone, and this means that they are at least partially delocalized. The effect that the soliton has on bond length means that it can be pinpointed to a limited degree on the chain, but not to the level of associating a soliton with a given point on the chain. The soliton may be considered as a distortion in spin centred around the point where the bond length alternation occurs, which in polyacetylene is where the two single –C– bonds are localized (Fig. 2.18). This is what we mean by saying that solitons are partially delocalized; the effect of the soliton might be over between 3 and 5 monomers. Solitons have an energy associated with them, and lie above the HOMO level, but below the LUMO. The soliton is located about midway up the band gap, which is a reflection upon the fact that the bond lengths either side of the centre of the soliton (i.e. two σ bonds are equal. The energy level associated with the soliton can only be partially filled, because, due to its spin, the soliton will be bound by Fermi–Dirac statistics.

Because solitons rely on the degeneracy caused by alternating bond lengths, they are rather specific to polyacetylene and related polymers. Polymers that have ring features along the backbone are not susceptible to this degeneracy, and so solitons are not created. If one, for example, looks at the structure of PPV (Fig. 3.3), alternating the bond lengths along the conjugation is difficult because the phenyl (benzene) ring is stable against such alternation.

2.4.5   Doping

Although the differences between charge transport in inorganic and organic semiconductors are significant, the existence of a band gap requires a little bit of help for the charge carriers to be able to give rise to conduction. In order to reduce the size of the band gap, the semiconductor must be doped by adding an impurity whose purpose is to shift the energy levels. The amount of impurity can reduce the band gap and can also, if a large amount is added, create metallic behaviour. However, the differences between organic and inorganic semiconductors are also significant where doping is concerned. In the case of organic electronics, redox chemistry is important, but this is not the case in inorganic semiconductors.

As for inorganic semiconductors, doping can be either *n-type* or *p-type*. The addition of group III atoms such as boron as an impurity into a group IV semiconductor such as silicon or germanium gives rise to *n-type* behaviour, because this group can donate an electron, and as such is often referred to as a *donor*. Similarly, group V elements such as phosphorus are sometimes known as *acceptors*, because they produce *p-type* behaviour, with holes as charge carriers. The electrons and holes give rise to new bands between the band gap. At room temperature these electrons will generally provide conduction, so the new energy level in which these electrons sit is close to the conduction band.

In organic semiconductors, n-type and p-type behaviour is created by chemical doping in which redox reactions play a large role. Redox means REDuction-OXidation, and it is worth including a few words about the meaning of these terms. Oxygen is an electronegative atom—a property that gives it a pivotal role in hydrogen bonding, and so can attract electrons from other atoms. Oxidation is therefore loosely described as the act of removing electrons from another element or molecule, and reduction is the opposite. It is easy to get confused here; surely reduction should reduce the number of electrons? This is not the case. Reducing agents or *reductants* are thus electron donors, and *oxidants* electron acceptors. A redox reaction is one in which both reduction and oxidation take place. An easy example is the formation of hydrochloric acid through reaction of hydrogen and chlorine: $H_2 + Cl_2 \rightarrow 2HCl$. Here the hydrogen is oxidized and the chlorine reduced.

We recall that polyacetylene obtained its remarkable Nobel Prize-winning properties through doping with iodine. Clearly, from the periodic table iodine will act as an acceptor. Iodine is an oxidative dopant, and when it is allowed to react with polyacetylene the anion I^- is created. The positive charge on the chain stabilizes the I^-, but the charge on the backbone and the negative charge on the I^- are spread out over a few monomers and so there is significant delocalization. Other oxidative dopants include $Fe(III)Cl_3$ and AsF_5, although the latter, while highly effective, tends to be discouraged due to environmental legislation. Iodine-doped polyacetylene and other synthetic metals using oxidative dopants are p-type, with holes as the majority carriers. This is because the hole cancels out the charge on the anion, given that the overall charge must remain zero.

Another means of doping in organic electrons relies on acid-base reactions. A good example of this kind of treatment is poly(3,4-ethylene dioxythiophene) (PEDOT), which is a synthetic metal, with the unfortunate property of being insoluble and therefore of limited use due to the difficulty in processing such a polymer. However, when synthesized in the presence of polystyrene sulfonic acid (PSS), it forms a highly conducting water-soluble complex. Here the acid acts as a dopant that counterbalances the positive charge on the PEDOT.

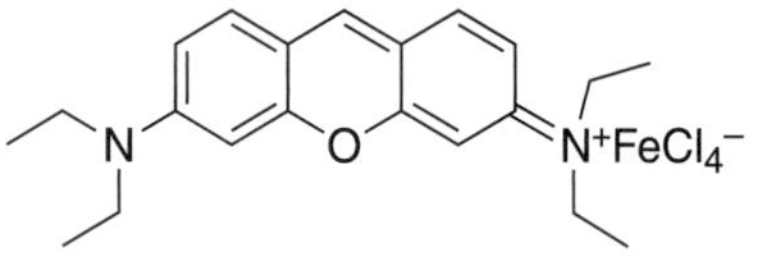

Fig. 2.19 Pyronin B is an organic dye. The fused aromatic (benzene) rings are typical of many dyes. Dyes have, by definition, well defined optoelectronic properties, so these molecules will have their own optical band gap. The presence of oxygen in the ring structure influences its n-type conduction behaviour (Section 3.7).

The *doping level* is defined as the ratio of the number of counterions to the number of monomers along the chain. If a chain contained 100 monomers and had 15 counter-ions associated with it, its doping level would be 0.15 or 15%. In fact such large doping levels, although usually undesired, are not uncommon. This is especially true of n-type dopants, which include metals such as lithium and caesium, and often need to be added in high concentrations for effective conduction. There are a variety of problems with this approach, such as the readiness of the dopants to diffuse during device operation. High dopant levels also change the chemical nature of the device and so are best avoided. As a result organic dyes such as pyronin B (Fig. 2.19) may be useful for n-type conduction.

Doping a semiconducting polymer allows the occupancy of electrons in bands close to the LUMO or holes in bands just above the HOMO. The resulting charge is sometimes called a radical cation (if the charge carrier is positive) or radical anion (if the charge carrier is negative), and in both cases these may be referred to as *polarons*. Doping does have an effect on the chemistry of the polymer chain, since the dopant causes the polymer to become charged. The electronic energy levels are therefore distorted and rearrange themselves in such a way as to minimize the cost of adding an electron or hole to the chain. This may be achieved by changing the nature of the bonds in the polymer. In polyacetylene, this is achieved simply because doping creates a soliton. In other polymers bond distortion may result in a new geometry such as the quinoid conformation (Section 3.1). An interesting consequence of this is that the vacuum level of semiconducting polymers does not represent the total energy cost of removing an electron. The polymer without the electron can relax into a new structure with different bond alternation, with the result of giving back some of the energy cost of removing the electron.

A consequence of p-type doping in polyacetylene is that a double bond is removed. This is stabilized by the addition of a soliton. Two of these solitons can combine on the same chain, if the doping level is high enough. In this case the solitons destroy each other in a process that might be considered the reverse of thermal soliton creation. The charges cannot of course destroy each other, because they are both the same, and will simply repel each other. The charges are sometimes known as charged solitons. Perhaps surprisingly they contain no spin, but this is due to the soliton being located in the middle of the band gap. The electron associated with a neutral (spin 1/2) soliton is removed to make a positively charged soliton, or another electron is added to create a negatively charged soliton. Either way, the net result is zero spin. Experimental evidence suggests that these charged solitons do indeed have no spin and may then contribute to conduction. Despite their importance to conduction in polyacetylene, solitons are much less important for conduction in other conjugated polymers, where polarons and bipolarons dominate charge transport.

Polarons, unlike solitons, are charged, but like solitons, are associated

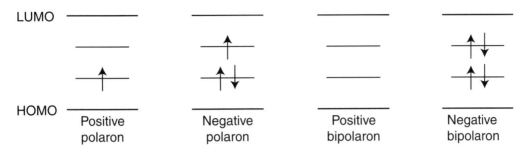

Fig. 2.20 Polarons and bipolarons are formed on energy levels within the band gap. The hole (positive) and electron (negative) polarons both have spin 1/2, and bipolarons have zero spin. Bipolarons have twice the electronic charge.

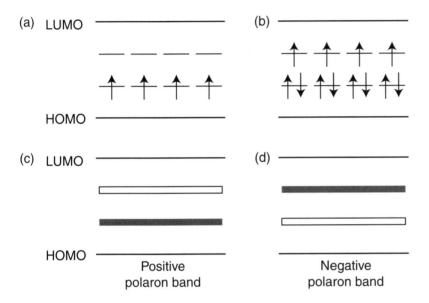

Fig. 2.21 If many polarons are located in neighbouring molecules (a and b), they interact with each other, and the energy levels are less well defined, meaning that they can form bands within the band gap (c and d).

with lattice distortions. Formally, a polaron is a neutral soliton-charged soliton pair, and generally they exist on chains with non-degenerate ground states (Section 3.1). Another important difference between polarons and solitons is that polarons are not associated with unpaired electrons, although they are located within the band gap. Polarons, therefore, have spin 1/2. A more complicated phenomenon occurs when the spins of separate polarons may combine to create a bipolaron. Here the charges, being alike, cannot cancel and a longer quinoid structure is formed with the charges at either extreme. A bipolaron therefore has twice the electronic charge and may be either positive or negative. Bipolarons have been determined spectroscopically, but remain something of a novelty because of their high-energy state. They are more likely to be formed when the charge density is high or in the presence of defects.

The polaron and bipolaron levels are located between the HOMO and LUMO. There are two such levels (Fig. 2.20). In the case of the negative polaron, the higher of these two levels can be considered to contain one charge, whereas the lower level is half-filled for a positive polaron.

Each of these has spin 1/2. In the case of the bipolarons, the negative bipolaron has all of the levels filled and the positive bipolaron none. These have zero spin. Some semiconducting polymers, especially those heavily doped with metals to give n-type behaviour, do not have such clear energy levels, and bands are spread out where the single polaron levels are formed. These are shown in Fig. 2.21.

We have presented *trans*-polyacetylene to describe how delocalization allows charges to move along the chain backbone, but this is not the only means by which charge transport can occur in polymeric semiconductors. Interchain *hopping* of polarons is a means by which effective conduction can be possible. We return to charge transport in more detail in Chapter 5 and hopping specifically in Section 5.2.

2.5 Further reading

The reader may wish to recall basic solid-state physics, and there are numerous textbooks in this area. The most popular is undoubtedly the book by Kittel (2004). The reader interested in understanding the origin of the band gap in polyacetylene would do well to read the original paper by Su, Schrieffer, and Heeger (1980), which is particularly clear (although one should note the erratum). This is also discussed in the first chapter of the graduate text by Heeger, Sariciftci, and Namdas (2010). The reader with an interest in more theoretical aspects of the electronic behaviour of polymers need only refer to the definitive text by Barford (2013), although a more concise approach is included in the review by Hoffmann et al. (1991).

2.6 Exercises

2.1. Consider a metal shaped into a cube of volume 0.125 m^3. How many free electrons are there with an energy of 2.46×10^{-35} J? Would this number change if the metal had a cuboid shape with sides 1 m $\times$ 0.5 m $\times$ 0.25 m?

2.2. The length of the C=C double bond in polyacetylene is 0.135 nm, and the dimer size is given as $2a = 0.280$ nm. What is the length of the C–C single bond?

The band gap of a polyacetylene sample consisting mostly of *cis*-polyacetylene is measured to be $E_g = 1.8$ eV. Consider such a polyacetylene sample with a conjugation length of 120 CH units having a density of states of 6.93 eV^{-1} at an energy of 1 eV above the Fermi energy. Using Su–Schrieffer–Heeger theory, what is the energy (within the HOMO or LUMO bands and relative to the Fermi energy) at which the density of states is a minimum? At what energies is the density of states a maximum?

What is the density of states 0.6 eV above the Fermi energy? In this question, the (one-electron) density of states is given as having units of eV^{-1}. Although the density of states can be measured (see Section 8.4), it is rarely reported with absolute values, and the (unitless as opposed to dimensionless) shape of the density of states as a function of energy is reported.

2.3. The band gap of *trans*-polyacetylene is $E_g = 1.5$ eV, and band size is $E_0 = 6.4$ eV.

(a) If an electron of energy E is defined by $ka = k_E a = 14\pi/30$, what is the value of E relative to the Fermi level?

(b) What is the value of $ka = k_{2E}a$ for an energy $2E$?

(c) What is the value of $k_E a$, below which $k_{2E}a$ becomes undefined? What is the value of E for such an electron?

3

Beyond polyacetylene

(a)

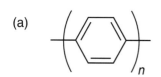

(b)

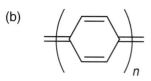

Fig. 3.1 (a) The chemical structure of underivatized PPP. The quinoid form is shown in (b).

Although we have tacitly accepted the need for other conjugated polymers besides polyacetylene, we have largely restricted ourselves to this polymer because of its simplicity. Polyacetylene has its limits that have been alluded to earlier in the book. Although we have acknowledged that the Durham route to synthesis of polyacetylene gave a massive boost to its usability, it remains a largely intractable polymer due to its insolubility. A convenient alternative to dissolution is to melt the polymer at elevated temperatures. Although polyacetylene thermally decomposes at high temperature ($> 400°C$), this is still below its melting temperature. In truth, polyacetylene is not particularly stable in air where it is prone to oxidation.

One solution that is commonly used in many semiconducting polymers is to derivatize the polymer. By this we mean making chemical modifications to increase the solubility of the material. In general rigid polymers such as polyacetylene and most other conjugated polymers are insoluble in most solvents. The addition of flexible side chains helps solubilize polymers, because the side chains are themselves soluble and, if long enough, will counter the insolubility of the main chain. Such a route is difficult in the case of polyacetylene, because any attempt to add side chains has a significant effect on the backbone, which causes it to lose its semiconducting properties. There are many other polymers which do have the possibility of being derivatized and therefore of being solution processable, and it is the purpose of this chapter to discuss their electronic properties.

The addition of side chains to aid solubility is an important means of increasing the applicability of polymers for semiconducting and opto-electronic purposes. However, there is much more that can be done to functionalize polymers. In this chapter we discuss individual homopolymers, but these need not be linear; they can be copolymerized with other monomers, grafted onto a main chain, and they can have defects imparted onto them. There are a multitude of modifications that can be made, depending on the final application, from changing the band gap in order to control the polymer's interaction with light to creating donor–acceptor block copolymers to create photovoltaic cells.

We have so far discussed polyacetylene, and we now turn to the addition of rings to the main chain, which aids greatly the ability to create solution processable conjugated polymers. The key polymers for this purpose include poly(p-phenylene) (PPP), poly(p-phenylene vinylene) (PPV), polythiophene, polypyrrole, and polyaniline (PANi).

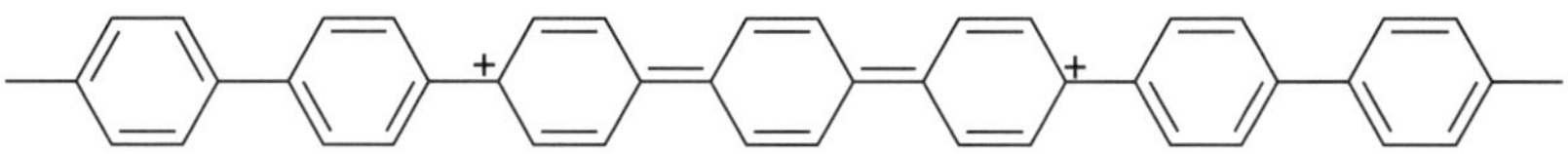

Fig. 3.2 The aromatic and quinoidal structures can coexist on the same chain. For example p-type PPP can contain bipolarons as shown here.

3.1 The quinoid conformation and PPP

The first aromatic conjugated polymer that we shall consider is poly(p-phenylene),[1] which is shown in Fig. 3.1. PPP is important because it is very stable (especially when undoped), can be used as an electrode, is able to emit blue light when undergoing electroluminescence, and, like other conjugated polymers, can also be used in gas sensing. PPP can be more easily processed than many similar polymers, because, for example, it can be fibre drawn.[2]

We mentioned in Section 2.4.5 that the quinoid conformation occurs due to bond distortion on the addition of electron or hole to the chain. The quinoid conformation is important since it breaks up the degeneracy of the polymer. The polymer ground state is non-degenerate, because any perturbation causes alteration of the bond lengths. The double bonds joining the phenyl rings in the quinoid conformation are shorter than those in the lowest energy state of PPP. In polyacetylene there are two bond lengths—one for the single bond and the other for the double bond. Alternating the bonds does not change the energy state of the polymer, and so it is degenerate. Alternating the bonding in aromatic polymers such as PPP clearly does have a significant effect, and raises the energy by a small amount (0.4 eV). Doping or photo-excitation are means that may be used to raise the energy state to that of the quinoid conformation, but this also requires overcoming an activation energy. Doped PPP is often associated with the formation of polarons and bipolarons bounding the quinoid form (Fig. 3.2).

PPP is a very simple model polymer to consider because it consists simply of aromatic rings bound together. This means that we can perform the same treatment that we did with polyacetylene to describe the molecular orbitals and simply the combination of the orbitals of the monomer. Charge transport in PPP is largely due to polarons. On doping, a polaron is created. Here, the polaron behaves differently to that of the polyacetylene case. In PPP doping induces a charge and a net spin, which delocalize over a few monomers. The length of delocalization can be determined because over this region the polymer has a quinoidal conformation. The distance of this delocalization remains constant and the charge and spin do not separate. For this reason polarons are responsible for charge transport in PPP, and many other aromatic conjugated polymers.

The energy bands of PPP must differ from polyacetylene due to their non-degeneracy. It is experimentally observed that if an electron is excited to the LUMO, the quinoid conformation forms. In fact the quinoid state is associated with polarons and so the energy of the quinoid confor-

[1] PPP can also be written as poly(*para*-phenylene)

[2] In a fibre drawing process, a nearly molten polymer is pulled or drawn into fibres from a given shape or preform.

mation may be given by a polaron state within the band gap (Fig. 2.20).

3.2 PPV

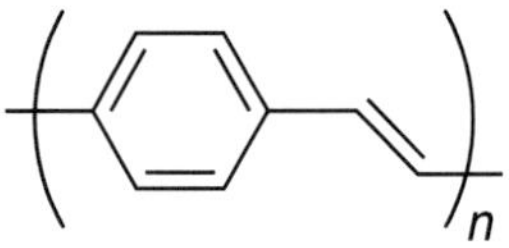

One of the most widely used semiconducting polymers is poly(p-phenylene vinylene), which is commonly used in the development of organic LEDs (OLEDs). The chemical structure of PPV is shown in Fig. 3.3. Although the polymer is insoluble, the use of substituents attached to the phenyl ring can aid solubility. Another advantage of adding functional groups along the chain is that they can be used to change the wavelength at which light is emitted by the polymer. At present, PPV derivatives can be used for yellow/orange to green emission, with a band gap that corresponds to light at wavelengths of between 590 and 515 nm.

PPV occupies a special position in the history of polymer electronics because it was the first for which electroluminescence was reported. The electroluminescence described in the first *Nature* paper was actually very poor (Section 10.1.2), but has been improved substantially over the years through derivatization. Unlike polyacetylene, which cannot be derivatized without breaking the conjugation, the processing and properties of PPV have been substantially improved. The first PPV was solution processed, i.e. a film of precursor polymer was created and then annealed (heated) to allow thermal conversion to PPV. This route is of limited applicability, and the use of derivatives has improved matters substantially, For every one photon emitted in the original PPV, more than 2000 electrons were injected; an internal electroluminescence efficiency of 5×10^{-4}. (Internal electroluminescence efficiency is defined in Section 4.12.) More recent results have shown that this has increased substantially, although the best results are typically $\sim 5\%$. Doping of PPV has been very effective, with conductivities upwards of 5×10^5 S m^{-1} reported.

Fig. 3.3 PPV consists of a phenyl ring and vinylene group.

3.3 Polythiophene

The chemical structure of polythiophene is shown in Fig. 3.4. Like PPV, polythiophene and its derivatives are usually doped with p-type materials with holes acting as majority charge carriers. (It is also possible to achieve n-type doping, but it is generally more difficult to get the required level of doping with an n-type dopant than with a p-type dopant.) Also, like PPV, polythiophenes are electroluminescent. The optical properties of polythiophene are particularly interesting with the doped polymer becoming blue compared with the undoped polymer, which is red. The level of doping governs the colour of the polymer.

As well as its optical properties, polythiophene is a popular polymer to work with for several reasons. It offers many possible routes to its synthesis. It has good environmentally stability in neutral and doped states, and it can be subjected to reversible redox switching. (The ability to reduce and then oxidize the polymer is useful in applications where

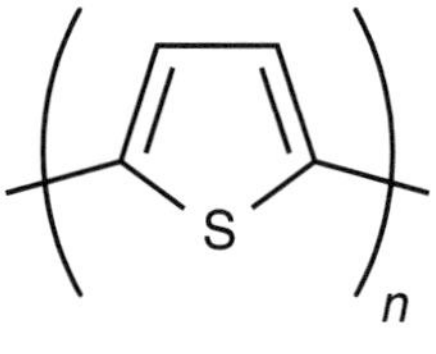

Fig. 3.4 The chemical structure of polythiophene.

switching may be useful, such as in electrochromic appliances, where the redox reaction changes the optical properties such as transparency or colour.) PEDOT, a polythiophene with commercial applications has an extremely high conductivity of up to 3×10^5 S m^{-1}. Even when modified by polymerizing in the presence of polystyrene sulfonic acid (to render it water soluble), conductivity is a perfectly respectable $\sim 10^4$ S m^{-1}.

3.4 Polypyrrole

Polypyrrole (Fig. 3.5) has a long history in polymer electronics. It showed promise as a good conductor in 1963 when the products of the pyrolysis of tetraiodopyrrole (C_4I_4NH, i.e. with an iodine atom attached to each of the four carbon atoms) was reported. The pyrolysis removes some of the iodine, promoting the polymerization of the pyrrole. Some iodine remains, and as for polyacetylene, dopes the remaining polymer. The resulting polymer exhibited conductivities of ~ 100 S m^{-1}, which is not metallic behaviour, and the polypyrrole was not unequivocally identified, but it is nevertheless interesting that this result preceded the chemical doping of polyacetylene by well over ten years.

Polypyrrole is not particularly suited to interesting optical properties; its importance is, like polyacetylene, due to its high conductivity. As such it has applications in areas where conducting polymers have traditionally been expected to be important, such as antistatic applications, or electromagnetic interference shielding.

3.5 Polyaniline

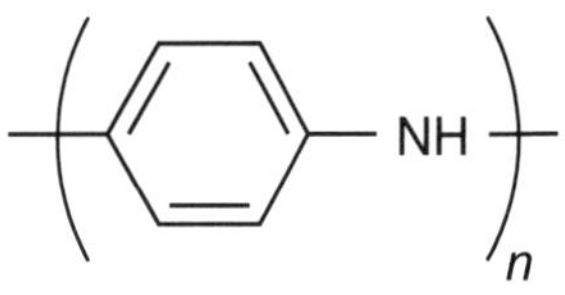

Like PPV, polypyrrole, and polyacetylene, polyaniline (PANi) also has historical importance, because it was reported—but not identified—nearly 150 years ago. PANi is in fact remarkably cheap to produce, which gives it a high level of importance in modern polymer electronics, with applications in anti-static packaging and electrochromic vision systems. The chemical structure of polyaniline is shown in Fig. 3.6. The qualities of polyaniline are perhaps not as good as those of, say PEDOT, which has better pH stability, for example, but one should never underestimate the importance of inexpensive manufacturing.

In its doped state PANi behaves as a synthetic metal, but it has other advantages. Its inexpensive production is not limited to synthesis. PANi is soluble in some organic solvents, it can be moulded, extruded (i.e. pushed or drawn through a former or die to create an elongated structure), and can also survive temperatures in air in excess of $200°C$, at least for short periods of time, and even greater temperatures before degradation ($> 400°C$) for some doped forms under inert atmosphere.

The parameter that defines the several different forms of polyaniline is its oxidation state. That shown in Fig. 3.6 is not oxidized. If we recall from Section 2.4.5, oxidation is the removal of electrons from a molecule. Because of the electronegativity of nitrogen, the electron from

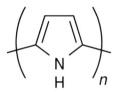

Fig. 3.5 The chemical structure of polypyrrole.

Fig. 3.6 The chemical structure of polyaniline.

a hydrogen atom is largely located on the nitrogen, and so removal of the hydrogen atom to form an imine bond is an oxidation process. This does not need to occur on every nitrogen atom along the chain, but there may be some degree of oxidation, between 0% and 100%. In the latter case we have pernigraniline, and in the former (i.e. no oxidation) we have the leucoemeraldine (Fig. 3.6). Pernigraniline and emeraldine (50% oxidation) are shown in Fig. 3.7.

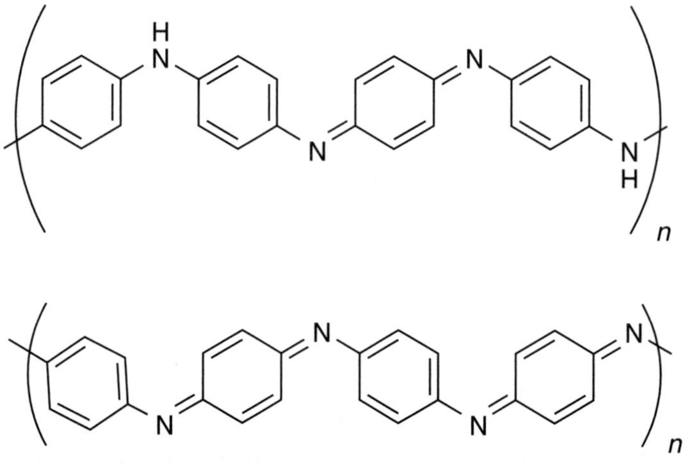

Fig. 3.7 (a) Polyaniline with an identical number of imine (=N–) and amine (–NH–) groups is known as emeraldine. There is no requirement in this, 50% oxidized, state that the bond order be as shown here, only that half of the amine groups have become imine groups. (b) Fully oxidized PANi is known as pernigraniline.

The emeraldine form of polyaniline is particularly interesting and important. It is important because this form is often synthesized, and interesting because of its doping. The emeraldine shown in Fig. 3.7a is known as emeraldine base, because the imine groups (=N–) can accept a proton, to form the emeraldine salt. This can form bipolarons, when the imine groups are adjacent, or polarons, when the imine groups are separated by amine units, as in Fig. 3.7. The bipolaron form is prone to lattice deformation, due to its asymmetry, and as such is less effective at conduction than the polaron form.

The synthesis of polyaniline is routine; aniline hydrochloride is reacted with ammonium peroxydisulfate. This reaction polymerizes the aniline in the form of the emeraldine salt, with chlorine counter-ions. Conversion to the base occurs in the presence of ammonium hydroxide, or another appropriate base. This process is reversible. There are other routes to synthesize PANi; the method reported in 1862 involved the electrolysis of aniline sulfate.

3.6 More on the limits to the conjugation of polymers

In considering polyacetylene, we discussed how solitons were formed at defects and kinks in the linear polymer. In fact, there are many limitations on the purity of conjugation, which can impede conduction, by acting as traps for example. Sometimes, however, defects can be intro-

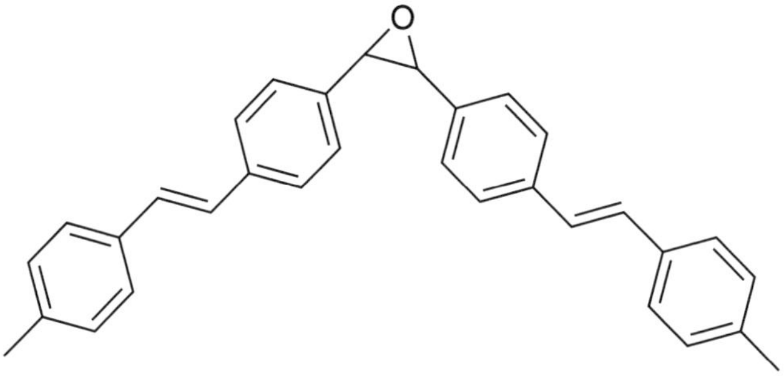

Fig. 3.8 PPV at high temperatures may be affected by the presence of oxygen in the atmosphere, which can react with unsaturated bonds, disrupting the conjugation.

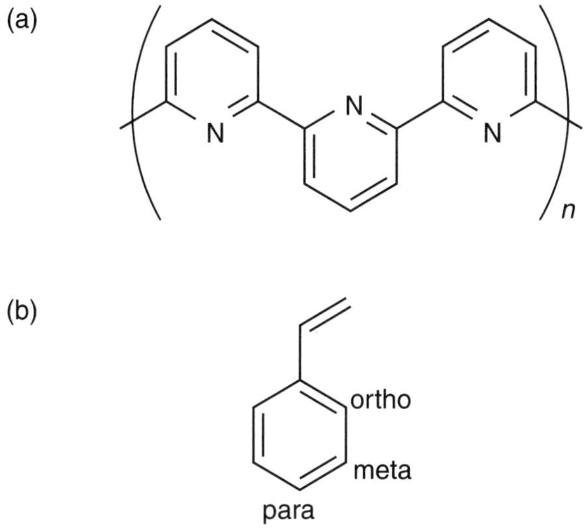

(a)

Fig. 3.9 Poly(2,6-pyridinediyl) is shown here as an example of a meta-linkage (a). Only para-linkages, such as those in PPV uphold conjugation. The other possible linkage in a ring is at the ortho-position, which is occupied by the nitrogen atoms in the present molecule. Ortho-links are rare due to stereochemical reasons. (Simply put, there is not enough space for the bond to occupy the ortho-position.) The ortho-position is always next to the bond linking the ring, the meta-position is next, and the para-position is opposite the link. These are pictured in the example for styrene (b)

(b)

duced for a specific purpose, such as controlling the band gap, which we shall discuss later in the book. In this chapter we have considered several polymers containing rings. Imperfect rings will spoil the conjugation, for example, by a so-called sp^3 defect. An sp^3 defect indicates the (not always unwanted) presence of an sp^3 orbital on a ring. Methane is a good example of an sp^3 orbital, which is marked by a tetrahedral bond conformation, and is a fully saturated bond. Such an orbital must break the conjugation. In order to picture this, one can add a hydrogen atom to benzene to see the effect of an sp^3 defect. The effect of this saturation is quite complex, because the sp^3 and sp^2 orbitals have different bond angles; the tetrahedral orbitals have bond angles of $109.5°$, whereas the trigonal planar sp^2 orbitals have angles of $120°$. These different bond angles put some strain on the structure, which will affect their bond lengths, and distort the molecule.

Unsaturated bonds are prone to oxidation, which restricts the stability of polymers for electronic and optoelectronic applications. The band gap in polyacetylene is adversely affected by oxidation, but most other semiconducting polymers can be similarly affected, with some worse than others. In Fig. 3.8 we show an example of how oxygen can introduce defects that break delocalization along the polymer.

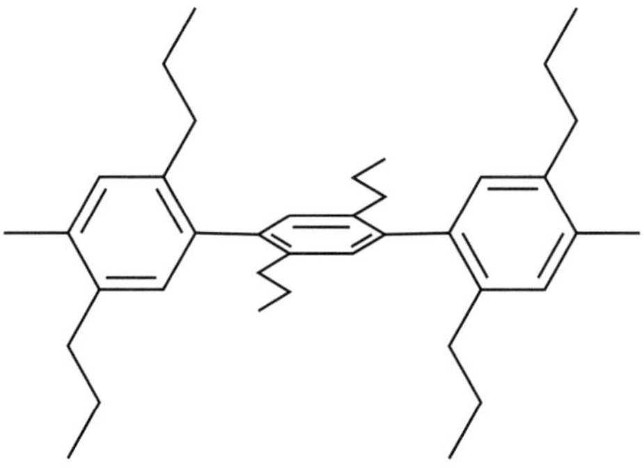

Fig. 3.10 PPP may be chemically modified by the addition of side chains along the molecule. However, too many of these can put strain on the backbone of the polymer causing rotation of the ring out of the plane that it normally occupies. Such a rotation impedes charge transport.

If we consider the structures of PPV shown in Figures 3.3 and 3.8, we see that the six-sided rings are all linked at the para-position (Fig. 3.9b). Such linkages are necessary for conjugation. Anything other than para-linkages limits the conjugation. In Fig. 3.9, poly(2,6-pyridinediyl) is shown. Here the meta-linkages mean that electronic delocalization along the polymer is limited to only a few monomers. In polythiophenes and polypyrroles this nomenclature is redundant, and the chain linkages must take place between the sulfur or NH; these are α-linkages. If the bonds are located on the other side of the five membered ring away from the sulfur or NH (β-linkage), the conjugation is broken.

Conjugated polymers are planar molecules, and any deviation from this would impede charge transport, along the chain. This is not normally a problem but it is very often necessary to solubilize the molecule by adding side chains. These side chains can allow solution processing of the polymer but also may cause stereochemical hindrance of the planarity; essentially the side chains can get in each other's way, with the resulting stress on the molecule forcing rotation along the backbone (Fig. 3.10).

3.7   Ladder polymers

One means of solving the problem of the steric hindrance of side chains is to reinforce the backbone. In Fig. 3.10, PPP is shown to be sterically hindered by side chains. However, a variant, ladder PPP (LPPP) may be forced into planarity in a ladder conformation, as shown in Fig. 3.11. In fact there are a variety of possible morphologies for ladder polymers, which may be either simple in the case of polyacene (Fig. 3.12), where the rings are fused together, or more complex in the case of BBL, which we describe below. A ladder polymer has the repeat unit attached to its neighbour by two separate bonds (*double-stranded*), whereas polymers such as polyacetylene or PPP can be said to be *single-stranded*. Note that the side chains shown in Figures 3.10 and 3.11 do not destroy the conjugation. The reader is invited to follow the single and double bonds in these figures (and also in Fig. 3.9a) to see how conjugation is upheld; where conjugation is retained, the bonding that describes polyacetylene

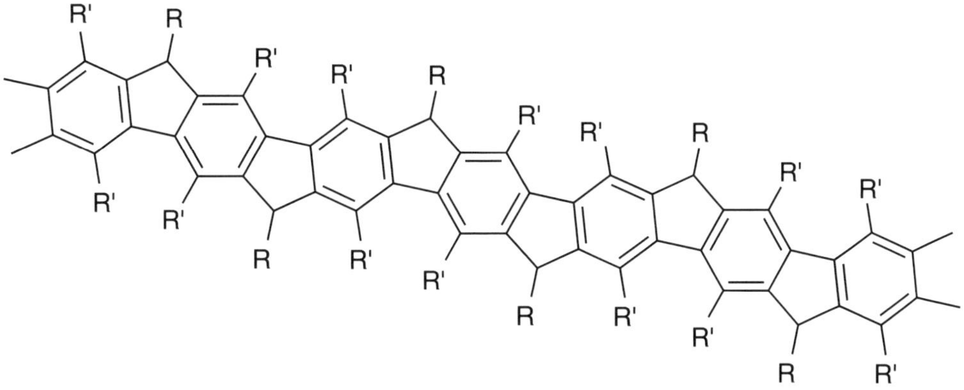

Fig. 3.11 Side groups R and R' can be added to PPP without steric problems by strengthening the backbone. In this example they remain planar, although there is some rotation of the rings relative to what would be expected for PPP.

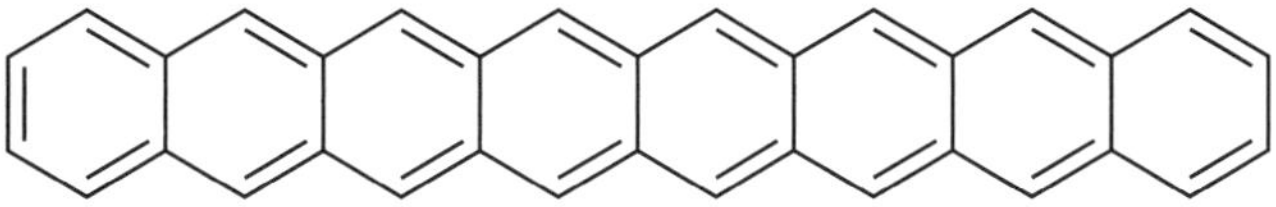

Fig. 3.12 Polyacene is formed of phenyl rings. However, there are important differences, because the ring does not lead to delocalized electrons around it. Each six-membered ring contains only two double bonds, and so the bond lengths are not the same.

is traced.

The simplest ladder polymer is polyacene. The motivation for the synthesis of such polymers is due to the possibility that they may have better conduction properties than, say, its nearest linear stranded polymer for comparison. In fact, it was once thought that some ladder polymers might have negligible band gaps giving them near-metallic behaviour. This has turned out not to be the case, and they are not too much different from some of the better alternatives and require doping to be electrically conducting. It is rather difficult to synthesize polyacene as a high polymer, with many hundreds of rings, but smaller molecules have found much use. The molecule with five rings, pentacene (shown in Fig. 3.13), has been particularly successful in transistor applications, due to its very high hole mobility.

Poly(benzobisimidazobenzophenanthroline), or more simply BBL, is another example of a ladder polymer and is shown in Fig. 3.14. This polymer is of interest because of its n-type behaviour. The polymer has shown promise in solar cells and also for electrochromic applications. Like most ladder polymers it has a very high glass transition (in this case above 500°C), and is very stable in air even at elevated temperatures, which is useful when device fabrication requires thermal processing steps. With these advantages, BBL has great promise in field-effect transistors (FETs). Most FETs have p-type behaviour, but BBL has outstanding qualities of stability and electron transport that it and materials based upon it have good prospects for commercial uses. BBL, one notes, is

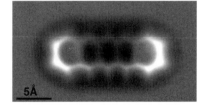

Fig. 3.13 This image of pentacene taken using high-resolution atomic force microscopy (Section 8.4.2) reveals the bonds present in the molecule. Taken from Gross et al. *Science* **325** 1110 (2009). Reprinted with permission from AAAS.

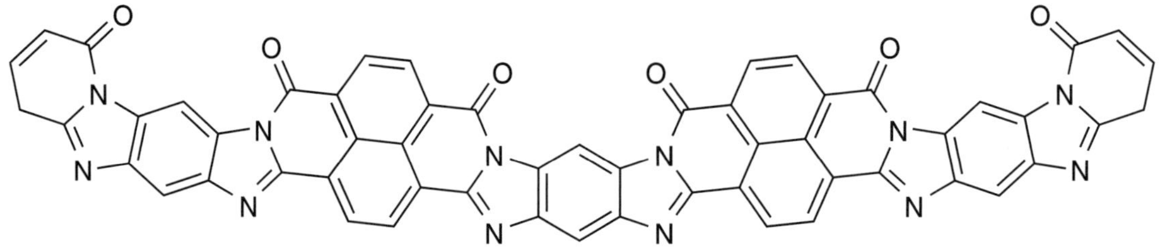

Fig. 3.14 BBL is a ladder polymer with a highly planar conformation. Rather than show one monomer unit, two full monomers are illustrated, because this reveals the structure with perhaps a little more clarity.

rich in oxygen in the side chains, and this improves electron mobility by withdrawing electrons from the conjugation, as was also noted for pyronin B (Fig. 2.19). The reader might find this counterintuitive, but if one recalls from introductory semiconductor courses, hole transport is best visualized as an absence of electrons. The common analogy is a row of people sitting on chairs with an empty chair in between. If each person moves to the left, it has the visual equivalent of an empty space (hole) moving to the right and we thus have hole transport as the motion of an absence of electrons. Similarly, if one removes electrons from the process, one starts to improve electron mobility; returning to our analogy, a prevalence of empty chairs means that we start to observe people moving left, which is our electron mobility. Oxygen is electronegative of course, so attracts π electrons from the conjugation.

BBL is also rich in nitrogen, which can often play a useful role in conjugated systems because it does not necessarily destroy conjugation. Nitrogen is electronegative, which means that it also can attract π electrons from the conjugation. This also raises the electron affinity of the molecule, which helps make it more air-stable. In the case of BBL, the nitrogen atoms that do not contribute to the conjugation can therefore play the same role as the oxygen.

3.8 Synthetic metals and low band gap polymers

Semiconducting polymers are often referred to as synthetic metals. However, we know that metallic behaviour in these polymers is only obtained through doping. The laws of physics have interrupted the quest for a polymer with no band gap, as we saw clearly in the case of polyacetylene. This does not mean that undoped zero band gap polymers cannot exist and efforts are continually being made to solve this problem. Even if the quest for a polymer with metallic behaviour were to prove insurmountable, it should be possible to lower the band gap to less than $k_\mathrm{B}T$, which is 0.025 eV at room temperature. It is generally understood, however, that any polymer with a band gap below 1.8 eV is a low band gap polymer. There are many examples of low band gap polymers in the literature, but there are few common threads chemically linking them and so we shall not treat them specifically here.

An important point about low band gap polymers is that their importance is not necessarily to get as close as possible to a synthetic metal, but to capitalize on their interaction with light. $E = 1.8$ eV corresponds to light of wavelength of $\lambda = hc/E = 690$ nm, which is red light. Lower band gaps correspond to the rest of the visible spectrum. We turn to the optoelectronic properties of conjugated polymers in the next chapter.

3.8.1 Sheet resistance

The quality of synthetic metals can be determined by their resistance. However, for many purposes in polymer electronics, the materials are used in films and current flow is in the plane of the film. For this reason the *sheet resistance* is often quoted. A good example of planar current flow is in FETs, whereby the flow from source to drain electrodes is measured. However, in the case of FETs, the material through which the current flows is a semiconductor and its motion is restricted to a channel defined by the source and the drain. Synthetic metals are applied in technologies, such as antistatic coatings, where a low resistance is required, because charge build-up must be prevented. Other areas that may be of importance in the future are in areas such as large area solar arrays, where metallic sheets can be used to collect charge to allow them to flow to electrodes, rather than have them flow through lower mobility semiconducting material. Even technologies as routine as a trackpad might find value in sheets of synthetic metals, if flexible technology is desirable in such applications. With these and other applications in mind the sheet resistance provides a useful contrast to bulk resistance for two-dimensional applications. A sheet resistance is given in the rather bizarre units of ohms per square ($\Omega/\square$), which has dimensions of ohms; the 'square' denotes that the resistance refers to a sheet resistance.

Unlike resistivity, sheet resistance is not an intrinsic quality of a material; multiplying the film thickness, d, by the resistivity, ρ, gives the sheet resistance,

$$R_\mathrm{s} = \rho d. \tag{3.1}$$

Nevertheless, the sheet resistance is often quoted, and values of $R_\mathrm{s} < 1$ k$\Omega/\square$ would indicate current market-leading performance for a material like PEDOT, with more usual values being a factor of a thousand worse. (By way of comparison, indium tin oxide, a common anode material, that we shall come to on numerous occasions in the remainder of this book, has a sheet resistance that can be lower than 10 $\Omega/\square$.) The reason why sheet resistance is a useful concept for performance comparison is because it is for two-dimensional flow, and as such the thickness is a less interesting parameter, although one would hesitate to dismiss it completely. Bulk resistance depends critically on all three dimensions, but sheets of conducting film can be cut up into different sizes, and the sheet resistance remains unchanged.

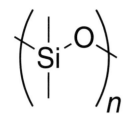

Fig. 3.15 The chemical formula for polydimethylsiloxane.

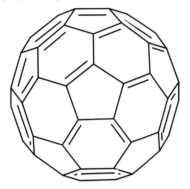

Fig. 3.16 The chemical structure of buckminsterfullerene (C_{60}).

3.9 Buckminsterfullerene, carbon nanotubes, and graphene

It is easy to forget that not all polymers have a carbon backbone. The valency of four is a prerequisite for polymerized structures; although nitrogen and oxygen can play an important role in polymer chains, nobody has yet created a polymer based on these molecules not incorporating a component with a valency of four. In the same way that carbon-based electronics is a viable alternative to silicon-based electronics, silicon-based polymers exist, and the classic example is polydimethylsiloxane (Fig. 3.15), which when crosslinked, is the elastomeric stamp used to create micro- and nanostructures in soft lithography.

So, we have established that polymers are not exclusively chains incorporating carbon atoms, but are all large molecule carbon-bonded species polymers? Specifically, are nanotubes and graphene polymers? They both contain bonds in two dimensions rather than the linear chain form to which we have become accustomed. Nevertheless, crosslinked polymers (gels and networks) retain credibility as polymers. The monomer would be a single carbon atom; there is no reason why this cannot be polymerized in other forms other than linear chains. Of course, the physical properties of nanotubes and graphene are rather dramatic, but then so were those of doped polyacetylene, and in its time, nylon was a considerable achievement. Maybe a better question is to ask whether or not these molecules belong in this book. The answer would be yes, if the book were two or three times as thick. Ultimately, the size of the research community involved in these remarkable carbon-based molecules is so large that it is a separate community in its own right. We cannot ignore these molecules, however, and their importance to the polymer community is such that they appear at later points in the book, so we shall satisfy ourselves with a brief introduction to them and some of their electronic and physical properties.

3.9.1 Buckminsterfullerene

The 'new' forms of carbon originate in the discovery at Rice University in Texas of C_{60} (Fig. 3.16) in 1985. Three of those involved in the original 1985 paper (Harry Kroto, Richard Smalley, and Robert Curl) were awarded the 1996 Nobel Prize for Chemistry. (Sir Harry Kroto is an alumnus of the University of Sheffield.) This molecule has a geodesic structure, and was named after the architect Buckminster Fuller, who created similar geodesic domes, if eleven orders of magnitude different in terms of dimensions. Each carbon atom is bound to three others (hence the suffix '-ene' to fullerene), which leaves a π electron. This π electron means that C_{60} is incorporated into many molecules as an acceptor (electron transporting) species. A good example is that of phenyl-C_{61}-butyric acid methyl ester (Fig. 9.11) which is particularly useful in organic photovoltaic cells. Buckminsterfullerene on its own is an insulator, but conducting and semiconducting properties can be obtained by incorporating metallic atoms into interstitial sites on the lattice, or trapped inside it.

3.9.2 Carbon nanotubes

The discovery of buckminsterfullerene was one of those discoveries that was worthy of a Nobel Prize independently of any obvious use or other importance. The idea that there is a third form (*allotrope*) of carbon besides diamond and graphite, and that this can be readily produced in combustion, but nonetheless remained undiscovered long after we were using personal computers is amazing. (Not that personal computers and C_{60} need have anything in common, but that something so fundamental can remain hidden while technology proceeds apace, is surprising.) The discovery triggered scientific interest and research into new forms of carbon quickly followed. Nanotubes were next, and their description in 1991 by Sumio Iijima of NEC is seen as the definitive paper, although they were observed in electron microscopy experiments and described much earlier than that—indeed, much earlier than the discovery of C_{60} for that matter.

Nanotubes exist in a variety of different forms. Their bond structure can be different depending on whether the central axis of the tube is parallel or perpendicular to the C–C bonds. In Fig. 3.17a the zigzag form is depicted, whereas the armchair form is shown in Fig. 3.17b. These can be further classified by the number of phenyl rings around the tube, so in Fig. 3.17a there are ten. This is known as a (10,0) structure, with the zero indicating that the rings are fused, which is always the case for zigzag structures. This perhaps rather opaque explanation for the nomenclature is clearer with the armchair structure. Here we show in Fig. 3.17b a (6,6) structure. There are six rings around the tube, each joined by a C–C bond (rather than being fused); the second six indicates that the C–C bonds joining the ring corresponds to another six rings a half a row above and below the first six. Nanotubes exist in forms other than armchair and zigzag, and these are generically known as *chiral* and classified by a *chiral vector*, $\mathbf{C}_h$,

$$\mathbf{C}_h = n\mathbf{a}_1 + m\mathbf{a}_2, \tag{3.2}$$

where $\mathbf{a}_1$ and $\mathbf{a}_2$ are unit vectors as shown in Fig. 3.18.[3] One can see from Fig. 3.18 that armchair nanotubes are those with $n = m$ and zigzag nanotubes, with $m = 0$. This information is sufficient for us to calculate the radius of the nanotube, r_n, which is given by

$$r_n = \frac{a}{2\pi}\sqrt{(m+n)^2 - mn}. \tag{3.3}$$

Furthermore, nanotubes can exist as multiwall and single wall structures. A multiwall structure is a series of concentric nanotubes, whereas a single walled nanotube is an isolated tube. Scanning electron microscope images of the two structures are shown in Fig. 3.19.

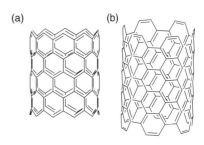

(a) **(b)**

Fig. 3.17 Chemical structure of (a) zigzag and (b) armchair carbon nanotubes.

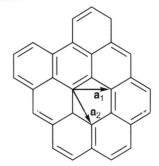

Fig. 3.18 Unit vectors $\mathbf{a}_1$ and $\mathbf{a}_2$ are used to define the orientation of nanotubes.

[3]The magnitude of these unit vectors is $a = 0.246$ nm, which the reader can check is consistent with the bond length between carbon atoms in phenyl rings of 0.142 nm.

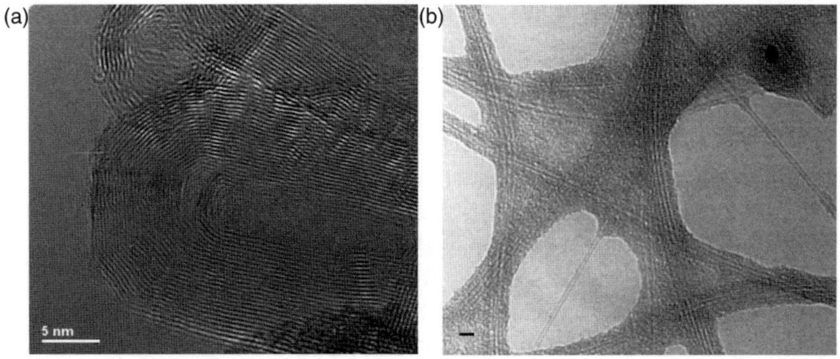

Fig. 3.19 Electron microscopy images of (a) Multiwalled and (b) single-walled nanotubes. The scale bar in both images is 5 nm. (a) is taken by Zabaeda Aslam and Rik Brydson (University of Leeds) and is from M. Grell 'Electronic and electro-optic molecular materials and devices' in *Nanoscale Science and Technology* edited by R. W. Kelsall, I. W. Hamley, and M. Geoghegan (Wiley, Chichester, 2005) and (b) is taken from D. S. Bethune *Physica B* **323** 90 (2002), with permission from Elsevier.

The electronic and mechanical properties of nanotubes are particularly noteworthy. As well as being very strong, with Young moduli of the order of TPa, they are also semiconductors. In fact the armchair configuration exhibits metallic conduction. The reader can trace the path that charge carriers would take along the nanotube, and note that the armchair configuration can take a path similar to *trans*-polyacetylene, whereas the zigzag form is that of *cis*-polyacetylene. The reader will note from Section 1.1 that only *trans*-polyacetylene behaved as a metal when doped. The Peierls instability (Section 2.4.4) is not possible with nanotubes, because the bonds must all be the same length, and so doping is generally not necessary. Nanotubes exhibit excellent hole-transporting properties, but can be made to transport electrons by doping with a metallic vapour.

3.9.3 Graphene

If we could unroll a carbon nanotube we should be left with a layer of graphene. Graphene is often thought of (not strictly correctly) as a single layer of graphite, and for the same reasons as for nanotubes, graphene has remarkable electronic properties. The simplicity of graphene, the production of which is easier than that of nanotubes, along with its electronic and mechanical properties were responsible for the award of the 2010 Nobel Prize for Physics to Andre Geim and Konstantin Novoselov, both of the University of Manchester. Graphene can be used as a basis for field-effect transistors, as can carbon nanotubes, but there are also possibilities for use in high-speed *ballistic* transistors. Like nanotubes, graphene can be categorized by chirality, although in the case of infinite two-dimensional sheets, it is meaningless to consider armchair and zigzag forms, except when considering their electronic properties.

3.10   Further reading

The reader requiring more on this area is best directed to more advanced texts such as the book by Heeger, Sariciftci, and Namdas (2010). Many chapters in the two volumes of the book edited by Hadziioannou and Malliaras will also be of considerable use to the advanced reader. The book by Blythe and Bloor (2005) is taken from perhaps a more materials engineering point of view, but its level is appropriate for readers of the present text.

3.11   Exercises

3.1. The image shown in Fig. 3.12 is a *cis* form. Draw the chemical structure of *trans*-polyacene.

3.2. (a) What is the molecular mass (in g/mol) of (i) pentacene, (ii) buckminsterfullerene, and of monomers of (iii) pernigraniline and (iv) emeraldine?

(b) If a given batch of P3HT (Fig. 6.23) has a molecular mass of 29.2 kg/mol, how many monomers are typically in a given polymer? How many polymer molecules of this molecular mass would be required so that it contained 1 mol of monomer, and what would be its volume? (P3HT has a density of 1.1 g/cc.)

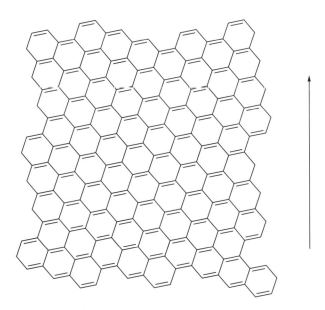

Fig. 3.20 An unrolled chiral nanotube. The axis of the nanotube is vertical, as indicated by the arrow.

3.3. (a) An unrolled nanotube is shown in Fig. 3.20. What are the values of n and m for the chiral vector, $\mathbf{C}_h$ (eqn 3.2)?

(b) Prove eqn 3.3.

4 Optoelectronic properties

Light is of course part of the electromagnetic spectrum. How light interacts with matter is what gives rise to the optical properties of materials, and the most simple of these is colour. Even colour is quite a complex property. The sky is blue for different reasons than gold being, well, gold. It is blue because long wavelengths are transmitted through the atmosphere. Short wavelengths are Rayleigh scattered, the intensity of which is inversely proportional to the wavelength of the light, explaining the dominance of blue. Gold has its yellowish colour due to the interaction of light with the free electrons that give it its metallic behaviour. These electrons absorb light over a large range of wavelengths in the blue region, and the resultant mix of reflected wavelengths give gold its yellow colour. Gold is unusual for a metal in that it is not silver in colour, and the reasons for this are due to the relativistic properties of its orbital electrons pulling them closer to the nucleus, permitting a transition and allowing absorption of blue. With other metals, similar transitions are in the ultraviolet and so visible light is not absorbed.

Semiconductors have a capacity to select light of given wavelengths either in absorption or in emission, with many important applications. The band gap is key to this selectivity, because the size of this band gap plays an important role in determining the absorption and especially emission of light. The ability of polymers to emit light at a specified wavelength is what makes polymer LEDs possible. The behaviour of molecules with a band gap is much more complex than absorbed energy causing light emission at a wavelength corresponding to the band gap. Not all polymers are particularly effective at emitting light; polyacetylene is a notable example of a conjugated polymer with poor optical properties.

The energy gap for electronic transitions can be understood in a similar manner as for simple diatomic molecules. In the case of a diatomic molecule, two electronic internuclear separation curves are schematized in Fig. 4.1. Here absorption of radiation will excite the molecule into a higher electronic level, with a greater probability of a transition occurring to a vibrational energy level whereby the internuclear separation does not change. This occurs at points where the electron is most likely to be located, that is where $\psi\psi*$ (i.e. the vibrational wave function multiplied by its conjugate) has a maximum. This is generally close to the edge of the potential energy curve; only the lowest vibrational state has a maximum probability in the middle of the vibrational energy level. The vibrationally excited state may then undergo radiationless transitions

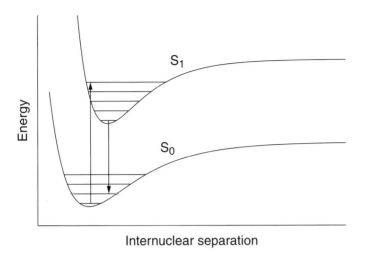

Fig. 4.1 Electronic transitions occur in both absorption and emission. The vibrational level to which the transition takes place is such that the internuclear separation does not change during the transition. After the transition, vibrational transitions can occur. Note that emission usually occurs from the lowest vibrational level; this is because the vibrational relaxation occurs faster than fluorescence, and is known as *Kasha's rule*.

down to the ground state, where an electronic transition may occur. Again, the transition is strongest closest to the electronic curve, where the wave function is largest. (These transitions are *vertical* and the idea that the internuclear separation does not change in an electronic transition is known as the *Franck–Condon principle*. This is easily understood if one remembers that the nuclei are very heavy compared to the electrons and therefore have little time to readjust after a transition.) In diatomic molecules, the energy spacing between the lowest vibrational transitions are equal. The principles described here equally apply to conjugated molecules, but internuclear separation is a redundant parameter, because there are more than two atoms present. Formally, one would consider a *configuration coordinate* rather than a separation.

To understand better the physical processes that occur when light interacts with materials, we need to understand all the relaxation phenomena that take place. This is best schematized in the Jabłoński diagram of energy states and relaxation processes.

4.1   The Jabłoński diagram

The Jabłoński diagram, named after the (Ukraine born) Polish physicist Aleksander Jabłoński presents energy levels and relaxation processes in a single diagram, an example of a which is shown in Fig. 4.2. The rate of the various transitions depend on whether or not they are forbidden. For example, triplet states cannot undergo radiative decay to singlet states because a change in the net spin is required. Spin–orbit coupling (Section 10.1.5) means that it is possible for a triplet state to radiate to the ground state. Being spin-forbidden, this is a weak transition, and so depletion of this state occurs more slowly han for an allowed transition. For this reason we use the word phosphorescence to describe these slow transitions, when a forbidden transition occurs. The time scales of radiation due to phosphorescence are generally accepted to be much

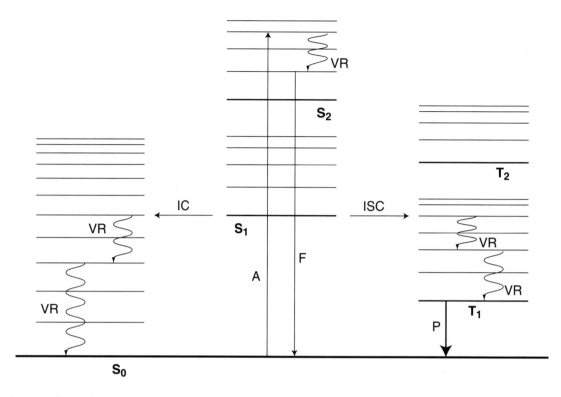

Fig. 4.2 A Jabłoński diagram showing ground and excited electronic and vibrational states of a molecule. Light is absorbed and excites an electron to a higher level (A). This may de-excite by fluorescence (F), but very often a radiationless vibrational relaxation transition (VR) occurs first. The fluorescence does not necessarily go to the ground state, but may end up at a vibrational level, where a radiationless transition to the ground state may then occur. Other transitions may also occur, such as internal conversion (IC), which is transition between energy levels of the same spin state, a transition from one spin state to another is known as an intersystem crossing (ISC), and is shown here from a singlet to a triplet state. The triplet state is generally long-lived and the radiation that eventually occurs is known as phosphorescence (P). Here the singlet spin states S_0, S_1, and S_2 are schematized, as are the triplet states T_1 and T_2.

greater than μs and can even occur on the order of seconds after initial excitation. By comparison, fluorescence generally takes place on the order of ns. For completeness, absorption is the quickest process, taking place with a time of the order of fs, while vibrational relaxations and internal conversion, being allowed transitions, occur in the same timeframe as fluorescence. However, when molecular potential energy curves overlap, it is relatively straightforward to have an intersystem crossing from singlet to triplet states. Intersystem crossings are also forbidden, but spin–orbit coupling again enables the transition. The intersystem crossing is generally faster than phosphorescence and one might consider it in the time domain of maybe ns to as long as ms. (The rate of phosphorescence can be increased by the addition of molecules containing heavy elements, and this is useful for generating white light, which we discuss in Section 10.1.5.) Intersystem crossings do not generally occur in the reverse direction; that is, a triplet state is less likely to form a singlet state by an intersystem crossing because triplet states are of lower energy than singlet states. In atomic physics one might remember this being due to Hund's rule; since the two electrons in a triplet state have the same spin, they must be separated in order to comply with the exclusion principle. This separation reduces Coulombic repulsion, lowering the energy compared to the singlet state.

4.2 Excitations in conjugated polymers

In conjugated polymers the π electrons are responsible for optical excitations and transitions are between the HOMO and LUMO. In fact the energy of a transition is a little less than that of the HOMO-LUMO gap because of the change in bond length that occurs when a molecule is excited. We have already seen this in Section 3.1 when we discussed transitions between the quinoid and aromatic forms of PPP. When an electron is excited into the LUMO, it experiences Coulombic attraction to the hole it has left behind. This attraction means that the electron and hole are paired, forming a quasi-particle known as an *exciton*, or sometimes as a polaron exciton. (Because the electron–hole binding energy is less than that of the free electron and hole it is not actually excited into the LUMO but rather just below it; if an electron were excited into the LUMO, it would be free of the corresponding hole. For the same reason, the hole has an energy just above the corresponding HOMO.) The band gap required to promote an electron to the LUMO is often differentiated from the energy gap between excitonic electron and hole by referring to it as the *electronic* gap, whereas the difference between the exciton and hole energies is sometimes known as the *optical* gap. It is generally the case that the optical gap is of lower energy than the electronic gap and lies within it.

The exciton as a quasi-particle may be treated in much the same way as other particles composed of two bound fermions such as the hydrogen atom. The exciton may be formed into both singlet and triplet states. The probability of this occurring under electroluminescence is currently a major theoretical and experimental problem. A triplet exciton can be formed when both the electron and hole have spin up or where they both have spin down. In both of these cases, they have total spin $S = 1$ because their spins are aligned and of value $1/2$. Quantum mechanically, we write both these states as $|1, 1\rangle$ for ($\uparrow\uparrow$) and $|1, -1\rangle$ for ($\downarrow\downarrow$). The third state with $S = 1$ is $|1, 0\rangle$. Here this is a linear superposition of both anti-parallel spin states, ($\uparrow\downarrow$) and ($\downarrow\uparrow$), which is denoted by $\frac{1}{\sqrt{2}}(\uparrow\downarrow + \downarrow\uparrow)$. The anti-symmetric linear superposition $|0, 0\rangle$ or $\frac{1}{\sqrt{2}}(\uparrow\downarrow - \downarrow\uparrow)$ corresponds to $S = 0$. There are three triplet states and one singlet state so, to a first approximation, the exciton is three times more likely to be found in a triplet state than in a singlet sate. There are two caveats to be noted here. Firstly, this is dependent on how the exciton is formed. If it is formed through excitation with light, triplet states are hardly likely to be formed at all, because a change in net spin is required, which can only be achieved through spin–orbit coupling and is very weak for photon excitation. (This is similar to the discussion in Section 4.1 of the decay of triplet energy levels to singlet states.) The second caveat is that when excitons can be formed statistically, they are still subject to different probabilities during formation. When electrons and holes first begin to interact, they may be on different polymer chains, which will change when their Coulombic interaction exceeds their thermal energy, whereupon a strongly bound excitonic state is formed on the same chain. If the process of transferring a singlet state on different chains to the same chain is considerably more efficient (i.e. faster) than that of the same process for the triplet state, and if the probability of an intersystem crossing of the triplet to the singlet state during this process is high enough (because of the slow transfer of the weakly bound triplet exciton to one chain), there will be a greater fraction of singlet states compared with the one in four that are statistically expected. In fact this is generally observed to be the case, which is fortunate for those involved in the development of electroluminescent devices, where triplet excitons can only decay to form photons radiatively through phosphorescence. These problems can be circumvented by *triplet harvesting*, and we turn to this in Section 10.1.5.

The binding strength of excitons is dependent on the band gap, but as we noted above, if the energies of the excitonic electron and hole corresponded to the LUMO and HOMO respectively, the exciton could not be bound; their potential energy would overcome any Coulombic attraction. However, if the exciton were located within the band gap then the electrostatic attraction would be large enough to stabilize the particle. So therefore why does the exciton not collapse to a point, as Coulomb's law would predict? In fact the exciton may be considered in the same way as a hydrogen atom, which is a bound electron and proton pair; here the exciton is a bound quasi-particle and the electron and the hole are stabilized by a balance between kinetic and potential energy. The kinetic energy is here in addition to the motion of the centre of the mass, and the potential energy is Coulombic.

The optical energy gap in organic molecules depends strongly on their length. We have discussed how the band gap decreases with increasing chain length in Section 2.4.1. This can be considered in a naive manner by treating the optical excitation following a particle-in-a-box argument, which was developed earlier in the book, leading to eqn (2.10). We can write eqn (2.10) in one dimension for a better comparison with (one-dimensional) chains of length L,

$$E_n = \frac{\hbar^2 \pi^2}{2m_e L^2} n^2. \tag{4.1}$$

An optical transition may take place between two energy states, and if we choose E_{n+1} and E_n, then the transition energy (band gap) may be written as

$$\Delta E = \frac{\hbar^2 \pi^2}{4m_e a^2 N}, \tag{4.2}$$

where we write the chain length L as $(2N - 1)\,a$. The assumption inherent in this calculation is that $N = n$, and this can be seen by counting the energy levels (solid lines) in Fig. 2.13, where n corresponds to the top of the HOMO and $n + 1$ to the bottom of the LUMO. Although a great simplification, this remarkably simple argument presents the starting point for an understanding of how longer molecules interact with light of longer wavelength than shorter molecules for the same electronic transition. To put eqn (4.2) into words, the excitation ΔE corresponds to light whose energy is inversely proportional to the conjugation length and hence the number of π bonds. The longer the molecule, the longer the wavelength it can couple to, and hence longer molecules (or specifically, molecules with longer conjugation lengths) have optical transitions at lower energies.

Excitons are not unique to polymers or indeed organic systems. Their inorganic counterparts are formed from the interaction with light in a similar manner to those in conjugated polymer systems. Like the polymeric version, inorganic excitons are also of a lower energy than the difference between the valence and conduction bands but the inorganic counterpart is much more mobile than those in polymer systems which are localized to regions of space defined by interfaces and defects. A polymeric exciton has typically a lifetime of several ns before it decays (usually by emission of light) during which time it can travel a few nm. Although excitons are here described as being due to optical absorption, they can also be formed when electrons and holes meet, as in electroluminescence.

Excitons play a crucial role in many technologies associated with polymer electronics. For example, light-emitting diodes emit light from the decay of an exciton. Conversely, photovoltaic applications require that excitons formed from the capture of photons dissociate into electrons and holes that can be efficiently collected at electrodes in the form of an electrical current. A key thread linking these two applications is the interaction of excitons with light. Light used to excite excitons need not be of a well-defined wavelength (this is not a resonance process), although it must of course be within the absorption band. Excitons subsequently formed are known as 'hot excitons' and decay non-radiatively to energy levels between the HOMO and LUMO.

On formation excitons are localized to within a few monomers on a chain. If they are very close together they are tightly bound (Frenkel excitons) and if they are further apart and thus weakly bound, they are known as Mott–Wannier excitons. These two names for excitons are limiting cases; Mott–Wannier excitons are weakly bound due to dielectric screening of their Coulombic interaction, which means that they are typically found in materials with a large dielectric constant and so are not normally relevant for organic systems, with a notable exception being carbon nanotubes. They may decay to their ground state either on a single molecule or via an interchain exciton. These interchain excitons exhibit efficient transport mechanisms, and occur either between chains stacked together or when the same chain is folded over. Interchain excitons are generally weakly bound. Because of the low conjugation length in these polymers, due to imperfections along the chain, or impurities, the excitons do not travel very far before they decay, radiatively or otherwise. This is in contrast to inorganic systems where excitons are more mobile and travel further within their more crystalline environments.

It is certainly true that the emission spectrum of an exciton is much narrower than its absorption spectrum. Whereas the absorption spectrum is broad due to, for example, the creation of *hot excitons*, the emission spectrum starts with excitons of a better defined energy with the electrons and holes sitting in their levels just below and just above the LUMO and HOMO respectively. This does not mean that the emission spectrum is one sharp peak corresponding to the electron–hole energy difference. *Inhomogeneous broadening* is responsible for broadening in both emission and absorption. The reasons for inhomogeneous broadening in both is due to disorder, which might for example be caused by the torsion of bonds in chains. Therefore impurities, bulk disorder, rotational (conformational) isomerism (known as *rotamers*) all contribute. The absorption peak is very asymmetrical, reflecting the fact that absorption is much more likely at high energy, but a long low energy tail is evidence of inhomogeneous broadening.

While the absorption spectrum is shifted to higher energy, the emission spectrum is correspondingly asymmetric around lower energies, or *redshifted*, due to a variety of possible factors. Excitons may for example become trapped at an imperfection or impurity. They may decay at such *traps*, or they may continue on when a significant jump in thermal energy allows it to escape the trap and continue. Despite the broadening, certain emission spectra may reveal *vibronic progression* where coupling between lattice vibrations (phonons) and the electronic state may be detected.

4.3   Exciton decay

As mentioned in the previous section, an exciton may decay radiatively or non-radiatively. Radiative decay is known as *luminescence*. There are a variety of forms of luminescence depending upon how the exciton was first formed. Photoluminescence refers to the radiative decay of an exciton formed by excitation with a photon, whereas electroluminescence is where an electric field is used to generate excitons; this is the process used in light-emitting diodes.

Processes leading to non-radiative decay can be understood with reference to the Jabłoński diagram shown in Fig. 4.2. Excitons of interest in most applications are generally formed in the singlet state, in which spins are anti-parallel with net spin $S = 0$. Excitons formed by photoluminescence, or by other forms of light generation such as optical pumping are uniquely formed in the singlet state. Nevertheless, these electrons may decay to triplet excitons by intersystem crossing, where they may decay radiatively, which is a phosphorescent process due to its being spin forbidden. It is more likely that these triplet excitons relax either vibrationally or through collisions or are dissociated yielding free electrons and holes. Dissociation is generally unlikely because it involves an energy barrier, but it may occur at an impurity or interface. All of the decay modes for triplet excitons are valid for singlet states, but it is recognized that the radiation rate constant of singlet states is much larger than that for triplet states. Another possibility is the creation of an *excited dimer*, which is often contracted to *excimer*,[1] which is an excitation-induced association, which also slows and redshifts the decay. An excited dimer is not specific to excitons, but does require that one of the components be excited in order to bind, in this case the exciton. An excited dimer is a resonance structure between different excited states, including charge transfer states, such as

$$AA^* \longleftrightarrow A^*A \longleftrightarrow A^+A^- \longleftrightarrow A^-A^+.$$

Excited dimers may also be formed between two different species A*B, and these are known as *excited state complexes* or *exciplexes*. We discuss both excited dimers and exciplexes in Section 4.9.

[1] Formally an excimer must be created between *identical* species, which is not the case in polymeric systems. We thus use the longer notation.

4.4 Photoluminescence

Most decay processes from excited states are exponential, and each excited state has a multiplicity of decay routes, meaning that the decay is generally the sum of a number of decaying exponentials. If we assume that the decay due to photoluminescence is a series of exponential terms and the rate constants control the decay, i.e. there are no branching ratios moderating the exponential (i.e. individual pre-factors to each exponential), then the photoluminescence intensity behaves as

$$I(t) = I_0 \exp\left(-kt\right). \tag{4.3}$$

Here, the total rate constant, k is additive and given by

$$k = k_{nr} + k_r, \tag{4.4}$$

where k_{nr} is the non-radiative and k_r, the radiative rate constant. Similarly, the time constant associated with the decay is given by $\tau = 1/k$ so

$$1/\tau = 1/\tau_{nr} + 1/\tau_r, \tag{4.5}$$

where the subscripts have their expecting meanings. It should therefore be obvious that

$$\tau = \frac{1}{k_{nr} + k_r}. \tag{4.6}$$

From these equations we can define the photoluminescence efficiency, Φ, by

$$\Phi = b\frac{k_r}{k_{nr} + k_r} = b\frac{\tau}{\tau_r}, \tag{4.7}$$

where b is the fraction of absorbed photons which produce singlet excitons. This equation is sometimes written as $\Phi = \frac{\tau}{\tau_r}$ because b is generally unity since only singlet states are expected to form in a photoexcitation process. Photoluminescence is generally strongly wavelength dependent, so Φ is usually stated as an integral over the entire emission spectrum.

4.5 Energy transfer

Despite the importance of photoluminescence as a loss process after photoexcitation, it is perhaps one of the least technologically useful. After all, photoluminescence is merely a means of replacing one photon with another of lower energy. The emission spectra also do not usually have a narrow distribution of energies either, and so there are few advantages in technologies whereby monochromation is important.

Other energy-loss processes include different means of energy transfer. For example, photons emitted by photoluminescence may be reabsorbed. A phenomenon such as luminescence reabsorption is more likely to be important in the solid phase because in solution there are likely to be fewer molecules available to absorb the emitted photon, unless of course, the solvent is capable of doing so. We do not consider luminescence reabsorption further except to point out that it can be considered an important loss process in some devices. Direct energy transfer by overlap of the electronic energy levels and, especially, dipolar interactions leading to Förster transfer are much more important, and we discuss these here.

Electronic overlap (usually referred to as *Dexter energy transfer*) is only important for direct neighbours where the excitation can simply tunnel from one molecule to another. The molecules are so close together that the excitons effectively do not know to which molecule they belong. The range of this interaction is consequently very small. Förster transfer, on the other hand, is of considerably longer range. Förster transfer is often known as FRET—Förster resonance energy transfer or NRET, meaning non-radiative energy transfer. Often FRET is used as a contraction of fluorescence resonance energy transfer, but this should be avoided because the process is non-radiative. The exciton is treated as an oscillating dipole, which can form a resonance on a neighbouring molecule. This rapid process (~ 10 ps) can occur over lengths as large as 5-10 nm. Dipole–dipole interaction energies scale as $1/r^6$, so the efficiency of the interaction also scales in this fashion. We write the Förster transfer efficiency as

$$E_{\mathrm{FT}} = \frac{1}{1 + (r/r_0)^6},\tag{4.8}$$

where r_0 is the Förster radius, defined as the distance whereby the energy transfer efficiency drops to half of its maximum value. After energy transfer the donor molecule relaxes back to its ground state non-radiatively, leaving the neighbouring molecule excited to either radiate or otherwise decay to its ground state.

4.6 Exciton dissociation

The singlet exciton formed by photoexcitation may simply be dissociated. To dissociate an exciton, either energy is required to overcome the exciton binding energy, or some impurity might form a trap at which the exciton can reduce energy by dissociation. If the exciton is to be dis-

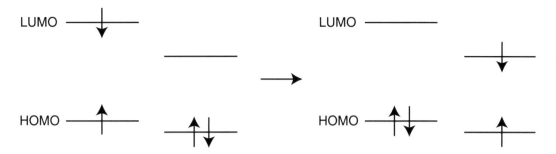

Fig. 4.3 If the exciton can reduce its overall energy by jumping from one molecule to another, it may do so. In this example, the reduction in energy of the electron jumping from one HOMO to another is easily enough to offset the increase in energy of the hole. Dependent upon the nature of the band gaps, in some case the hole, rather than the electron, will reduce its energy, or both the hole and electron may reduce their energy in undergoing such a process.

sociated in the absence of defects or other interfaces, an external source of energy would be required. Here it is necessary to separate the electron and hole by a distance r_C so that their electrostatic (Coulombic) attraction is equal to their thermal energy, $3k_BT/2$. In this case

$$r_C = \frac{e^2}{6\pi\epsilon_r\epsilon_0 k_B T}, \tag{4.9}$$

where ϵ_r is the dielectric constant and ϵ_0 is the electric constant (permittivity of free space). When the electron and hole are separated further than r_C, thermal energy is enough to keep them apart. An electric field would be one means of achieving this, since the number of monomers separating them must increase in an electric field. Of course, in many cases these excitons may simply be reformed by the recombination of excitons and holes, and so these competing effects need to be considered. Recombination is perhaps less likely at a trap because the trap has already lowered the energy of the components, meaning an energy barrier has to be overcome for recombination to take place. When the electron and hole recombine at a distance greater than r_C, the process is known as bimolecular recombination. This unusual terminology is historical, because for small molecules it would not be possible in such cases to have the electron and hole located on the same molecule. This bimolecular recombination need not take place with the same hole and electron, and may form with other free carriers. The other recombination process, which occurs for separations smaller than r_C, is known as geminate recombination.

Excitons that are formed from charges on different chains are known sometimes as polaron pairs. These interchain excitons have a comparatively larger dipole moment than excitons formed from polarons located on the same chain. Because the polarons that make up the exciton are usually further apart than those on the same chain, they are easier to separate because their distance apart is closer to r_C.

Whether an exciton is formed on the same chain or on different chains, it is known as a *charge transfer complex*, which may either recombine (radiatively or otherwise) in a decay to the ground state, or dissociate.

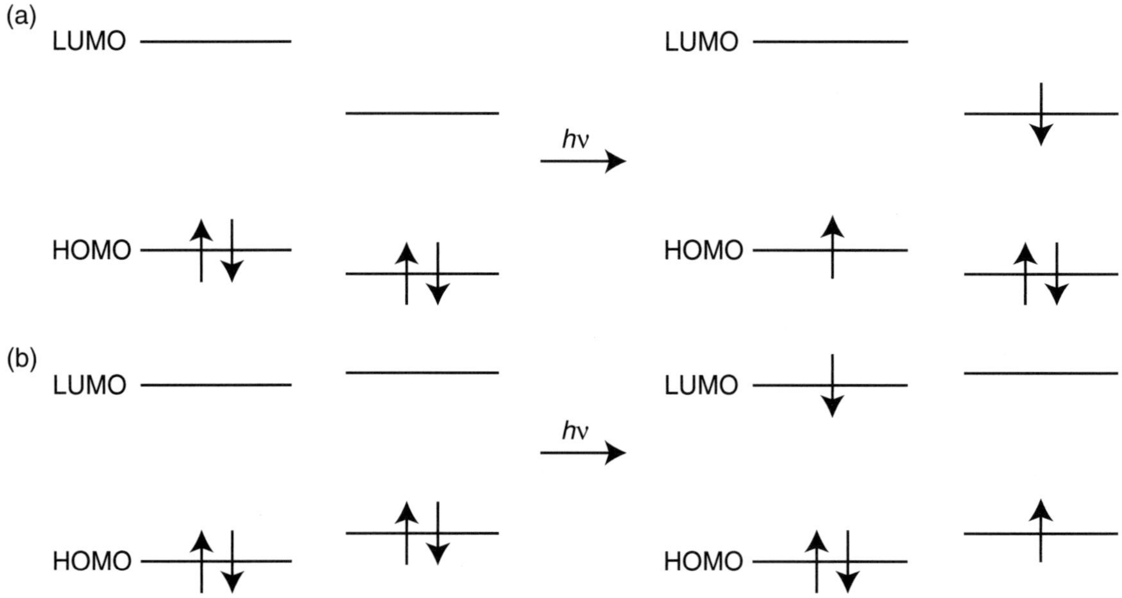

Fig. 4.4 (a) The donor (left-hand species) absorbs the incident photon, the energy of which is used to transfer an electron to the LUMO of the acceptor. This sensitization process involves a charge transfer complex being formed without the formation of an exciton on the donor. (b) The equivalent direct process to transfer a hole to the acceptor is also possible and occurs by exciting an electron in the HOMO of the acceptor to the LUMO of the donor.

Hot excitons have an energy above that of the charge transfer complex and may dissociate without it being formed; i.e. the exciton might directly separate into an electron and a hole; the excess energy means that the exciton is unstable.

The loss of excitons due to energy transfer is understood simply on the basis of the reduction in energy when an electron and hole swap places in neighbouring (but different) molecules. This is illustrated in Fig. 4.3. The ability of excitons to jump from one species to another has a number of advantages. Because this is a very fast process (several ps), it will dominate over fluorescence decay. It is probable that the decay of the exciton on the second species will be through fluorescence emission. A particularly important use of exciton transfer is in devices whereby the active molecule has a low absorption coefficient. Absorption can take place in a different molecule, which uses this exciton transfer process to excite the active species. A good example of this process is the blending of PPV with poly(vinyl carbazole). The PPV captures the photon, creating an exciton, which can then be transferred to the poly(vinyl carbazole) for use in applications such as xerography. Another important use is in the dye sensitization of photocells, whereby the sensitivity of the cell can be optimized in a given wavelength range by the addition of an appropriate absorber. There are other uses, for example in the control of the wavelength emission of the device, or in the deliberate creation of energy traps in a device. Such energy traps have a variety of uses, one of which might include the suppression of charge migration in devices that depend upon the control of photorefractive behaviour, for example in holographic materials.

4.7   Photoconduction

The illumination of a semiconductor can enhance conduction by increasing the quantity of mobile charge carriers. Illumination adds energy to the system, which may be used to excite electrons to the LUMO. This is achieved usually with the help of a sensitizer, which is a dye molecule that absorbs light from a region of the spectrum that cannot be usefully used to this purpose by the semiconductor. The sensitizer is then able to use some of that energy to create electrons and holes, which are in turn injected into the main material. In its strictest sense, dye sensitization does not involve an exciton being formed on the donor molecule, but rather a process known as charge transfer absorption. The donor absorbs incident radiation, but, rather than forming an exciton, an electron from the HOMO of the donor is excited to the LUMO of the acceptor (Fig.4.4a). Because the energy levels of the acceptor lie lower than those of the donor, a lower-energy photon is required for this process. The redshift in the absorption spectrum is a signature of charge transfer absorption processes and can therefore be used to identify it in spectroscopic experiments. It is of course possible that the dye is excited up to its LUMO forming an exciton, which lowers its energy by transferring that electron to its neighbouring host molecule (Fig. 4.3).

In order that dye sensitization work most effectively the absorption must be redshifted relative to the band gap of the host. In this way light will only be absorbed by the dye and polymer and not the polymer alone. This is important; if the polymer were to absorb the light it is possible that the light would be re-emitted at a longer wavelength, so we require a dye that can aid the conversion of light to charges. This does not necessarily mean that the band gap of the dye should be smaller than that of the host, because the excitation process is such that the electron from the HOMO of the dye is directly promoted to the LUMO of the host, and the LUMO of the dye need play no role in the process. The electron in the polymer LUMO may now contribute to charge transport whereas the hole remaining in the dye is trapped.

For hole carrying materials, the HOMO of the dye must be at a lower energy than that of its polymeric host. In the direct process, absorption of light promotes an electron from the HOMO of the host molecule to the LUMO of the dye (Fig.4.4b). This is again a form of charge transfer absorption and is equivalent to a hole jumping from the dye to the polymeric semiconductor.

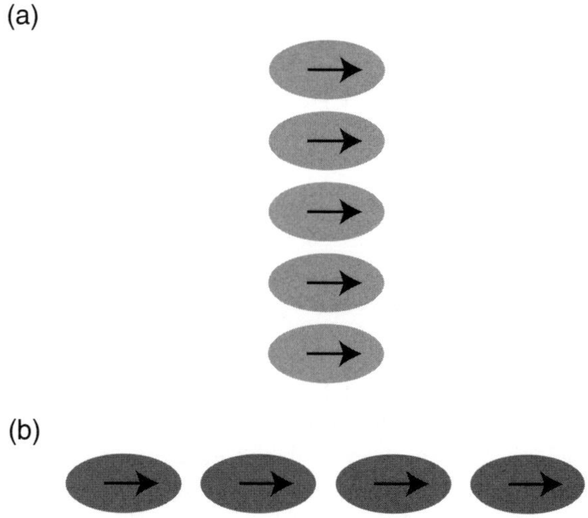

(a)

(b)

Fig. 4.5 (a) H-aggregates are formed with their transition dipole moments parallel, whereas for J-aggregates (b) these are arranged end-on. The arrows indicate the direction of the dipoles, and the shading indicates the shape of the molecules.

[2]There is no dipole moment in a π-conjugated system; if there were electrons could not move freely along the chain backbone. Excitation by light creates a *transition* dipole, which allows for dipolar intermolecular interactions. A transition dipole should not be confused with the standard dipole moment, but it does have units of Cm.

4.8 Aggregates

Aggregates exist in solution when molecules are placed in a poor solvent. When the molecules are in low concentration, or in a marginal solvent, smaller aggregates may occur. We are here interested in small aggregates with interesting optical properties, known as H- or J-aggregates. Which is which depends on the alignment of their transition dipoles.[2] J-aggregates align with dipoles head-to-tail and H-aggregates align parallel. Both of these are schematized in Fig. 4.5.

In the solid state, molecules generally (but not always) tend to align and form crystalline structures. They align because intermolecular forces can best be accommodated by moving them into positions whereby their energies can be reduced. Of course there is an entropic price to pay in this regard, but this is usually not prohibitive. In semiconducting polymers, this can be due to the stacking of aromatic rings (π-stacking), which is illustrated in different contexts in Figures 4.8 and 7.5. Dipole moments are also important in affecting ordering; for example, van der Waals solids exist because dipoles (permanent or induced) align other dipoles in neighbouring atoms or molecules. In a solution these factors remain important, but crystalline solids are not formed—rather, small aggregates of molecules.

4.8.1 Aggregates of small molecules

The formal physical description of H- and J-aggregates is really only accurate for small-molecule behaviour, but the respective blue- and red-shifts in their optical lines have parallels in the optical properties of polymers, and so we must introduce these aggregates for small molecules and polymers separately.

The two different structures shown in Fig. 4.5 have different energies. The parallel structure in Fig. 4.5a is of a higher energy than that in

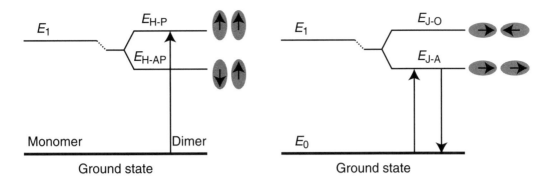

Fig. 4.6 The energy levels of monomers are split when they are formed from dimers, with a higher energy state corresponding to charged segments being together, and a lower energy state when adjacent charges are opposite. The midpoint of these two energy levels ($0.5(E_{H-P} + E_{H-AP})$ and $0.5(E_{J-O} + E_{J-A})$) is lower than that of the isolated molecules (E_1). If a dimer is excited with parallel dipoles, an exciton is formed between the two molecules that cannot decay because there is no route to the lower energy state, hence a blueshift in the absorption spectrum. If the molecules align end-to-end, then absorption into the higher energy state E_{J-O} is suppressed, but absorption and emission can take place into or from E_{J-A}, The ground state is labelled E_0.

Fig. 4.5b. This is due to the fact that the dipoles are aligned in the head-to-tail structure (Fig. 4.5b) so that the positive end of the dipole meets the negative end of its neighbour, must be a lower energy state than having the ends of the same charge aligned, as in Fig. 4.5a. If we now excite these states with light, the parallel structure has an absorption that is blueshifted with respect to that of the small molecule, and does not emit. The head-to-tail structure has a redshift in its absorption spectrum relative to the isolated molecule, and this excited state can fluorescence down to its ground state. The energy levels are shown in Fig. 4.6.

The absorption of light in both H- and J-aggregates is associated with the formation of excitons. To understand the redshift in the J-aggregates, we return to Fig. 2.13, which shows the band gap decrease in size as more alkene units are added. The alignment of molecules in a J-aggregate can be seen as being equivalent to this, and thus the conjugation length is increased. An increase in conjugation length corresponds to a redshift (Section 4.2). The parallel with conjugation length exists for interchain excitons, as must be the case for those participating in excited aggregates. There is no reason why, when excited, these molecules cannot fluoresce, and this is indeed observed.

H-aggregates exhibit the opposite behaviour to that of J-aggregates. The alignment of the molecules forces a shorter wavelength exciton than the individual isolated molecules. In the absence of other interactions, H-aggregates would simply separate and so their electronic states are associated with anti-bonding orbitals, and are consequently blueshifted with respect to the individual molecules. Such an exciton cannot fluoresce easily due to Kasha's rule (Fig. 4.1), which tells us that an H-aggregate (or indeed any electronically excited species) must reach the lowest energy state of that electronic level before fluorescing. To achieve this there must be a switch from parallel dipoles to anti-parallel dipoles, which is clearly forbidden. Absorption in both J- and H-aggregates is not only optically shifted, but produces absorption lines that are considerably sharper than those for the isolated molecules.

4.8.2 Polymer aggregates

Although one can calculate the dipole moment for a semiconducting polymer, it is of little relevance to aggregates.[3] Here the conjugation length is relevant, and this length is over only a few monomers and partly reflects the stiffness of the molecule. For a bond angle θ, a monomer dipole moment I_0, and a conjugation length covering N_c monomers, the dipole moment will be to a first approximation $N_c I_0 \cos (0.5 (\pi - \theta))$. (What constitutes the bond angle is shown in Fig. 7.2.) The conjugation length can be taken as the dipole in a polymeric J-aggregate. Should two polymers come together in solution, their co-location may be stabilized by two elements of conjugation aligning (i.e. two dipoles) aligning in a J-aggregate, as schematized in Fig. 4.7. Such a dipolar alignment cannot extend the conjugation, because these are two separate chains. However, it can support an interchain exciton, and because the length over which this exciton exists is larger than one conjugation length, the J-aggregate is redshifted.

H-aggregates are likely to form due to packing advantages, such as those that would give rise to semi-crystalline structures of polymers (see, for example, Fig. 7.5). These structures are relatively stable, and polythiophenes are a particular class of conjugated polymer for which such aggregates can readily form.

Polymer aggregates are less well behaved than small-molecule H- and J-aggregates; the coupling between the dipole and the exciton is not always complete, so the blueshift of an H-aggregate may not always be observed or may be accompanied with redshifts in the absorption spectrum. Such phenomena are quite complicated and may involve a contribution due to phonons, and is well beyond the level of this book.

[3]The dipole moment of a polymer is the vector sum of the dipole moments of its monomers. This will average to zero for large polymers in no external field, but the magnitude of the dipole moment will be given by $I = \sqrt{N}I_0$, where there are N monomers, each with dipole moment of magnitude I_0. This is the standard *random walk* calculation and will be derived in a different context in eqn (7.2).

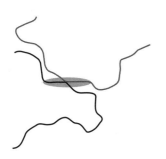

Fig. 4.7 A polymeric J-aggregate may occur when two regions, each capable of band transport, of different chains align to allow an exciton to form (shaded region), which can help to stabilize the aggregate.

4.9 Excited dimers

It is possible for the absorption of light to excite a metastable aggregate of two conjugated molecules. Such an excited state is an excited dimer when the two molecules are the same, or exciplex when they differ. There is only a limited attraction between individual molecules in the ground state, and these are generally dispersive, i.e. due to long-range van der Waals interactions. (True excimers are actually repulsive in their ground state.) Other possible causes of intermolecular attraction are the geometric effects causing crystallization. Electronic interactions are certainly prohibited because the orbitals are filled, and the like charges therefore repel, as they do when any two molecules are brought together.

The excited state interaction is stabilized because both polymers can share the excitation. To achieve this, π-stacking generally needs to oc-

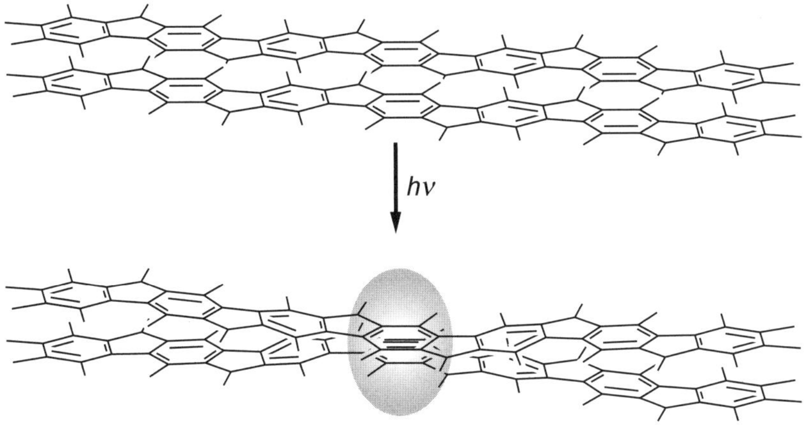

Fig. 4.8 Excited dimers form between conjugated polymers under the dual absorption of light. Ladder PPP, being a rigid planar polymer, is a good candidate for emission due to such a process. Its highly rigid structure keeps the dimer stable, but its emission, when it occurs, is redshifted. LPPP normally emits in the blue region of the electromagnetic spectrum, but its excited dimer emission emits yellow light. (Experimental evidence for excited dimer emission in LPPP may be complicated by emission due to defects.) The groups R and RR' (see Fig. 3.11) are omitted for clarity.

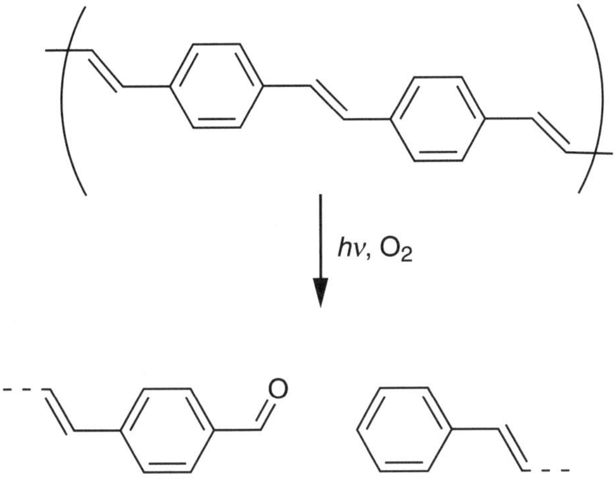

Fig. 4.9 Chain scission can occur when PPV is exposed to UV light. In conjugated molecules the double bond is usually the point at which this occurs, since all other bonds are more stable. In this example, ambient O_2 reacts with the conjugated polymer to break the bond, with oxygen incorporated in one end to form an aldehyde group (RCHO, where R is standard notation for a side group). Another possible outcome of photo-induced oxidation of PPV is the inclusion of a ketone group (Fig. 3.8), which still breaks the conjugation.

cur, as shown in Fig. 4.8. One of the two polymers absorbs light leading it into an excited state, which can relax down to a lower energy state.

4.10 Photodegradation

As we have seen, light absorption leads to interesting affects, many of which can be exploited for practical use. The degradation of polymers due to light absorption is also a common effect and is usually detrimental to what scientists and engineers are trying to achieve, and avoiding it often requires a great amount of effort. (Sometimes, however, it can be useful, for example some synthetic methods might take advantage of photochemical crosslinking.) The most common photodegradation route is via oxidation, which is therefore avoided by working in an oxygen-free atmosphere, such as in vacuum or in nitrogen or an inert gas. Such alternatives, while manageable in many cases, limit the applicability of any device and raise its cost. Oxygen may break the conjugation through chain scission (Fig. 4.9) but may also have other effects such as chemical

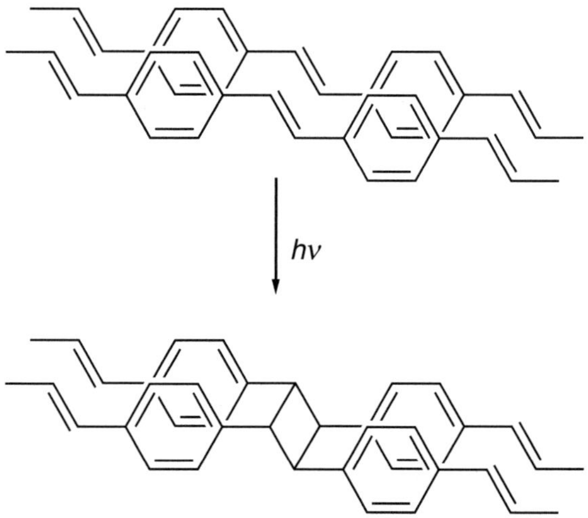

Fig. 4.10 Photodimerization is quite possible in conjugated polymers where chain stacking allows conjugated bonds to break, forming interchain bonds. This *photocyclodimerization* reaction is common in polymeric crystals.

dimerization (Fig. 4.10), which is a form of crosslinking reaction.

The effects of these reactions are such that the conjugation is broken, but in the case of intermolecular interactions, the solubility of the polymer will also be affected. The extent to which photochemical degradation occurs depends on the different polymer. Some polymers are very photostable, and others less so. No polymer is entirely immune to photochemical degradation; even stable saturated molecules will degrade if the wavelength is short enough; after all, photons bring energy into the system, and this energy can break bonds. Oxidation may occur in small amounts, but even this will have a significant effect on the optical properties of the polymer. Any effect that reduces the conjugation length induces a blueshift for the reasons discussed in Section 4.2. Oxidized regions are very effective traps and can dissociate excitons very easily, simply because the highly electronegative oxygen will take the electron from the exciton. Oxidation therefore decreases the luminescence efficiency of the polymer.

4.11   A simple device

Here we introduce the phenomenon of electroluminescence, which is used in light-emitting diodes. Before starting this discussion, it is worthwhile to consider the basic construction of a device in its simplest terms. To operate a device, one needs two electrodes which are connected to a power supply. The active layer is between the electrodes and there are many different forms that it can take, not necessarily being a single discreet layer. Single layers of one component are possible, but are less effective because they generally do not display the required transport behaviour and are less efficient. For both photovoltaic devices and LEDs, electrons and holes need to both travel through the medium, and this is best if there is one component available for hole transport, and another for electron transport. Nevertheless, in Fig. 4.11 we show the electronic

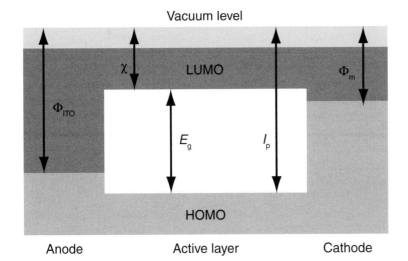

Fig. 4.11 The energy-level structure of a single layer device. The anode is usually (but not always) made from indium tin oxide, which has a large work function Φ_{ITO}. The cathode is a low work function (Φ_{m}) metal. The active layer is characterized by its LUMO, HOMO, band gap (E_{g}), electron affinity, χ, and ionization potential, I_{p}.

structure of a single-layer device.

The anode is used for hole injection for LED operation. In the case of photovoltaic behaviour, the anode is where holes are collected, but its properties should be the same. Indium tin oxide (ITO) is a commonly used anode because it has good transparency to visible radiation. The reader will appreciate the importance of letting light in for a solar cell, or allowing light out for a LED. (It is not a requirement that the anode specifically be transparent, merely that one of the electrodes is. Given the suitability of ITO, this is generally the anode of choice, allowing some flexibility for the cathode.) The cathode is usually made of calcium or aluminium, and provides electron injection in LEDs and electron collection in photovoltaic devices.

The cathode and the anode must have different properties to be useful for electron and hole injection respectively. Since the hole is travelling in the HOMO, a large work function is necessary. A large work function prevents electrons being sent into the LUMO, but it should not usually be too difficult for holes to tunnel into the active layer, because the energy barrier is small. Similarly, the cathode must have a low work function. Here, electron injection is important, and these electrons must make it into the LUMO, so a cathode Fermi level close to the LUMO allows tunnelling of electrons from the cathode into the LUMO.

The active layer should have the relevant transport properties for electrons and holes, and is characterized by the difference between the LUMO and HOMO (its band gap, E_{g}), its ionization potential, I_{p}, and its electron affinity, χ. The electron affinity represents the decrease in energy in going from a neutral element to a singly negatively charged anion. No element repels electrons, i.e. has a negative electron affinity, although those with full outer electron orbitals have affinities that may be taken as zero. For these reasons, all molecules will also have electron affinities above zero. For semiconducting materials, the electron affinity must be limited by the LUMO, since putting an extra electron in

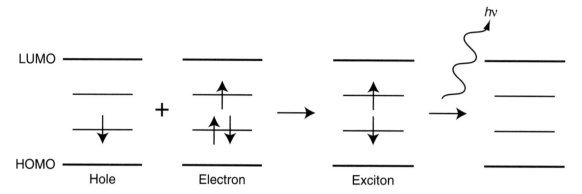

Fig. 4.12 In electroluminescence a hole polaron and an electron polaron move under the action of an electric field to combine and form an exciton. This exciton radiatively decays to its ground state.

the HOMO is not possible, because it is thermodynamically filled, and there are no states available in the band gap. The ionization potential is therefore given by $I_p = \chi + E_g$. Note that the ionization potential and the work function are not the same. The work function is a bulk quantity related to the Fermi level, whereas the ionization energy is a molecular quantity. The work function is the energy required to raise an electron from the Fermi level to the vacuum levels, whereas the ionization potential is the energy required to remove the outermost electron to infinity. There is no point in describing a work function for an undoped semiconductor or insulator, because there are generally no electrons at the Fermi level, since this is located between the valence (HOMO) and conduction (LUMO) bands. (The reader with a strong background in physics may recall that the work function is also dependent upon the Miller indices of the surface orientation.)

4.12   Electroluminescence

Electroluminescence is the emission of light in response to an electric current passing through a material. As mentioned previously, electroluminescence is the phenomenon harnessed for application in light emitting diodes. An electric field is applied across an active material transporting holes and electrons in opposite directions. These meet at some point within the active layer, combining to form an exciton, which may decay radiatively. The decay processes of the excitons thus formed are identical to those that may occur in photoluminescence discussed in Section 4.4. This should not be surprising because the exciton, being merely a bound electron–hole pair should be independent of its origin.

Under an applied electric field, positive and negative polarons are injected at the anode and cathode respectively. They may meet within the active layer, or they may simply travel through to the other electrode. Should they meet and interact, then an exciton will be formed. This is most likely to occur when they are on the same polymer chain. The energy levels for this process are shown schematically in Fig. 4.12.

The technological challenge is to increase the electroluminescence efficiency, i.e. to emit as many photons as possible per unit of power dissipated in the device. Physically, we consider this as per electron and hole injected into the device. To this end it is instructive to consider the loss processes in electroluminescence. The first is stated in the preceding paragraph, which is the possibility that the electron and hole do not meet; they simply carry on to the other electrodes or are otherwise lost. Clearly the efficiency of photoluminescence is proportional to the probability that the electron and hole recombine, η_1, where $1 - \eta_1$ is the probability that they do not meet. Should they meet, we learnt in Section 4.3 that triplet states are spin forbidden to decay via fluorescence emission to their ground states. We therefore need to take into account that the electroluminescence efficiency is proportional to the probability of forming a singlet state, η_2. Finally, we need to acknowledge that not all singlet excitons decay radiatively, and this means that we should take into account the quantum efficiency for (singlet) photon emission. This quantum efficiency has already been discussed in Section 4.4 and may be given by $\eta_3 = \tau/\tau_r$. Putting all of these probabilities together, we have the (internal) electroluminescence efficiency (internal quantum efficiency) as

$$\eta_{\mathrm{EL}} = \eta_1 \eta_2 \eta_3. \tag{4.10}$$

Improving η_{EL} requires great thought being given to the design of the device. It is not possible to improve η_2 because this is set by physics. Generally, $\eta_2 \approx 0.25$ because this represents the statistical probability of an exciton being formed in the singlet state rather than triplet state. However, as discussed in Section 4.2, it is often the case that $\eta_2 > 0.25$ because of the competition between the transfer of a triplet exciton to be localized on one chain with the likelihood of an intersystem crossing occurring to the singlet state. Experiments reporting the size of η_2 are unreliable because it is particularly difficult to extract the different components of η_{EL} and there remains controversy in this subject; many researchers do not accept that there are significant departures from a purely statistical value of $\eta_2 = 0.25$. It should be sufficient to say, that device design is not an appropriate parameter for increasing η_2 but it might be that a different choice of material or active component might help. The same applies to η_3; there is not much a device scientist can do to improve this efficiency other than to change the active component from which the luminescence arises. This leaves η_1 as the efficiency where most improvements can be made. How this is done will be described later in this book (Sections 8.2 and 10.1). The qualification *internal* to the product of the efficiencies is an important one. For example, the light must travel out of the device, possibly through transparent electrodes or some other window, both of which will have their own absorbances, limiting the practical efficiency of the device. The internal electroluminescence efficiency is essentially a scientific problem and improving it requires control of the physics. The external electroluminescence efficiency (or external quantum efficiency, EQE) requires not just control of the physics, but also the more practical, engineering, problems such as the windows, and ensuring that light is emitted in the right direction (*out-coupling*). Although it is largely an engineering matter, out-coupling is an area of great interest and importance with many research groups putting significant effort into this aspect of improving performance.

4.13   Further reading

The physics behind excitations leading to optical behaviour is covered in many texts. All the reader could ever want to know about the basic principles underlying the Jabłoński diagram and related photophysics, as well as the experimentation involved in the measurement of optical properties is covered by Lakowicz (1999). This book covers fluorescence generally, so for questions specific to conjugated polymers the reader is directed to review articles on the subject such as those by Bässler (1997) and Scholes and Rumbles (2006) on excitons, or Akcelrud (2003) and Friend et al. (1999) on electroluminescence. Photophysics is covered by Rothburg (2007). Much of the material in this chapter is given a strong theoretical treatment in the book by Barford (2013), The material here is appropriate for those with a good understanding of the basics of theoretical physics.

4.14   Exercises

4.1. The band gap in eqn (4.2) is an approximation based on $n = N$ for very large N and its derivation requires a binomial expansion. Retaining the assumption $n = N$, what is ΔE when the next correction (expansion) term in N is included?

4.2. The energy-level structure of a simple single-layer device is shown in Fig. 4.11. Show that the Fermi level of the semiconducting layer can be given by $\epsilon_F = (I_p - \chi)/2$.

4.3. Given that the dielectric constant of poly(9,9-dioctylfluorene-*alt*-benzothiadiazole) (F8BT, Fig. 6.2) is 3.2, calculate the maximum number of monomers associated with an exciton at 320 K assuming (a) a density of 0.92 g/cc, or (b) a bond length of 0.142 nm and a bond angle of 120°. (These values are appropriate for benzene rather than F8BT.) Do the alkyl side chains (C_8H_{17}) in this polymer stop the formation of interchain excitons?

The answers that you should obtain for the maximum number of monomers associated with an exciton are generally much larger than the results of experiments. What factors might be responsible for limiting the size of an exciton to perhaps one or two monomers?

Charge transport

5

How electrons and holes move within polymers is of considerable importance for the design of all-polymer transistors (Chapter 9) and optoelectronics devices (Chapter 10). In this chapter we shall consider the basics of charge transport within polymer devices, and discuss the two main mechanisms of charge transport: hopping and band transport. We also consider the way that doping affects charge transport, as well as the behaviour of traps.

It is an important consideration that polymer crystallinity is never even close to 100% (Section 7.2), except in certain rather artificial and largely unimportant environments. In a normal metallic conductor, charge transport is much less affected by grain boundaries than might be the case in polymers. In an inorganic semiconductor, charge transport is similarly less affected. It is also for this reason that the diffusion lengths of excitons in inorganic materials is somewhat greater than in polymers. The morphology of polymeric materials therefore plays an important role in charge transport. We shall discuss the physics of polymers later in this book, but we need to be aware of the important issues at this point. Conjugated polymers are stiff by their nature; the double bond conveys a degree of stiffness to the molecule that is lacking in saturated polymers. Stiffness is an aid to packing and this therefore means that many conjugated polymers are semi-crystalline, which also has consequences for charge transport.

To quantify charge transport in polymeric materials we need to consider a rather simple situation whereby a material is injected with either holes at the anode or electrons at the cathode. These are injected without losses into the material at any desired density, controlled by the bias voltage. In real situations there are complicating factors; materials may be mixed, and both holes and electrons may be injected into the device. Nevertheless, to understand transport processes, such complications must be eliminated.

The form of charge transport most accessible to explanation in conjugated polymers may be considered to be intrachain transport, because this is the starting point for our understanding of the properties of polyacetylene, for example. This would be an underestimation of the complexities of the subject. It is not only that polymer chains have a finite length, but also that there are other mechanisms of transport that play a significant role. Some polymers, for example the polycarbazoles, which can have non-conjugated backbones, have a strong interchain charge transport mechanism known as *hopping*. (Of course, hopping can take

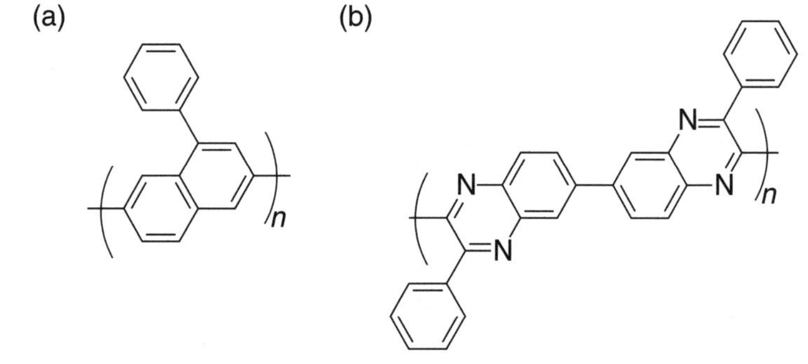

(a)　　　　　　　　　　　　　　(b)

Fig. 5.1 Two examples of n-type semiconducting polymers: (a) poly(2,6-(4-phenylquinoline)) and (b) poly(phenyl quinoxaline).

place along the chain too, and such intrachain hopping may well be significant in the polycarbazoles and other similar polymers.) Interchain hopping is facilitated in semi-crystalline materials where close-packed neighbouring chains are available. However, grain boundaries and impurities also exist and impede transport. Fortunately, charges can cross grain boundaries, and this illustrates the challenges in understanding the complexity of such morphology-dependent charge transport.

There are other matters to be overcome when one starts to think about optoelectronic devices, which we shall discuss in Chapter 10. Charge transport should also be considered along with charge injection. For example, in polymeric LEDs one needs to convert hole and electron polarons into excitons to generate the light. By applying a voltage across an active layer, we ensure only that a current will flow. Ideally this will be of equal amounts of electrons and holes, because this will maximize electroluminescent efficiency, but it may be that close to 100% of the current is due to holes. Clearly, in such circumstances, excitons cannot be generated, and the device will not work. This is not only a matter of choosing the right active layer, but of the careful selection of electrodes, a problem that was briefly discussed in Section 4.11. Similar issues are relevant for photovoltaic devices, whereby electrons and holes are generated from the exciton created by the incident light. If electron mobility were negligible in the device, then only half the possible current could be generated, because only the holes would contribute. However, the problem is worse than that because static electron polarons would generate a charge imbalance which can only be removed by annihilation of a hole, thus further limiting the current. If we can inject the requisite density of holes and electrons into the sample, then their uniform transport requires *ambipolar* behaviour,[1] an important property of devices, which was also mentioned in Section 4.11. Unfortunately, ambipolar transport is not easily achieved and is a challenge in the field of polymer electronics. Presently, most devices are dominated by hole transport, which is generally much more efficient than electron transport. For this reason we need to consider what determines whether a polymeric semiconductor exhibits electron or hole transporting behaviour. Good electron and hole balance for optoelectronic devices can, however, be achieved by device engineering tricks, such as mixing two materials.

[1]Ambipolar diffusion requires movement of positive and negative charges at the same rate. In organic electronics an ambipolar material is one in which both holes and electrons have significant mobilities, and the requirement that they be the same is generally relaxed.

Although increasingly good electron transporting polymers are being observed in polymeric semiconductors, hole-transporting materials tend to have better mobility and stability. The reasons for the relative quality of hole transporting materials compared to electron transporting materials are phenomenological and not rigorously understood. Electron traps are one means of inhibiting electron mobility, and the presence of oxygen is a good means of trapping electrons. Oxygen is a powerful means of mopping up electrons due to its high electronegativity, and it is present on many impurities that might get into the device. Water is, for example, rather difficult to remove from materials. Heating is one option, but that might degrade some materials; vacuum treatment is also effective at removing water and can be used. (The boiling point is of course the temperature at which the vapour pressure of the liquid is equal to the atmospheric vapour pressure, and so by lowering the environmental pressure, one also lowers the boiling point; a pumped vacuum chamber therefore provides a means of eliminating water and other liquid contaminants from a device.) Unfortunately, the use of vacuum treatment can add significantly to processing costs and one has to perform further processing to ensure that after removal from the vacuum atmospheric water or other contaminants do not return. Good electron-transporting devices can be made by using materials having a low lying (deep) LUMO, and these are increasing in number. To this end, a LUMO that is more than 4 eV is generally regarded as air-stable, because –OH groups can easily form traps to disrupt electron transport. A deep-lying LUMO (high electron affinity) leaves electrons significantly below the vacuum level and thus the energy provided by hydroxyl (–OH) and similar traps is insufficient to interact with the electron from the polymer. The classic n-type polymer is the ladder polymer BBL, shown in Fig. 3.14. Other nitrogen-containing polymers also have good possibilities for use as electron transporting materials, for example, polyquinolines and polyquinoxalines (Fig. 5.1). All of these polymers have high electron affinities, which is very useful for the efficient injection of electrons into devices, such as LEDs, as well as raising the energy barrier to oxidation. Recent developments have added better solubility to the properties of air-stable electron transporting polymers, and an example is shown in Fig. 9.8.

Another possible explanation has been proposed for electron transport being less prevalent than hole transport is the suggestion that holes and electrons have asymmetric properties; mobile electrons are associated with the the LUMO and holes with the HOMO. There is no clear reason why these should result in symmetric properties and equivalent charge transport. It has been postulated that the LUMO, to which the electron polaron is associated, is more localized than the HOMO. Consequently, electron transport is less efficient than hole transport. (Localization differentiates inorganic materials, which have delocalized charge carriers, and consequently more wave-like motion. The more localized charges are, the more hopping contributes to the transport properties, and the less efficient the mobility becomes.) Experiments on highly crystalline materials do however show that high electron mobilities are indeed possible, and so most would ascribe traps as the reason why electron transport is generally not as good in organic materials as hole transport.

5.1   Band transport

We have introduced many of the issues that give rise to the complexity of charge transport processes in polymers. However, we can treat these problems by considering band transport as exclusively for intrachain motion and hopping for both intrachain and interchain transport.

Band transport is the mechanism that we should expect when one considers the idea of conjugation. The wave function of the charge is affected by an applied field and so the polaron travels along the chain in a direction defined by the field. There are limits to how well band transport can contribute to conduction in polymeric samples as discussed above. Even if a pure polymeric crystal were synthetically possible there would be limitations on band transport. Firstly, there is no reason to assume that the chain is aligned with the field; a polaron on a chain aligned perpendicularly to the applied field cannot contribute to band transport without hopping onto another chain. Even without such trivial arguments, a perfect crystal will have polaron transport disrupted by phonons. These lattice vibrations increase with temperature, and so it follows that the resistivity of the material does too. Given that polarons can be scattered by phonons, there is a temperature dependence on the mobility, and experimentally the carrier mobility scales with temperature as $\mu \propto T^{-\nu}$, where $1 < \nu < 3$. In fact, the Einstein relation can be used to show that the mobility of a polaron is expected to be inversely related to temperature

$$\mu = \frac{De}{k_{\mathrm{B}}T} \tag{5.1}$$

provided that the diffusion coefficient D of the charge carriers has itself no complicated temperature dependence. The diffusion coefficient can be given by $D = v\lambda_{\mathrm{m}}$, where λ_{m} is a mean free path and v a drift velocity. The problem here lies with λ_{m}, which is also temperature dependent; at higher temperatures more phonon modes (shorter wavelength lattice vibrations) are excited, which decreases the mean free path of the polarons before scattering. In a non-metallic crystal, at temperatures above what is known the Debye temperature, the number of phonons increases linearly with temperature, which, all other things being equal, will give rise to a diffusion coefficient inversely proportional to temperature and a mobility which scales as $\mu \propto T^{-2}$. Here we are starting to make assumptions which may or may not be valid. At this point it is sufficient to note that the temperature dependence of the mobility is expected to be somewhat greater than T^{-1}, i.e. $\nu > 1$.

This temperature behaviour of band transport is more effectively tested on small organic materials which can form perfect crystals, such as the polyacenes (Section 3.7) rather than true polymeric materials. In polymeric systems, the presence of traps complicates the temperature dependence of carrier mobility considerably, and they are thus poor candidates for such experiments. Even in small-molecule organic crystals, there remains debate over the true contribution of band transport to conduction, because of reports of rather small carrier mean free paths in these crystals that are smaller than would be expected even accounting for phonon scattering.

The presence of traps in a material do not preclude the existence of a band transport mechanism, for one can modify the mobility to take this into account. Here one assumes that the charges move under an applied electric field. When they are caught in a trap, they require a certain amount of thermal energy to break out, which is expected to follow an Arrhenius-type rate behaviour. The modified mobility is thus given by

$$\mu = \mu_0 \exp\left(\frac{-\Delta\epsilon_t}{k_B T}\right), \tag{5.2}$$

where $-\Delta\epsilon_t$ is the depth of the trap, which usually indicates that the energy of the trap lies somewhere in the band gap, and μ_0 is a constant that depends on the degeneracy of the trap states (i.e. how many states there are with a trap energy $\Delta\epsilon_t$ below the LUMO for an electron, or above the HOMO for a hole). Clearly only polarons in traps within an energy around or smaller than $k_B T$ from the LUMO (or HOMO, depending on the nature of the charge carrier) have a significant chance of being released quickly enough to contribute to charge transport. This modified band transport mechanism is known as multiple trapping and release band transport. In some ways it is a useful contribution to understanding charge transport, but it is also a little disingenuous in that it can confuse the nature of the transport mechanism. For example, there is no guarantee that the polaron, on being released from a trap, might not find itself on a different chain, and this multiple trapping and release model does not differentiate between the two. In fact the temperature dependence of the behaviour can be construed as similar to hopping anyway, so this model muddies the distinction between the two mechanisms.

5.2 Hopping

Hopping (*thermally activated tunnelling*) is a process by which charge carriers 'jump' from one monomer to another (in a quantum-mechanical tunnelling process) rather than travelling coherently (Fig. 5.2). Although it cannot be denied that band transport is an important charge transport mechanism in both polymeric semiconductors and their small-molecule organic counterparts, it is clear that hopping mechanisms are much more important in contributing to the carrier mobility. For conjugated polymers it is likely that a mix of the two occur, since a pure hopping mechanism is also unlikely. One should note that hopping can also occur in intrachain transport, even for conjugated polymers. A kink or defect in a conjugated polymer will give rise to the opportunity for

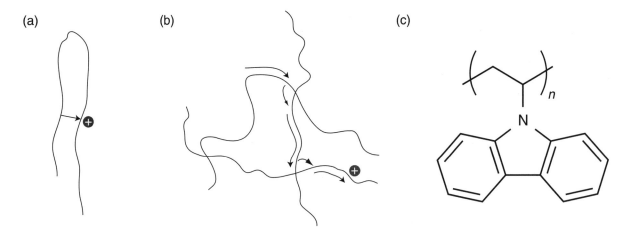

Fig. 5.2 One way of differentiating hopping transport from coherent (band) transport is that the charge carrier does not need to travel from one monomer to the next. This does not preclude intrachain transport, which may occur by a hopping mechanism, especially when there are defects in the chain (a). In conjugated polymers a mixture of coherent transport and hopping is possible, with the charge carrier hopping from one chain to another (b). Some polymers such as poly(vinyl carbazole) (c) can only exhibit hopping transport because they have no conjugated backbone.

charge to move from one part of a chain to another without travelling along the chain (Fig. 5.2a). Similarly, some non-conjugated materials enjoy good semiconducting behaviour with charge transport being exclusively due to hopping. Poly(vinyl carbazole), often denoted by PVK, is perhaps the simplest example of a polymer where hopping transport occurs exclusively (Fig. 5.2c). Transport in such a polymer may be between adjacent chains, or it may be along the same chain either by jumping from one part to the other as in Fig. 5.2a, but it can also occur along the chain, mimicking the behaviour of coherent band transport. Of course, the carrier cannot travel coherently because the chain is not conjugated, but the (in this case) hole can move between the aromatic rings without involving the chain backbone in its motion.

Hopping transport does not give rise to mobilities as large as those for pure band transport. One contributing factor for the lower mobility is the need for an activation energy in the hopping process. In the case of PVK, delocalization occurs on and between the two aromatic rings, and for a hole to leave one and move to another carbazole unit it needs enough energy to leave the first. For this reason, hopping transport becomes more efficient at higher temperatures, in contrast to band transport which we have seen (Section 5.1) decreases the mobility through phonon scattering. The simplest way to couple an electronic motion with phonons is to recognize that phonons contribute energy to the system and therefore change the energy levels of the charge carrier. A hole on a carbazole unit, for example, is localized in its HOMO. When a phonon arrives it increases the energy of the hole, which gives it the opportunity to lower its energy by moving to a nearby carbazole. The likelihood of hopping taking place will depend on the energy difference between the two sites, as well as the distance between them. A simple theoretical formalism for this hopping mechanism was proposed in 1960 by Allen Miller and Elihu Abrahams for transport in doped inorganic semiconductors, but there is no fundamental reason why this model cannot be equally applied to organic materials. In order for the Miller–Abrahams model to be applied to polymeric systems, we must treat each monomer as essentially isolated from its neighbour, even if they exist on the same chain. This must be the case for the polycarbazoles, because the monomers are connected through the non-conjugated backbone. The hopping rate of this Miller–Abrahams model is given by

$$\gamma = \gamma_0 f_{\mathrm{FD}}(\epsilon_i)(1 - f_{\mathrm{FD}}(\epsilon_j)) \exp\left(-\frac{\epsilon_j - \epsilon_i}{k_{\mathrm{B}}T}\right) \exp\left(\frac{-2R_{ij}}{a}\right), \qquad (5.3)$$

where R_{ij} is the inter-site separation, ϵ_i and ϵ_j are the respective energies of the donor and acceptor sites, a is a distance constant known as the localization radius, γ_0 is a constant, and the function f_{FD} represents the Fermi–Dirac probability distribution (eqn 2.15). The two exponentials are to be expected; it is not surprising that the hopping rate should decrease exponentially with increasing energy of the acceptor state and distance apart. The exponent containing $\epsilon_j - \epsilon_i$ is a simple Boltzmann factor, whilst the exponential containing R_{ij} is a measure of the degree of overlap between the electronic wave functions. If the donor energy level is greater than the acceptor, i.e. if $\epsilon_i > \epsilon_j$, then this exponential is replaced by unity, so that

$$\gamma = \gamma_0 f_{\mathrm{FD}}(\epsilon_i)(1 - f_{\mathrm{FD}}(\epsilon_j)) \exp\left(\frac{-2R_{ij}}{a}\right). \qquad (5.4)$$

The Fermi–Dirac probability distributions represent the probability that there is a charge carrier available in the donor site available to hop to the acceptor and the probability that there is a state available in the acceptor to take that charge carrier.

The Miller-Abrahams model is important because it links distance and energy to the hopping rate. However, it is difficult to apply this model to real systems because of the disorder inherent in non-crystalline materials which means that it is not possible to relate R_{ij} to a position on a (crystalline) lattice. Similarly, it is problematic to define the energy states ϵ_i and ϵ_j given that these are affected by phonons. To account for the spread in energies and distances, Heinz Bässler and colleagues included a Gaussian density of states in the energy levels (as well as a degree of position disorder) that can also be assumed to be Gaussian.

The Gaussian density of states may then be given by

$$g(\epsilon) = \frac{N}{(2\pi\sigma_g)^2} \exp\left(\frac{-\epsilon^2}{2\sigma_g^2}\right), \qquad (5.5)$$

where N is the total number of states and σ_g is the width of this Gaussian density of states. This Gaussian density of states is of a different form to the density of states shown in Fig. 2.3 because we are considering entirely different carrier distributions. In the free electron model of metals all electrons are available for conduction, and one allocates them energies dependent upon the availability of states at a given electronic wave vector starting from the lowest available wave vector until all electrons have been considered. In the case of this hopping mechanism we need to think carefully about the differences between this mechanism and the coherent transport mechanism. Coherent (band) transport represents the simple physics that we should like to be able to apply to semiconducting polymers, because the resultant band theory has many parallels with inorganic materials. For example, we can readily compare the density of states of an inorganic semiconductor (Fig. 2.7) to a perfect conjugated chain with a density of states that may behave as shown in Fig. 2.16. Here the HOMO and the LUMO levels are defined either side of the band gap. Hopping, however, does not lend itself to such lattices. The wavelengths of the charge carriers is determined primarily by defects in the structure and the availability of locations to where they can hop. The charge carrier has a wave function that is essentially confined within a limited lattice because of disorder in the surrounding medium. The number of available states increases for similar reasons as that in crystals; one starts with a limited number of low-energy states which increases due to increasing degeneracy as one goes to smaller wavelengths. However, at the smallest wavelengths, which we equate to a distance between monomers, there remains disorder inhibiting a sharp cut-off in a density of states distribution. Similarly, because we are not considering perfect crystals there will be a distribution in nearest neighbour distances, which means some carriers will experience different maximum wave vectors to others. These two effects—the spread in the HOMO and the positional disorder—are referred to as *diagonal* and *off-diagonal* disorder and are characterized by (Gaussian) departures from a mean HOMO by σ_g or position by Σ. The resultant density of states therefore differs from that discussed in Section 2.4.

To express the effect of these different parameters, Bässler proposed a carrier mobility given by

$$\mu\left(E, T\right) = \mu_0 \exp\left(-\frac{2}{3}\left(\frac{\sigma_g}{k_B T}\right)^2\right) \exp\left(\left(\left(\frac{\sigma_g}{k_B T}\right)^2 - \Sigma^2\right)\sqrt{E}\right),$$

(5.6)

where E is the electric field and μ_0 is a constant. If one recalls the definition of carrier mobility from eqn (2.3), one imagines it to have a simple inverse dependence on E. We see clearly from eqn (5.6) that this is not the case. One should note also that the LUMO is affected by this energetic disorder in the same way as the HOMO, and this is included in eqn (5.6). For low fields, the second exponential can be neglected, and the temperature dependence is controlled by first term so that logarithmic plots of mobility as a function of the square of inverse temperature should be linear. For many materials this is the case, and an example is shown in Fig. 5.3.

An interesting consequence of eqn 5.6 is that the mobility can decrease with *increasing* field should Σ^2 be large enough, i.e. $\Sigma > \sigma_g/k_B T$. Indeed

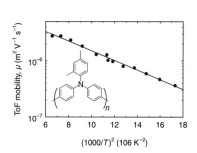

Fig. 5.3 Mobility data as a function of the square of inverse temperature obtained using the time-of-flight method for hole transport in a polytriarylamine (inset). The data are for temperatures between 240 and 390 K. These data are taken from Veres et al. *Adv. Funct. Mater.* **13** 199 (2003).

there are experimental data that demonstrate this to be true. The effect of increasing disorder becomes smaller for the most disordered systems. To account for that, Σ^2 is usually not allowed to exceed 2.25 because any more disorder is not seen as causing further decreases in mobility. This might seem rather *ad hoc*, and indeed the model originated in computer simulations, but it explains rather well mobility due to hopping in disordered systems. Nevertheless, the model is not perfect and fails, for example, to account for polarons (i.e. the charge carriers) affecting their environment or for experimental observations that mobility can increase with increasing carrier density.

5.3 Space charge

We have seen that charge carriers move in a semiconducting polymer by either coherent transport or by the hopping of electrons or holes along the chain, or from one chain to another. We know that traps will affect the movement of charges, and we shall turn to this below in Section 5.3.2. However, the nature of the material and its purity are not the only factors controlling charge transport. If we apply electrodes across a conducting medium, a current will flow from one electrode to the other. However, the current has a simple dependence on the applied electric field if the charge density within the medium is small, in which case the current density is given by eqn (2.1). In such a case, the charges present in the medium do not experience an environment affected by each other, but can be treated as isolated charges under the influence of an electric field, E. The voltage between the electrodes can therefore be treated as a parallel plate capacitor (or a capacitor of whichever geometry is appropriate) with a capacitance given by $C = Q_e/V$, where Q_e is the charge on the electrodes and $E = V/d$, where d is the plate separation. The capacitance of a parallel plate capacitor of plate area A is $C = \epsilon_r \epsilon_0 A/d$, so the minimal presence of charge carriers in the medium is to simply treat the medium as a dielectric of constant ϵ_r. This is therefore of little interest if we are considering conducting materials, so we need to look at the case where there is a substantial presence of charge carriers in the medium.

The criterion by which we do not neglect the charge carriers in the medium is if their total charge Q is comparable to Q_e. Under these circumstances, the applied electric field is screened by neighbouring charges and its effect is consequently lower. Because these neighbouring charges are closer to a given test charge than the charges on the plates, the electrostatic force (and consequently the field) exerted on a test charge by the neighbours is greater than that by the plates. The current drawn between the electrodes is therefore limited by these space charges, and is known as the *space charge limited current* (SCLC), and is an important concept in practical devices such as light-emitting diodes. The SCLC density (which is a charge flux) is given by

$$j = \frac{9\epsilon_r \epsilon_0 \mu V^2}{8d^3},$$ (5.7)

where μ is the free charge mobility.[2] This equation is valid when the movement of the charges is dominated by electrostatic effects (i.e. the applied electric field) rather than simple thermal diffusion. The importance of this result, which is often known as *Child's law* or sometimes the *Mott–Gurney law*, is that it is independent of the injection barriers at the interface between active layer and electrodes that were first introduced in Section 4.11, but rather the bulk properties of the medium (i.e. ϵ_r and μ) and the applied voltage V and layer thickness, d. In fact, this SCLC regime provides a possible means to determine μ for particular materials, because the mobility is the gradient of a plot of j as a function of V^2.

Although we present a formal proof of Child's law in Section 5.3.1 below, a simple argument can be used to explain its features. We suppose that a given charge will take a time τ to travel from one electrode to the other. The charge of interest should be Q_e, which means that $j = Q_e/A\tau$. Given that the charges travel distance d in time τ (by our definition of τ), we can use eqn (2.3) and the capacitance of the medium to obtain $j = Q_e \mu E/Ad = \epsilon_r \epsilon_0 \mu V^2/d^3$.

5.3.1   Proof of Child's law

A more precise derivation of the SCLC density between parallel electrodes is obtained by first solving Poisson's equation

$$\nabla^2 V = -\frac{\rho}{\epsilon_r \epsilon_0},$$ (5.8)

where ρ is the charge density in the medium. In electromagnetism, Poisson's equation is a simple restatement of Gauss's law, which is (in one dimension)

$$\frac{\mathrm{d}E}{\mathrm{d}x} = \frac{\rho}{\epsilon_r \epsilon_0}.$$ (5.9)

It is helpful to make the substitution

$$\rho = e\left(n - \bar{n}\right),$$ (5.10)

where n is the number density of free charge carriers in the medium and $\bar{n}$ is n in the absence of an applied field. We shall use a method proposed by Murray Lampert in 1956 which requires the following reduced variables:

$$w_{\mathrm{L}} = \frac{e^2 \bar{n}^2 \mu x}{\epsilon_r \epsilon_0 j},$$ (5.11)

$$u_{\mathrm{L}} = \frac{e\bar{n}\mu E}{j},$$ (5.12)

[2]It is inherent in the derivation of Child's law that the mobility is independent of applied field. For situations where this is not so (eqn 5.6), the theory requires modification.

and

$$v_L = \frac{e^3 \bar{n}^3 \mu^2 V}{\epsilon_r \epsilon_0 j^2},$$ (5.13)

where x is a distance from one electrode in the direction of the other. Using eqns (2.2) and (2.3), it is readily shown that $u_L = \bar{n}/n$, so

$$\frac{du_L}{dw_L} = \frac{1}{u_L} - 1.$$ (5.14)

We note that we can also obtain v_L from

$$v_L = \int_0^u u_L \frac{dw_L}{du_L} dx = \int_0^u \frac{u_L^2}{1 - u_L} du_L,$$ (5.15)

although readers might prefer to convince themselves of this result by replacing the reduced variables in the integrand explicitly. In this case, one should note that $du_L/dw_L = (E/e\bar{n}) \, dE/dx$ and so

$$v_L = \int_0^{u_L} u_L \frac{e\bar{n}\mu E}{j} \frac{e\bar{n}}{\epsilon_r \epsilon_0} \frac{dx}{dE} \frac{e\bar{n}\mu}{j} dE = \int_0^{u_L} \frac{e^3 \bar{n}^3 \mu^2}{\epsilon_r \epsilon_0 j^2} E \, dx,$$ (5.16)

where we use $V = \int E \, dx$. We need to require $u_L = 0$ at $w_L = 0$ as our boundary condition and noting that eqn (5.15) is a standard integral, we obtain

$$v_L = -\frac{1}{2} u_L^2 - u_L - \ln(1 - u_L).$$ (5.17)

We can also integrate eqn (5.14) to obtain

$$w_L = -u_L - \ln(1 - u_L).$$ (5.18)

Unfortunately, there is no analytical solution to recover j from eqns (5.17) and (5.18), but we can make approximations. In the SCLC regime, we have large j, so we can assume $w_L \ll 1$. This approximation allows us to expand $\ln(1 - u_L)$ as a Maclaurin series (i.e. a Taylor expansion about zero), which gives

$$w_L = \frac{u_L^2}{2}$$ (5.19)

and

$$v_L = \frac{u_L^3}{3},$$ (5.20)

which can be combined to give

$$v_L = \frac{2 w_L^{3/2}}{3}.$$ (5.21)

We simply substitute for v_L and w_L from eqns (5.13) and (5.11) into eqn (5.21), and rearrange for j to obtain Child's law (eqn 5.7).

5.3.2 Traps

The derivation of the SCLC density in Section 5.3.1 ignores the presence of traps, which act as regions where charge is stored. Traps reduce j, the SCLC density, because they help to screen any electric field across the material. Although charge carriers trapped in defects or at the interface between two different species do not move from one electrode to the other, and so do not contribute to the SCLC, they do affect their surroundings, i.e. they contribute to the space charge. One can therefore replace μ in Child's law (eqn 5.7) by an effective mobility, μ_{eff} given by

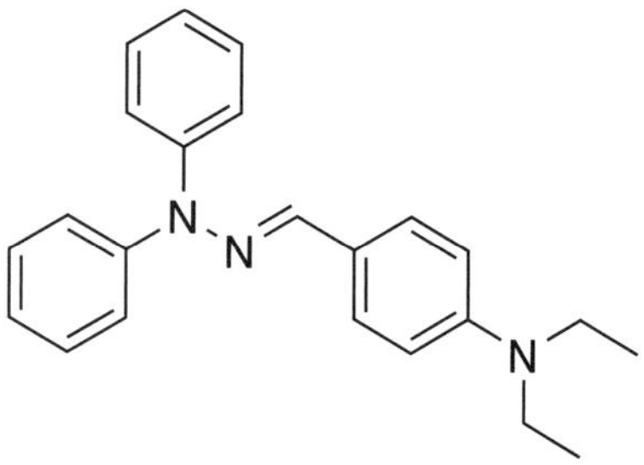

$$\mu_{\text{eff}} = \frac{\bar{n} - n_{\text{trap}}}{\bar{n}} \mu, \tag{5.22}$$

where n_{trap} is the density of trapped charge.

Traps are often categorized as either *shallow* or *deep*. A shallow trap has a depth of $\sim k_{\text{B}}T$, meaning that the charge will escape the trap with its thermal motion. When transport is primarily due to hopping, it is difficult to say when hopping ends and motion in traps begin since both have a thermal activation energy barrier. Charges may escape deep traps if the applied field is large enough. The field, after all, affects the energy landscape, and by applying an electrostatic force on the trapped charge Eq, it may raise its energy above the depth of the trap and in so doing allow it to escape. Clearly then, the greater E is, the more charges that may be mobile, which in turn raises μ_{eff}. Child's law (eqn 5.7) will fail in such circumstances with $j \propto V^{\xi}$ where ξ can be very large indeed, with values $\xi > 10$ having been reported, although this is likely to be observed over a small range of V. The small range in V can often be explained as being caused by the traps having a similar chemical nature.

Although traps are generally unpredictable, one can introduce known quantities of impurities into a material in order to control the behaviour of a device. The logic is simply that if charge carriers are trapped in isolated impurities, then the addition of enough impurity in a controlled manner should allow charges to move between impurities. As an example, PVK (Fig. 5.2c) is a polymer with a high ionization potential where hopping transport is dominant. The high ionization potential means that it takes a lot of energy to remove an electron, which is essentially the case for a polymer exhibiting hole transport. Of course, when hole transport is taking place, large energies are not required, because when a hole hops from one site to the other, the energy cost in arriving at a new site is paid by the energy gained in leaving one site behind. This is not energy that necessarily has to be bought at one point to be paid back later; although an activation energy does exist in hopping transport, the overlap of the wave functions facilitates the movement of the charges from one site to another. If this hole finds itself in a trap of lower ionization energy, it will lower its overall energy for the same reasons; the energy gain behind the PVK site it has left behind is more than the energy cost of moving to the trap. This hole will then be trapped until it acquires enough thermal energy to escape. Such traps can therefore ruin the performance of devices. However, by adding a low-ionization

Fig. 5.4 4-(diethyl amino)benz-aldehyde diphenylhydrazone is an example of a low-ionization energy molecule that efficiently traps holes.

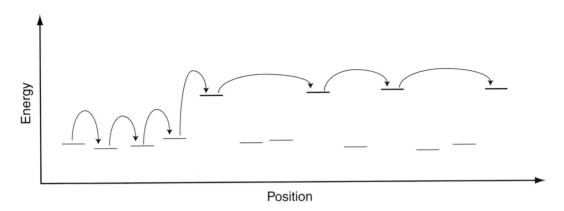

Fig. 5.5 Schematic representation of holes hopping from one site to another. If an impurity is present, these holes can be trapped. However, if there are many such traps, e.g. by introducing a controlled impurity into the system, then a high mobility will be recovered because the holes will hop from trap to trap. Traps are indicated by energy levels marked as bold lines. In this diagram the energy refers to the electron energy, so holes jumping 'up' in energy to a trap is not an increase in energy of the system, because when the hole leaves for a trap of lower ionization potential, the site it has left behind recovers its electron, which because of the difference in ionization potentials means a net reduction in energy for the system. Note also that the sites do not have equal energies, but rather a spread in energy.

energy impurity such as 4-(diethylamino)benzaldehyde diphenylhydrazone (DEH) shown in Fig. 5.4, the holes can travel from one DEH to a neighbouring DEH (Fig. 5.5). This means that as small amounts of DEH are added, hole mobility decreases due to its effective function as a trap, but as its concentration increases, this decrease is arrested, before mobility increases with increasing DEH concentration.

5.4   Time-of-flight

The development of new materials requires of course that they be well characterized. Similarly, even well-understood materials need to be tested, as part of quality control. Arguably the most important property of a polymeric semiconductor is its charge mobility. The two most popular means of measuring mobilities are the time-of-flight (ToF) method and field-effect transistors. Transistors are sufficiently important to be be discussed fully in Chapter 9. The ToF method (Fig. 5.6) involves a laser pulse of short duration to create charges in the material. The charges will move under an applied electric field to their respective electrodes. The time it takes for these charges to arrive at the electrodes is used to calculate the device mobility.

If one wishes to measure hole mobility, the laser pulse will photogenerate charges near the anode. Electrons formed in the pulse will be removed immediately by the anode leaving just the holes of interest. Ideally the laser pulse would be of infinitesimal duration so that the holes generated would leave the anode at the same time. The time that

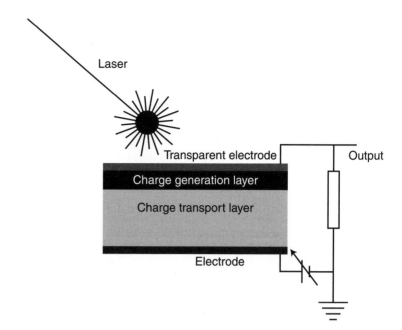

Fig. 5.6 The time-of-flight experiment requires a short laser pulse to generate charges in a layer near one electrode. The charges travel through the material of interest, which is typically a few microns in thickness, to the other electrode. The output (*transient*) is sent to an oscilloscope and recorded on a computer.

the holes arrive at the cathode would then be given by

$$\tau = \frac{d}{E\mu},$$
(5.23)

where μ is the hole mobility, E is the applied electric field, and d is the layer thickness. Of course this is simply a rearrangement of eqn (2.3), so the idea is really very simple. Of course, the laser pulse is not of infinitesimal duration, so not all holes start at the same time, and each hole has a different journey to the cathode, experiencing different numbers of traps, which broaden the distribution of holes arriving at the electrode compared to that created by the laser pulse, as schematized in Fig. 5.7.

In ToF experiments the sample should be of sufficient thickness such that the initial width of the charge carrier distribution is small compared to the thickness of the sample. This initial width is dependent on the penetration depth of the laser into the sample. This is often not useful because one might have limited material or one might desire (as is usually the case) to fabricate a thin device. In such cases one can use thin dye layers at the surface to capture the light. This dye-sensitizing layer is an example of the photoconduction described in Section 4.7. It is important in the sensitization process that the injection of charge carriers into the host material be fast. If this injection is a slow process compared to the drift time in the sample, then interpretation of the results will be awkward. The contacts that linking the sample and electrodes of the ToF instrument should not cause the injection of other carriers into the device. Clearly, it is desirable only for the originating from the initial pulse to be mobile within the sample.

In a perfect experiment, the width of the charge carrier distribution

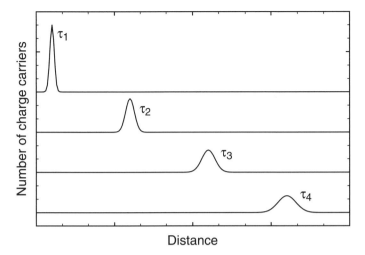

Fig. 5.7 Charge carriers are generated by a laser pulse and shortly after formation formation (τ_1) they have a rather narrow velocity distribution. Under the influence of an applied electric field the charges drift towards the relevant electrode, with the width of the distribution broadening as the charges get closer to the electrode, as shown at times τ_2, τ_3, and τ_4.

would be controlled solely by the temperature; at higher temperatures the thermal velocity distribution governed by the Maxwell–Boltzmann is broader than at lower temperatures. The Maxwell–Boltzmann distribution in one dimension (because here we shall neglect thermal motion orthogonal to the direction of the applied electric field) is given by

$$p\left(v_x\right) = \sqrt{\frac{m_q}{2\pi k_\mathrm{B}T}} \exp\left(\frac{m_q v_x^2}{2k_\mathrm{B}T}\right), \qquad (5.24)$$

where m_q is the mass of the charge carrier, and v_x its speed in the direction of the applied field. Of course the carrier kinetic energy is related to applied voltage by $eV = m_q v_x^2/2$ so one can deduce that this velocity distribution has a width that increases in time with $\sqrt{k_\mathrm{B}T/eV}$. This might not be obvious from the Maxwellian distribution because it contains no indication that the velocity distribution should be dependent on time, but one can equate the kinetic energy to the electrical energy to give

$$\tau^2 = \frac{m_q x^2}{2eV}, \qquad (5.25)$$

where x is the distance travelled over the transit time τ. The significance of this is that, as τ increases, so does x, and consequently the width of the distribution at x increases also. This is different from a standard Maxwell–Boltzmann velocity distribution for an ideal gas, because in that case the average position of a gas molecule does not change.

Actual ToF data are shown in Fig. 5.8, and it is clear that the results are somewhat different from the rather idealized situation discussed above. The complications to the ToF measurement are very often due to thickness effects. If d is small, then the centre of the charge carrier velocity distribution can become rather ill-defined and the distribution itself becomes asymmetric. This is known as *dispersive* transport and requires much more sophisticated modelling to explain the charge carrier velocity distribution.

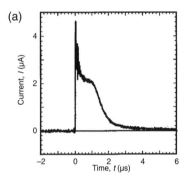

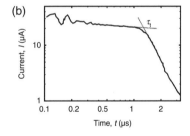

Fig. 5.8 (a) Time-of-flight data for a poly(9,9-dioctylfluorene) sample. (b) The transit time, τ_t is obtained from the intercept of the two broken lines in the double logarithmic plot of the data. Taken and adapted from Redecker et al. *Appl. Phys. Lett.* **74** 1400 (1999).

Fig. 5.9 The metal-semiconductor junction is schematized for (a) the metal and undoped semiconductor (pristine polymer), (b) metal and p-doped semiconducting polymer, and (c) metal and n-type semiconductor. The left-hand images show the separating materials, and the right-hand side after contact. In each case it is assumed that there are enough charge carriers to equalize the Fermi levels. χ is the electron affinity of the semiconductor, I_p the ionization potential of the semiconductor, Φ_m the work function of the metal, and ϵ_F the Fermi level for either the semiconductor or the metal. There is no band bending for the pristine polymer because there are no mobile charge carriers in that material. In the case of the p- and n-type materials holes and electrons may flow from the semiconductor into the metal as well as *vice versa*. In the example shown in (c), the Fermi level of the metal remains greater than that of the semiconductor, so net electron flow will be from the metal to the semiconductor but for the p-type semiconductor, charge flow can be in both directions. In both (b) and (c) mobile charges within the semiconductor respond to the potential difference caused by the equalization of the Fermi levels by forming a depletion zone at the interface. It is here that band bending occurs.

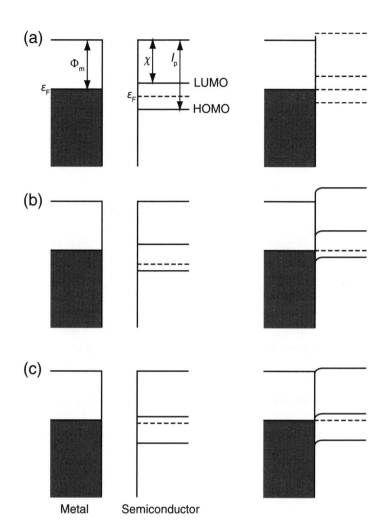

5.5   Injection

5.5.1   Band bending

If two metals are brought into contact, their Fermi levels are equalized. Free electrons in the metal of the larger Fermi level are able to lower the overall energy of the system by moving into the other metal. This continues until the two Fermi levels are the same, or equivalently, when the chemical potential is continuous and equal across the boundary. In a polymer device, metal contacts are still used, although there are a few exceptions when synthetic metals such as PEDOT can be used. Under no applied field, electrons from the metal cannot escape into the LUMO because these are generally higher energy states.

In a doped semiconducting system, the flow of charge at the metal–semiconductor interface is important. Here, there are sufficiently mobile charge carriers to move across the interface to equalize the two Fermi levels, assuming an adequate level of doping. Although the Fermi levels can be equalized, the band gap, which is a property of the electronic structure of the polymer, must remain unchanged in magnitude. (Although the Fermi level is a property of any material, it is a reflection of the number of mobile carriers in that material and so can be altered by the addition or removal of charge carriers.) For an n-type semiconductor the HOMO and LUMO levels are both lowered (by the same amount) to accommodate this change in Fermi level, and for a p-type semiconductor they are both raised. The equalization of the Fermi levels is associated with a potential difference across the boundary, even at equilibrium. This potential difference is accompanied by an electric field, whose effect is to repel further charges from crossing the interface. The electric field has a magnitude ϵ_b/l_d, where l_d is a measure of the distance over which it crosses (typically a few nm) and ϵ_b is the height of the barrier, given by

$$\epsilon_b = I_p - \Phi_m \qquad (5.26)$$

for a p-type device, where Φ_m is the work function of the metal. If the work function is large enough ($\Phi_m > I_p$), there is no barrier, and carriers do not have to tunnel across the boundary. In a device, this condition means more efficient carrier injection is possible. For an n-type semiconductor, the injection barrier is given by

$$\epsilon_b = \Phi_m - \chi, \qquad (5.27)$$

where χ is the electron affinity of the semiconductor. This barrier inhibiting the further injection of charges is known as a *depletion layer* (Fig. 5.9). This band bending phenomenon is not limited to metal–semiconductor interfaces, but also occurs at the interface between two different semiconductors (heterojunctions). Although these effects are particularly important for inorganic semiconductors, they will only be significant in organic semiconductors that have a high level of doping.

5.5.2   Electron and hole injection under an applied electric field

The application of an electric field across a polymeric (or indeed any) semiconductor will affect the bands; the HOMO and LUMO will not remain at a constant energy, but will vary as a function of distance from the electrodes. A reverse bias will put a potential barrier across the device by forcing the electrons to raise their energy to travel across to the anode, with the same being true for holes to the cathode. Carrier injection will not be possible (or will at least be severely impeded) under these circumstances. A forward bias, however, will allow charge transport across the active layer. If that bias is large enough, holes or electrons (or both) will be able to tunnel through the barrier to be injected into the semiconducting polymer. The larger the field across the device, the shorter the tunnelling distance becomes for charge injection. An alternative but less important injection route is when the barrier is

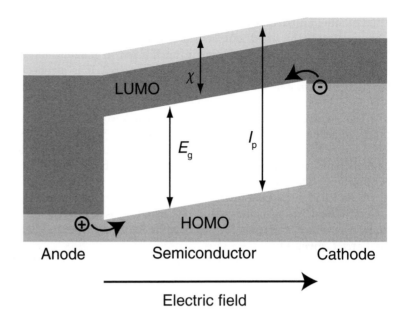

Fig. 5.10 Modification of Fig. 4.11 showing how the energy levels of the polymer are modified under an applied electric field. Electrons are injected at the cathode, and holes at the anode. If the applied field is large enough, charges can be injected either by tunnelling through the barrier (or by thermal injection above it). The applied electric field changes the shape of the HOMO and LUMO because these have to accommodate the Fermi levels of the electrodes. The change in the vacuum levels to accommodate the potential difference across the device is not shown, but is discussed in Section 10.1.3.

small enough for significant charge injection due to the thermal energy of the carriers. This is shown schematically in Fig. 5.10. Note that we have not shown any band bending in Fig. 5.10. The applied field across the device will be enhanced in the region where band-bending occurs, which can improve injection by increasing the field where it is needed the most. For most devices this is not an important issue; experiments show that Child's law (eqn 5.7) is often obeyed, demonstrating that band bending effects do not play a significant role in charge injection in many polymeric semiconducting materials.

The applied electric field moves the vacuum levels without changing the work functions of the materials, and is discussed in Section 10.1.3. Note that in the conducting state shown here, there can be no equalization of the Fermi levels because of the effect of the bias voltage. The Fermi levels would equalize under zero bias and this is important in photodetectors. Fig. 5.10 is applicable to devices operating under forward bias such as light-emitting diodes.

5.5.3 Electrodes

For a device to work as part of a circuit, charges that are injected at one end must be allowed to leave at the other, even in an LED, since not all charges are lost to the creation of photons. If current does not easily flow then the voltages used to drive the device need to be rather high, which is not desirable. Electrodes are therefore very important in a good device, because the choice of electrodes is the key to good charge injection. Ideally, one would have *Ohmic contacts*, where no charge injection barrier exists at the interface between the electrode and the active layer.

Anodes

ITO, a blend of In_2O_3 and SnO_2, is possibly the most important anode in polymer electronics because it is transparent, which is clearly useful for optoelectronic devices. Metals are not transparent; at the beginning of Chapter 4 we briefly discussed the colour of metals. The interaction of the free electrons with light is too strong. ITO is thus transparent only in thin films; in its solid bulk form it has a yellowish colour.

There are disadvantages of ITO, which should be accounted for when it is being used. Important problems include its inflexibility, which means that foldable electronic newspapers will require something other than ITO to become reality. Also, when a voltage is applied across the device, resistive (Joule) heating can occur, which raises the local temperature such that indium can migrate from the anode and into the active layer. (The ITO can be decomposed by high electric fields present during operation, making indium migration possible.) This is readily prevented by a separate PEDOT:PSS layer separating the ITO from the active layer, although the acidic nature of the PEDOT:PSS complex does allow etching of the ITO itself, which lets indium migrate into the PEDOT:PSS layer. From a practical point of view, deposition of ITO layers requires the creation of a vacuum, which adds to the cost of the device. (ITO itself is not cheap to purchase and the search for replacements to ITO is a vigorous area of research as indium becomes harder to source.) However, deposition procedures are no less awkward with cathode electrodes, although polymeric anodes can be used. Despite these disadvantages, ITO remains the anode of choice for most requirements. It is very moisture-resistant and has very good stability.

PEDOT:PSS films are an ideal complement to ITO anodes, because they have similar work functions (both are ~ 5 eV) and so injection barriers are not a problem. In some cases, PEDOT:PSS can be used as an anode without ITO; after all, doped PEDOT is a synthetic metal. PEDOT:PSS also has poor absorption of visible radiation, which is a prerequisite if it is being used with ITO. When PEDOT:PSS films are included in the device, the interface with the active layer becomes smoother than would be the case with an ITO layer alone. Smooth interfaces are a common occurrence with solvent cast films.

Cathodes

The cathode requires a low work function metal in order to facilitate electron injection. Metals with low work function are generally very prone to oxidation. This is not difficult to understand, as a material is oxidized if it loses an electron. If the work function is small, the energy barrier to taking that electron is small too. So a compromise has to be reached. Calcium has a relatively small work function (2.9 eV) and is commonly used, but to protect it, it needs to be encapsulated and so is often capped by another layer, such as silver.

5.5.4   Transport across the barrier

The size of the barrier of a device under a positive bias (Fig. 5.10) controls the injection of holes and electrons into the device. The injection across this barrier can be divided into regimes of small and large electric fields across the device. For low fields, the Schottky equation is appropriate to describe the emission,

$$j_{\mathrm{S}} = AT^2 \exp\left(\frac{-\left(\Delta W - W_{\mathrm{S}}\left(E\right)\right)}{k_{\mathrm{B}}T}\right),\tag{5.28}$$

where $A = 1.20 \times 10^6 \ \mathrm{Am^{-2}K^{-2}}$, and the T^2 term is dependent on the density of states scaling with the square root of carrier energy, which is valid only in the free electron model. In any case, the exponential term dominates this temperature dependence and it is therefore difficult to test experimentally the T^2 term in eqn (5.28). ΔW is the injection (Schottky) barrier, which is lowered by an amount $W_{\mathrm{S}}\left(E\right)$, given by

$$W_{\mathrm{S}}\left(E\right) = e\sqrt{\frac{eE}{4\pi\epsilon_{\mathrm{r}}\epsilon_0}}.\tag{5.29}$$

A proof of the reduction in the energy barrier is shown in Appendix A. The term $W_{\mathrm{S}}\left(E\right)$ differentiates eqn (5.28) from the Richardson–Dushman equation for thermionic emission, which is recovered when $W_{\mathrm{S}}\left(E\right) = 0$.

For larger fields a tunnelling current exists given by

$$j_{\mathrm{FN}} = \frac{BV^2}{d^2\Delta V}\exp\left(-\kappa\frac{d\Delta V^{3/2}}{V}\right),\tag{5.30}$$

where B and κ are constants, and V is the bias voltage across the device of thickness d. Equation (5.30) is known as the Fowler–Nordheim equation and is mechanistically different from the Schottky equation (eqn 5.28) in representing a tunnelling current. The electric field distorts the HOMO and LUMO by allowing the charges to tunnel through the barrier. This is the more common mode of injection in devices, and is useful in explaining behaviour, even in experimental verification is not always successful due to the injection current being compensated by a reverse current.

When $\Delta V = 0$, there is no injection barrier and injection is Ohmic. The Fowler–Nordheim equation is not valid under these circumstances because no tunnelling occurs, and so we do not have the situation whereby $j_{\mathrm{FN}} = \infty$. In practice, eqn (5.30) is valid for $\Delta V \gtrsim 0.3$ eV. Ultimately the barrier height limit of ΔV amounts to a consideration of whether or not charge transport or charge injection is the rate limiting process. For small ΔV, injection is not a major obstacle to device performance. However, this is a rule of thumb for typical devices. $\Delta V = 0.3$ eV is an order of magnitude greater than thermal values, and so, for small bias voltages, tunnelling will still be important.

5.6 Further reading

There are a few reviews covering the material contained with this chapter. The article by Kaiser (2001), whilst not recent, has a large breadth of coverage of the area. The chapters by Grell (2005) and Blom et al. (2007) summarize the material in this chapter nicely.

5.7 Exercises

5.1. Explain why carrier mobility during band transport might have the form $\mu \propto T^{-2}$ at high temperatures. How might this be modified by a large presence of traps?

5.2. Write down an equation for the number of carriers at a distance x and a time t from the initial pulse in a time-of-flight experiment, if the carrier velocity is constant.

Two samples are measured in a time-of-flight experiment with an applied electric field $E = 10^7$ V/m. The carrier mobility of the first sample is $\mu_1 = 10^{-9}$ m^2 V^{-1} s^{-1}; what is the mobility, μ_2, for the other, if the maximum in the carrier distribution having travelled $x = 10$ μm is half that of the first? An identical number of carriers were created in the initial pulse for both samples. You may assume that the carrier velocity is constant in both samples.

5.3. The T^2 term in eqn (5.28) is not expected to be particularly relevant in polymer or small-molecule organic electronics. Discuss the physical circumstances under which we should be able to test whether or not this T^2 term should exist.

5.4. A potential difference of $V = 6$ V is applied across a semiconducting polymer layer of thickness $d = 135$ nm. It is known to have a dielectric constant of $\epsilon_r = 5.1$. The transport properties of two doped samples are studied; in one case a carrier (hole) density of $n_1 = 1.7 \times 10^{23}$ m^{-3} is determined, whilst in the other a hole carrier density, $n_2 = 2.6 \times 10^{20}$ m^{-3} is obtained. The hole mobilities are respectively $\mu_1 = 2.2 \times 10^{-7}$ m^{2}V^{-1}s^{-1} and $\mu_2 = 8.7 \times 10^{-11}$ m^{2}V^{-1}s^{-1}. Calculate the corresponding fluxes, j_1 and j_2.

5.5. A time-of-flight experiment involves the sample in parallel with a resistor (Fig. 5.6). Because it is a semiconductor, the sample can be treated as a capacitor of areal capacitance, C. If we treat the sample (of area, $A = 5 \times 10^{-5}$ m^2), including charge generating layer, as a capacitor of dielectric constant, $\epsilon_r = 3$ and thickness 10 μm, what controls the maximum value of the resistance, R if the transit time of the carriers is $\tau = 20$ μs? What factors would dictate a lower limit for R?

6 Synthesis and macromolecular design

The properties of synthetic polymers have been known to depend to a great degree on their microstructure. Microstructure is a broad term explaining the material properties such as hardness, ductility, density, resistance to wear and so on. Very often these properties may be determined by nanostructure, or even macrostructure, but in all cases it will depend on the processing conditions. Examples of different microstructures would be fibres, crystallites, or amorphous structures. The chemical structure of the polymer determines to what form the microstructures may evolve; the actual structure is a result of the thermodynamic and hydrodynamic process occurring during solidification from solution or melt. Polymer materials are very sensitive towards their thermal history; for example, many structures might depend on whether the polymer was allowed to evolve in any of the semi-crystalline or liquid crystalline phases during processing.

Conjugated polymers and other organic conducting and semiconducting materials are no exception; in fact many, if not all, of their conducting, semiconducting, and optoelectronic properties will be very sensitive to the state of order from the molecular level upwards. Indeed, the performance of optoelectronic devices based on organic conducting and semiconducting materials will depend on the structural organization at length scales spanning eight orders of magnitude. At the nanometre and sub-nanometre scale, traps can affect charge and excitonic transport, but we still require devices that can be used by the consumer which means large lateral length scales of the order of centimetres and greater.

Thus, one can exploit the versatility of recent developments in polymer synthesis to control the primary (by which we mean chemical) structure of conjugated polymers producing soluble and processable macromolecules with tailored electronic properties and controlled material structure.

Historically conjugated polymers were synthesized in an insoluble and intractable form long before their remarkable conductive properties were recognized. In 1958, Giulio Natta and colleagues, then working at the Politecnico di Milano, mixed a titanium isopropoxide, $Ti(OCH(CH_3)_2)_4$, catalyst with triethyl aluminium $(Al(C_2H_5)_3)$ to create the conditions whereby acetylene could be polymerized. This created an insoluble and generally intractable black powder but was nevertheless a landmark in the synthesis of conducting polymers. A few years before the work of

Natta, Karl Ziegler found that he could use similar catalysts to produce high-density polyethylene, and in so doing develop a synthetic route that is still in use today. Natta improved the method further and so the Ziegler–Natta catalyst bears both of their names. For this pioneering work, both Natta and Ziegler were awarded the 1963 Nobel Prize for Chemistry.

In this chapter we shall discuss the chemistry of semiconducting polymers. This material might be considered as inappropriate for some physics courses, and so this chapter can be omitted. Nevertheless, it should be accessible to physics students, and we trust that students from other disciplines will also find it useful.

6.1   Polymerization

There are many mechanisms for synthesizing polymers, of which the two most popular are *polycondensation* (condensation polymerization) and *addition* reactions. Addition polymerization involves growing a polymer chain by the direct addition of monomers. Such reactions are *initiated*, then *propagate* before being *terminated*. The propagation step generally involves the addition of the monomer onto the growing chain without the production of a small molecule. Such a propagation reaction can be written as $M^*_n + M \rightarrow M^*_{n+1}$. The propagation step is often difficult to control, and very rapid, and so the polymerization requires a separate termination step. (Addition polymerizations without a termination step are known as *living* polymerizations. These have the advantage that they can create polymers with very narrow molecular weight distributions.) Not all addition polymerizations are awkward to control, and these are usually prefixed '*controlled*', but will not be discussed further here.

In polymer electronics, the polymers are more often than not synthesized by condensation chemistry, although we shall describe exceptions in Sections 6.4.2, 6.7.2 and 6.9 for the syntheses of polyacetylene, polythiophenes and polypyrrole respectively. In condensation polymerization, monomers react with each other and the growing chain, eliminating molecular fragments or small molecules during the elementary reaction steps. An example of a condensation reaction can be summarized by $A+B \rightarrow C+D$, where A, B, C, and D refer to relevant molecules (B and C should not be taken as boron and carbon), although formally this is a description of the broader concept of step-growth polymerization, of which polycondensation reactions form a subset. In condensation reactions the reaction takes place step-by-step, which is why such reactions are very commonly known as step-growth polymerizations. It is also possible in a condensation reaction for other polymers to join the growing polymer rather than just monomers. This is a key difference with addition polymerizations, which will react only with monomer units. Addition polymerization falls into the category of polymerization or chain-growth reactions, and may be described by the form $M_n + MR \rightarrow M_{n+1} + R$.

To achieve a condensation polymerization, the monomer must have the required functionality to polymerize. If the reaction were to occur in a bath of monomers with only one reactive group, the products could be no more complex than dimers, because once that functional group has reacted, there is nothing left on the monomer to continue the polymerization. So, monomers should generally have two functional groups. The functional groups that we shall see the most are probably halides, but carboxylic acid groups are another example that are used, because these can often react with alcohols or amines, which afford great flexibility in the chemistry. As well as the correct monomer functional groups these reactions are catalysed. Catalysts are not going to aid the propagation of an addition reaction, although they are often useful to speed up the initial stages of the reaction. However, catalysts are crucial in all forms of condensation reactions. This is because each time it is necessary to add to the growing chain, a new reaction takes place; the growing chain in a condensation reaction is usually a stable molecule because all of its bonds are satisfied.

6.1.1   Carothers equation

Condensation reactions generally take place in a 'soup' of components, consisting of solvents, catalysts, and monomers. These are very often complemented by one or more of co-catalysts, ligands, pre-catalysts, and other species with a role in the chemistry. The key components from a mathematical consideration of the process are the monomers and the growing polymers. The effectiveness of a polymerization is therefore defined by the *extent of reaction*, p, which is given by

$$p(t) = \frac{n_0 - n(t)}{n_0},\tag{6.1}$$

where $n(t)$ is the number of (unreacted) monomers present at time, t and n_0 is the number of monomers at the start of the polymerization, $t = 0$.

The *degree of polymerization* or *polymerization index* is an important concept because it is related to the polymer molecular weight by the total molecular weight of the polymer divided by the molecular weight of the monomer. For example, polyacetylene, with the chemical formula $(C_2H_2)_n$ has a monomer molecular weight of 26 g mol^{-1}. A polyacetylene sample with a total molecular weight of 2600 g mol^{-1} therefore has a polymerization index $N = 100$. We can relate a time-dependent polymerization index, $X_n(t)$, to the number of monomers by

$$X_n(t) = \frac{n_0}{n(t)}.\tag{6.2}$$

which can readily be related to the extent of reaction from eqn (6.1) by

$$X_n(t) = \frac{1}{1 - p(t)}.\tag{6.3}$$

This equation (6.3) is known as the *Carothers equation* after Wallace Carothers, a pioneer in polycondensation reactions, who was best known for his work at DuPont, where he invented nylon in 1935. One can easily calculate the degree of polymerization knowing the extent of polymerization using the Carothers equation, but unfortunately, polymer chemistry is not so easy. Clearly, with a half the monomers reacted, $X_n = 2$. This however is only an average; specifically, it is the *number average* degree of polymerization, which allows us to define a number average molecular weight by

$$\overline{M}_n = \frac{\sum\limits_{i=1}^{\infty} n_i M_i}{\sum\limits_{i=1}^{\infty} n_i}, \tag{6.4}$$

where M_i is the molar mass of the chain with polymerization index n_i. Ultimately, to have a reasonable polymeric sample, one needs values of $p > 0.99$, where we use $p = p(\infty)$.

Although the number average molecular weight is an important parameter, it does not give any feel for how *disperse* the sample is. The *dispersity* of a given polymeric sample must therefore involve some knowledge of another means of describing the polymer molecular weight in order to give some value of the width of the distribution of the molar masses. The required parameter is known as the weight average molecular weight. The numerical fraction of molecules of a polymerization index given by eqn (6.2) is not necessarily the same as the mass fraction of polymers of a given chain length. The mass fraction is unsurprisingly the mass of polymers with a polymerization index i divided by the total mass of polymer, or

$$w_i = \frac{n_i M_i}{\sum\limits_{i=1}^{\infty} n_i M_i}. \tag{6.5}$$

The *weight average* molecular weight is given by

$$\overline{M}_w = \sum\limits_{i=1}^{\infty} w_i M_i, \tag{6.6}$$

which, using eqn (6.5) gives

$$\overline{M}_w = \frac{\sum\limits_{i=1}^{\infty} n_i M_i^2}{\sum\limits_{i=1}^{\infty} n_i M_i}. \tag{6.7}$$

The *dispersity index*, D_M, of a polymeric sample is then defined as the ratio $\overline{M}_w / \overline{M}_n$. Values of $D_M < 1.2$ can reasonably said to be *uniform*, whereas $\overline{M}_w / \overline{M}_n > 1.5$ should be considered non-uniform.[1] It is not possible for the dispersity index to be less than unity.

For polycondensation reactions, the dispersity is given by

$$D_M = 1 + p, \tag{6.8}$$

[1] Uniform and non-uniform polymers are often known as monodisperse and polydisperse respectively. These terms are no longer recommended by the International Union of Pure and Applied Chemistry (IUPAC). Similarly, 'dispersity index' is recommended as a replacement for 'polydispersity index'.

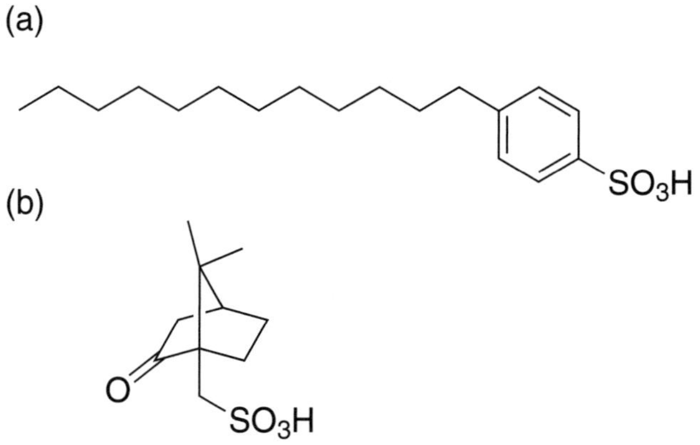

(a)

(b)

Fig. 6.1 Dodecylbenzenesulfonic acid (a) and camphor sulfonic acid (b) are amphiphilic molecules that have been used to dope polyaniline. The SO_3H group is ionized in water, which makes it hydrophilic. The remainder of these two molecules are hydrophobic. In the case of polyaniline they also confer solubility in some organic solvents.

and this is proved in Appendix B. We see from eqn (6.8) that the dispersity is limited to $1 \leq D_M \leq 2$.

6.2 Macromolecular design

The importance of polymer chemistry cannot be understated. If one polymer is not fit for a specific purpose, there is plenty of scope to make another one. There are a variety of different issues that molecular design can resolve, and we briefly address these in turn.

6.2.1 Solubility

Conjugated polymers are relatively rigid molecules; the double bonds do not encourage flexibility, and alternating single and double bonds generally have high persistence lengths (Section 7.1) and have consequently limited solubility; the large persistence lengths allow these chains to efficiently aggregate in solution. In order to circumvent this problem, alkyl side chains are often added to polymers to improve their solubility, but without breaking the conjugation. (Side chains do, however, limit the conjugation length simply because they improve the flexibility of the polymer.) These are flexible in solution, and so the polymer can dissolve efficiently. In the solid state the side chains often bring another length scale into play, and can improve packing. Polyfluorenes (Section 6.3), polyphenylenes (Sections 6.3, 6.5, and 6.6), and polythiophenes (Section 6.7) particularly benefit from this possibility. Nevertheless, it is not necessary to design side chains into the chemistry. Clever synthesis can result in the creation of a tractable polymer from a precursor, such as in the Durham route to polyacetylene (Section 6.4.2), or alternatively, synthesis *in situ*, by for example, electrochemistry (Section 6.9).

6.2.2 Doping

Clearly, doping is necessary to achieve high conductivities, as can be seen from the examples in Table 2.1. However, doping also provides some more subtle benefits, such as compatibilization with other materials and solvents, which can aid processing. An example of this is for the case of polyaniline, which can be doped with sulfonic acids to aid solution in some aromatic solvents such as *m*-cresol. These molecules are *amphiphiles*, which means that they have hydrophobic and hydrophilic parts (Figure 6.1). Apart from improved solubility and conductivity (polyaniline films have been shown to have conductivities greater than 10 kS/m), such doping allows for better blending with non-conductive materials, e.g. polymers such as polystyrene and poly(methyl methacrylate) (PMMA). Blending will reduce conductivity considerably, but often the conductivity is good enough, for example in antistatic coatings. (Another example of polymer additives to improve performance is discussed in Section 9.5.1.)

Perhaps a more prominent example where polymer doping is important is the PSS-doped PEDOT complex, introduced in Chapter 1 and Section 3.4, which is discussed in more detail in Section 6.7.3. PEDOT is the most successful conducting polymer, and its doping with PSS has enabled its commercial success.

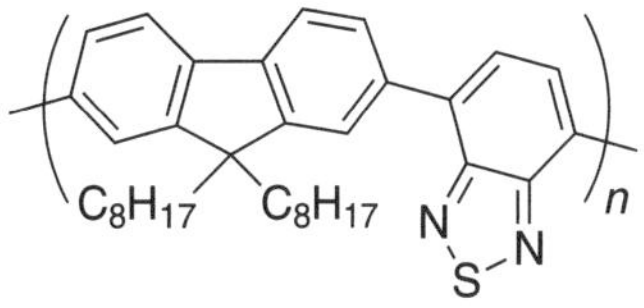

6.2.3 Control of the band gap

Many of the cross-coupling routes that are described in Section 6.3 allow the creation of *alternating copolymers*. These are polymers for which two covalently-joined species become the repeating unit. If we create an alternating copolymer of two distinct conjugated monomers, we can control the electronic properties (without spoiling the conjugation), essentially because the electronic properties of one modifies those of the other, and *vice versa*. An example of this shown in Fig. 6.2 is poly(9,9-dioctylfluorene-*alt*-benzothiadiazole) (F8BT), which contains an electron-rich (donor) and electron-poor (acceptor) unit to control the band gap. The 9,9-dioctylfluorene has a relatively large band gap (it emits in the blue), and acts as a donor. The acceptor unit (benzothiadiazole) modulates this and the alternating copolymer emits in the green. The benzothiadiazole unit also gives the copolymer reasonable electron mobility, which is another example of how the band gap may be controlled.

The use of electronegative substituents such as nitrogen or oxygen has the effect of controlling the location of electrons within the polymer, which permits better electron transport. In a rather hand-waving argument, this is because there are fewer mobile electrons present which means their motion is notable in comparison to holes. This argument has already been presented in Section 3.7. In the case of F8BT the nitrogen atom of the benzothiadiazole has this effect, but BBL (Fig. 3.14) is a more spectacular example. Ultimately, nitrogen and oxygen both

Fig. 6.2 Poly(9,9-dioctylfluorene-*alt*-benzothiadiazole) (F8BT) is an alternating copolymer made up of a donor (9,9-dioctylfluorene, left) and an acceptor species (benzothiadiazole, right).

function as electron withdrawing groups, which helps electron transport. Furthermore, electron withdrawing groups generally cause a reduction in the LUMO level, which make the molecule more stable in air by increasing the energy to remove electrons (electron affinity) from the LUMO.

Control of the band gap can also be rooted in physics, rather than chemistry. For example, we showed in Fig. 2.13 how increasing complexity of alkenes increases the number and density of the energy levels in the molecule. A consequence of this is a reduction in the band gap, and the same applies to conjugation. An increased conjugation length therefore causes a smaller band gap. Of course, in polymers this is a small effect, but it can be significant if there are many traps and defects present.

Although the presence of traps decreases the conjugation length these cannot be removed simply by improving the purification of the polymer, because the structure of the molecule can affect its stability, and consequently its conjugation length; an example is shown in Fig. 3.10. Increasing rigidity in the molecule, such as in ladder polymers (Section 3.7) can help. Saturated polymers are not rigid, and so having double bonds along the main chain backbone increases rigidity and the conjugation length. Quinoid polymers (e.g. Fig. 3.1b) will generally have smaller band gaps than non-quinoid polymers.

Of course, it is sometimes useful to limit conjugation to tailor the band gap. In such cases one can add substituents into the chain to break the conjugation. For example, a regularly repeated oxygen in the chain backbone will act as a spacer limiting the conjugation length, which, for example, may aid in the emission of shorter wavelength photons, when this is necessary. Another strategy would be to hang the conjugated units from a saturated polymer main chain.

As a final note, the reader should be aware that design strategies involving the conjugation length are very often based upon the assumption that (intrachain) band transport is the only mechanism by which charges can move. Of course, hopping is also a major contributor to electronic transport, and one should not discount interchain band transport either. The predominance of band transport in molecular design strategies is arguably due to band transport being easier to understand than the other mechanisms. If a chemist can understand band transport but not hopping, then that chemist is not going to synthesize polymers according to the principles of hopping transport. Of course, it would help if physicists properly understood hopping transport too...

6.2.4 Charge transport requirements

The better the charge transport, the more applications for which the polymer can be used. The design principles are mostly unsurprising. Rigid and planar chains with few defects are well-suited to good charge transport. This helps the overlap of orbitals such as π-stacking. Excellent crystallinity serves these purposes too by removing defects and helping the stacking of chains, facilitating inter-chain charge transport.

Given crystallinity, the orientation of the polymers is important too; it is preferred that the chains crystallize along the direction of charge transport, facilitating band transport towards the electrodes.

Better electron and hole transport is controlled by the electron affinity and ionization potential respectively. A small ionization potential and large electron affinity are also desirable for efficient injection. A large electron affinity keeps the electrons away from the vacuum level, limiting their removal due to oxidation.

6.2.5 Improved optoelectronic behaviour

If one wishes to design a polymer with suitable luminescent properties, then one needs to make sure that the chromophores are not subjected to interference with their environment or with each other.

Chromophores must not be able to easily vibrate or rotate. Multiple internal degrees of freedom allow chromophores to interact with their environment through, for example, collisions, and dissipate their energy in non-radiative transitions. Chromophores can also interact with each other (interchain effects), which can, for example, form excimers, also affecting their behaviour in undesirable ways.

In some cases, one may prefer amorphous optoelectronic materials, because these will not contain grain boundaries. Of course, amorphous materials will have worse transport behaviour than their crystalline counterparts, but grain boundaries can scatter light if they are of length scales commensurate with the wavelength of the photons. Clearly this optoelectronic behaviour needs to be balanced with charge transport, and compromises like these are common in designing materials for specific purposes.

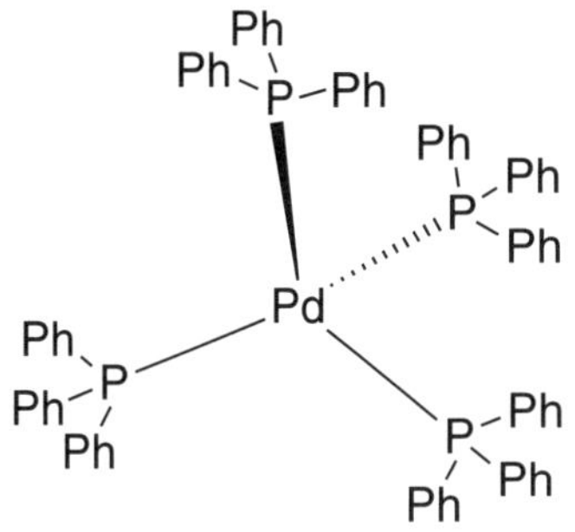

6.3 Coupling and cross-coupling reactions

One of the most frequently employed methods of creating C–C bonds between two unsaturated molecules is what is known as a *cross-coupling* reaction. This is a branch of chemistry where chemical moieties are reacted in the presence of a metallic catalyst. Cross-coupling means that two different molecules are reacted to form a third species. (Coupling reactions are very similar to cross-coupling reactions but only one kind of molecule is reacted.) A common catalyst for both coupling and cross-coupling reactions is $Pd[P(C_6H_5)_3]_4$, known as tetrakis(triphenylphosphine)palladium(0), often denoted TPP palladium. The chemical structure of TPP palladium is shown in Fig. 6.3. The main reason that palladium is commercially mined is because of its predominant role as a catalyst. Here palladium binds to four triphenylphosphine molecules; these bonds are known as *coordination* bonds and the *coordination number* of TPP palladium is 4. The ligands are *monodentate* because each one has only one bond to the metal. The importance of *palladium-catalysed cross-couplings in organic synthesis* was highlighted by the award of the 2010 Nobel Prize for Chemistry. The contributions of two of the recipients,

Fig. 6.3 Tetrakis(triphenylphosphine) palladium(0) is a common catalyst for coupling reactions. In this diagram the phenyl rings have been replaced by the symbol Ph, which is common practice in organic chemistry, especially when, as is the case here, space constraints preclude the drawing of aromatic rings.

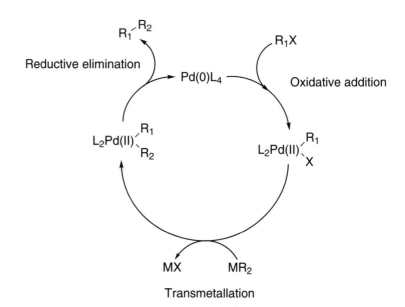

Fig. 6.4 The cross-coupling reaction proceeds in three steps. Firstly, oxidative addition breaks up an aryl or vinyl (R_1) halide (X) with the insertion of the metal within its bonds. This means that halogen and carbon cannot 'share' electrons within a normal covalent bond, and these must be donated by the palladium catalyst, consequently increasing the oxidation state of the molecule by 2. Transmetallation is a process for the exchange of the halogen on the palladium complex for another organic reactant. The reaction is complete with the reductive elimination of the palladium complex, regenerating the Pd(0) catalyst.

Akira Suzuki and Richard Heck, are described in Sections 6.3.2 and 6.3.6.

The cross-coupling reaction consists of three steps, as shown in Fig. 6.4. Metals such as palladium can complex to more molecules in a process known as *oxidative addition*, which is the insertion of a metal complex into a covalent bond. Because a covalent bond has been interrupted, two electrons are removed from the metal which increases its *oxidation state* by 2. By this process TPP palladium will go from an oxidation state of zero, which is what the (0) refers to in its explicit name, to an oxidation state of (2). The IUPAC defines oxidative addition as $L_nM^m + XY \rightarrow L_nM^{m+2}(XY)$, where for TPP palladium $n = 4$ and L is triphenylphosphine. (The symbol L denotes ligand, and M refers to metal.) For these kinds of reactions XY will be an aryl or vinyl halide. (An aryl halide is a molecule with a halogen bonded to a conjugated ring, such as a phenyl or thiophene ring, and a vinyl halide is an alkene (C_nH_{2n}) with one or more halogen atoms replacing hydrogen atoms.)

The cross-coupling reaction proceeds by complexing two components of the reaction onto the catalyst before ejecting the catalyst. The first stage therefore is the attachment of one of the reactants to the palladium. The next stage is called transmetallation, where another organometallic complex reacts with the palladium complex to remove the halide, which is swapped with the organic compound originally complexed with the metal. The transmetallation step is key because it brings the two reactants together. Once this has been achieved, it is necessary to remove the palladium using a process known as *reductive elimination*. Reductive elimination is easy to understand being simply the opposite process to oxidative addition. This process recovers the palladium complex in its original form, which is a necessary criterion for a catalyst; it must be left untouched by the chemical reaction it facilitates. It might seem strange to a non-chemist that one can start with an oxidative addition reaction, which one can suppose is exothermic and end with the opposite (reductive elimination) reaction, which must not be endothermic. We are therefore required to choose the reactants carefully so that all stages proceed as rapidly as possible. Organometallic chemists have studied reactivities for many years and the correct choices are well known. Ultimately, if the R_1–R_2 bond is stronger than the bond R_1–X, there is a good chance that there is transmetallation step that would allow for the reaction to proceed to completion. Of course, to reduce chemistry to this naive level of simplicity misses important details, but following energies is never a bad starting point.

In practice these reactions are generally 'single pot' syntheses. This means that all the reaction components are present together and the important reactions are to take place sequentially. The risk with all single pot syntheses is that there are unnecessary and unwanted side reactions which create contaminants. In the case of polymerizations these can result in chemical impurities in the chain; rather than synthesizing a pure polymer, one would end up with a random (statistical) copolymer of the desired material and a contaminant. Other unwanted phenomena could include branching or crosslinking reactions. These reactions generally occur in solution, but this can be changed depending on the target polymer. Of course most semiconducting polymers do not easily dissolve, so solubilizing side chains are often included. Side chains enable complete solution processing to take place, but that is not an option for many other polymers which may be grown in the solid state, perhaps on the catalyst. In such cases, the monomer may be introduced in the form of a gas, and quite large polymers may be synthesized on the catalyst. This is an example of *heterogeneous catalysis*, which refers to a chemistry where the catalyst is a different phase of matter to the reactants. (Phase separated liquids also may constitute a heterogeneous catalysis, but this is not of great importance in semiconducting polymer synthesis. The key point is that a catalyst surface exists on which a polymer can grow.) For solution phase chemistry organic polar solvents are popular, with dimethyl sulfoxide (DMSO), which has the chemical formula $(CH_3)_2SO$ being a common solvent for many of the reactions described below.

Although we shall concentrate on the reaction chemistry here, a good laboratory procedure is more than using pure and well chosen reagents and performing the reactions in a carefully controlled environment. A certain degree of pragmatism is also necessary; part of the popularity of Suzuki coupling (Section 6.3.2) is due to its use of low toxicity materials. Another important aspect of the chemistry not covered here is the necessity to clean up the materials afterwards. In the cross-coupling reactions described here, metals are invariably used, mainly in the catalyst. If these remain after purification, their effect can be seen in device performance. For example, in field-effect transistors, a current may flow when the transistor is nominally 'off' because the metal provides charge

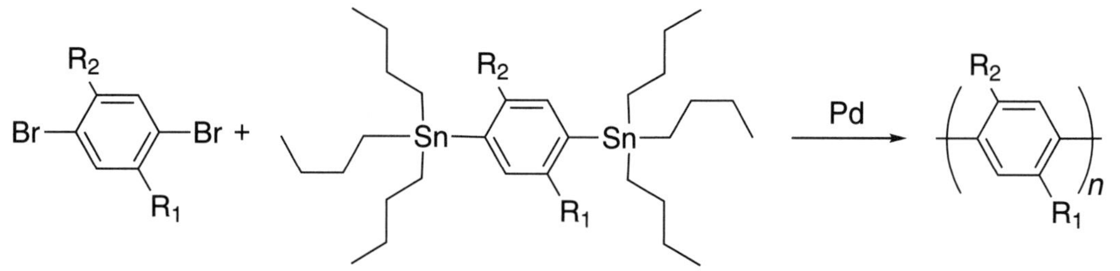

Fig. 6.5 The Stille reaction is a fairly standard route to PPP. The solubilizing side chains R_1 and R_2 are usually identical. Here, the tributyl stannane mentioned in Section 6.3.1 is shown for the transmetallation step, but other stannanes are also applicable. In this figure we have omitted other reaction products and for convenience have shown the final reaction product as the polymer.

carriers that can contribute a current under any bias.

There are a wide variety of coupling reactions and these are classified largely by the metal used for the transmetallation reaction. Some of these will be discussed below, but this discussion is by no means complete because there are other possibilities. Important examples that we shall not discuss here due to their relatively minor contribution to the synthesis of conducting and semiconducting polymers include, for example, Hiyama and Negishi coupling reactions. Ei-ichi Negishi will presumably accept this apparent snub, also having a share (with Heck and Suzuki) of the 2010 Nobel Prize for Chemistry.

6.3.1   Stille coupling

Stille coupling is perhaps the generic coupling reaction because it takes place under neutral conditions. The process therefore is as described above, with an oxidative addition stage followed by transmetallation, and then reductive elimination. Here the transmetallation process requires an organostannane to add the second reactant to the palladium complex. Stannane is tin hydride (SnH_4), but this chemistry is commonly achieved using a tributyl stannane, $(C_4H_9)_3SnR_2$. This reaction is certainly not mild since the organostannanes are very toxic. The use of stannanes on a large scale is difficult on safety grounds, and so other coupling reactions such as Suzuki coupling (Section 6.3.2) are more popular, at least industrially. Despite these weaknesses, Stille reactions are very versatile because they work with a large variety of functional groups. Synthesis of a variety of polythiophenes, poly(p-phenylene)s (Fig. 6.5) and poly(phenylene vinylene)s, amongst others, can be readily performed using Stille coupling.

Stille coupling was named after John Kenneth Stille, who, with his colleague David Milstein, invented the technique in 1977. Tragically, Stille was to perish twelve years later at the age of 59 when a commercial flight on which he was a passenger crash-landed in Iowa.

6.3.2 Suzuki coupling

The Suzuki coupling method, first described in 1979 by Norio Miyaura, Kinji Yamada, and Akira Suzuki of Hokkaido University, involves the use of boronic acids or other organoboranes in the transmetallation process. PPP (Section 6.5) and polyfluorenes can readily be synthesized using Suzuki coupling. Boronic acid has the general form $R–B(OH)_2$, where R is an appropriate group, which very often contains a derivative of an unsaturated group. The chemistry of Suzuki coupling is more complicated than that of the Stille method described above, because, for example, it requires the use of a base to create an intermediate product between the oxidative addition process and the organoboronic transmetallation stage. The purpose of the intermediate stage in which the base is added is to create a better controlled reaction route for the boronic acid. In general organoboranes are quite reactive and it is well worth including an intermediate stage so that the rate coefficient of the transmetallation is enhanced to reduce the contribution of significant unwanted side-reactions. A simple intermediate reaction could be $L_nPdR_1X + NaOH \rightarrow L_nPdR_1OH + NaX$. At the same time the base will attack and activate the boronic acid, which will improve the speed and efficiency of the transmetallation step.

Despite its apparent complexity, Suzuki coupling is quite popular from a large-scale synthesis point of view because it involves compounds of relatively low toxicity as well as being rather easy to remove unwanted side products.

6.3.3 Kumada coupling

The Kumada coupling reaction was developed in 1972 independently by two groups, although it is named after Makoto Kumada, the leader of one of them. In the synthesis of conjugated polymers, its biggest contribution has been in creating polythiophenes. Here, what are known as Grignard reagents were used for the transmetallation step. Grignard reagents, which have been used since the early twentieth century and are named after the 1912 Nobel chemistry laureate François Grignard, are based around the metal magnesium. In these reactions nickel catalysts may replace the palladium catalysts that are common in other coupling reactions. (An important Nickel catalyst is 1,3-bis(diphenylphosphino)-propane nickel(II) chloride, shown in Fig. 6.6.) Grignard reagents are readily commercially available and there is a wide variety of choice, which makes the method still used. The Kumada reaction has the advantage of taking place at room temperature, which gives it an advantage over some other routes to polythiophenes, which require cryogenic conditions. However, these reagents are very basic and very reactive and so the efficiency of the reaction is limited unless functional groups that are unaffected by Grignard reagents are used. A particular problem is atmospheric humidity, which often requires sealed reaction vessels to prevent moisture ingress.

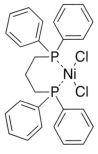

Fig. 6.6 1,3-Bis(diphenylphosphino) propane nickel(II) chloride (which is usually abbreviated to Ni(dPPP)Cl2, and is shown above) is a typical nickel catalyst for Kumada coupling reactions. Formally, this molecule is a precatalyst; the two chlorine atoms are lost to make the catalyst 1,3-bis(diphenylphosphino)propane nickel(0), which is the molecule regenerated at the end of the reaction.

Fig. 6.7 The Yamamoto coupling reaction is commonly used for the polymerization of fluorenes (a) and phenylenes (b). The monomers are most commonly brominated, but other halides may be used. Here the monomers are shown with alkyl side chains to aid solubility. An advantage of this method is the ease with which copolymers can be synthesized, which might include a copolymer of the two monomers shown here.

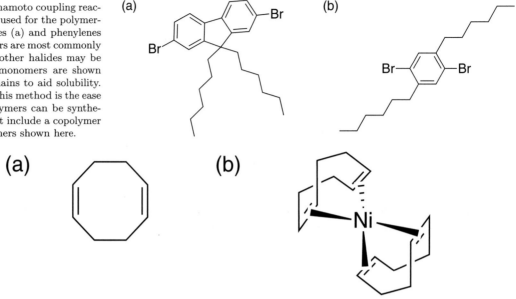

Fig. 6.8 The catalytic effect of nickel in Yamamoto coupling (and other such reactions) is effective only if bound to a ligand, such as 1,5-cyclooctadiene (a). 1,5-cyclooctadiene is a diene, which means it has two double bonds. The *cyclo* means it is a ring, and the *octa* indicates eight carbon atoms. The '1' and '5' specify the diene positions on the ring. The complex with nickel (b) is what is known as a *chelate*, which refers to a compound in which one or more metals are trapped by one or more organic molecules. These metals are held in place by four coordination bonds binding it to the double bonds of the 1,5-cyclooctadiene. Note how the bonds with nickel instigate a shape change in the complex.

6.3.4 Yamamoto coupling

Yamamoto coupling involves a nickel reagent to facilitate coupling reactions involving aryl halides. A couple of examples of monomers used for Yamamoto coupling are shown in Fig. 6.7. The nickel needs to be complexed with a ligand to be effective as a catalyst. This is a common procedure and Yamamoto coupling proves a classic example with the use of 1,5-cyclooctadiene as the ligand, as described in Fig. 6.8. The reaction proceeds with an oxidative addition stage involving a halogenated (X) monomer (R)

$$2RX + Ni(0) \rightarrow 2RNi(II)X,$$

but does not involve a transmetallation step, but rather a process known as *disproportionation*. Here both reduction and oxidation occur as a species is broken into two components. In the case of Yamamoto coupling disproportionation proceeds by

$$2RNi(II)X \rightarrow Ni(II)R_2 + Ni(II)X_2.$$

In this disproportionation process, X is a halogen, so $Ni(II)X_2$ is reduced and $Ni(II)R_2$ oxidized.[2] Finally, reductive elimination occurs through

$$Ni(II)X_2 \rightarrow Ni(0) + R_2.$$

[2] Disproportionation is a form of redox reaction, and does not change the net oxidation state.

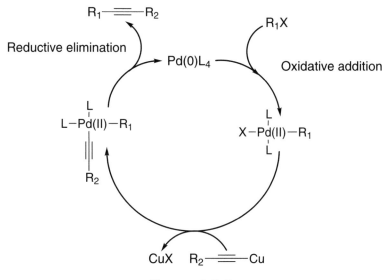

Fig. 6.9 The Sonogashira cross-coupling reaction is an excellent example of the cross-coupling process. X, is typically iodine, although other halogens can be used. Isomerization takes place as part of the transmetallation step. The co-catalyst is a copper halide, the halide of which is removed in preparation for the transmetallation step in a reaction with a third species, R_3, which is not shown.

The reaction can proceed with different monomers to contain alternating copolymers. This reaction consumes half of the nickel to form $Ni(II)X_2$. For this reason nickel is not strictly a catalyst, and the quantities of nickel used determine the extent of reaction possible, because half of the remaining $Ni(0)$ reagent takes place in the next reaction step, and so on. Since the halides remain on the end of the growing polymer, the reaction can continue. The rate limiting factor for many Yamamoto coupling polymerizations is the solubility of the polymer. To achieve PPP this way would be limited to between 10 and 15 *p*-phenylene units, although the use of monomers with solubilizing side chains can improve matters.

6.3.5 Sonogashira coupling

Clearly organometallic chemistry was thriving in Japan in the 1970s. In Osaka in 1975, Kenkichi Sonogashira and Nobue Hagihara reported a method which bears the former's name, using copper for the transmetallation. In this case copper can be labelled as a co-catalyst, since the copper halide will be regenerated during the reaction. The reaction process is shown in Fig. 6.9. This reaction is very useful for producing acetylenes, especially those with phenyl rings contained within them (aryl acetylenes). The mild reaction conditions make this a popular choice for producing such compounds.

6.3.6 The Heck reaction

The Heck reaction is an important chemical route for the synthesis of different forms of PPV and is different from the other coupling reactions described here in that it does not involve a transmetallation step. In

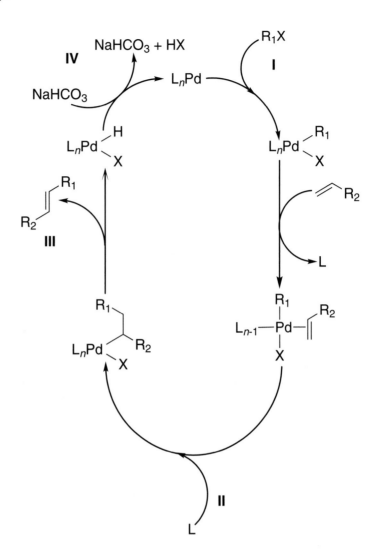

Fig. 6.10 The Heck reaction begins with an oxidative addition step (**I**). The second group (R$_2$) is not added by a transmetallation step but is typically connected to an appropriate organic group (e.g. the vinyl group shown here) that displaces a ligand on the product of the oxidative addition step. An immediate and rapid (migratory) insertion reaction (**II**) puts the ligand back in place, breaking the double bond in the process. The desired reaction product is then eliminated in a process known as beta hydride elimination (**III**), which occurs before regeneration (reductive elimination) of the catalyst (**IV**) takes place.

the Heck reaction, the important (rate limiting step) is oxidative addition; the speed of the oxidative addition reaction defines the efficiency and speed of the entire coupling reaction. The entire process of the Heck reaction is shown in Fig. 6.10. Like many of the coupling reactions described above, a palladium catalyst is used, and TPP palladium is typical. The next step is rather complicated and is shown in Fig. 6.10 as two steps. The group R$_2$ is not complexed to a metal for a transmetallation step but is covalently bonded here to a vinyl group which will interact in such a way with the palladium complex to first displace one of the ligands. The complex that is thus formed is known as a π complex; the reader will note that the bond from the palladium does not end at a carbon, but rather in the middle of the vinyl group. This reflects a movement of electrons around the complex, but because it involves the π-electrons of the double bond there are changes in the chemistry of the vinyl group such as an increase in bond length. At the end of this process, the ligand returns to the complex as the final step (*ligand addition*) in this *migratory insertion* reaction. A migratory insertion reaction involves one part of the complex (the vinyl R_2 component) reacting with another part (R_1), with the removal of R_2 allowing the ligand to return to the process. This migratory insertion can only occur if the two species are next to each other on the complex; technically, they must be *cisoidal*. The final reaction product is obtained by a process known as beta hydride elimination. This is not as complicated as it sounds. The α- and β-positions on the complex are specified in Fig. 6.11; the hydrogen from the β-position takes the place of the bond at the α-position, eliminating the desired (final) reaction product from the complex. The final stage of the reaction is the reductive elimination step, which regenerates the catalyst. To achieve this a proton acceptor (base) is required, and examples of those commonly used are $NaHCO_3$ or triethylamine.

The Heck reaction was named after the American chemist Richard Heck, of the University of Delaware. Given the contribution of the Japanese to cross-coupling chemistry, it will come as no surprise that a Japanese group (led by Tsutomu Mizoroki at the Tokyo Institute of Technology) simultaneously developed this form of palladium catalysed coupling reaction.

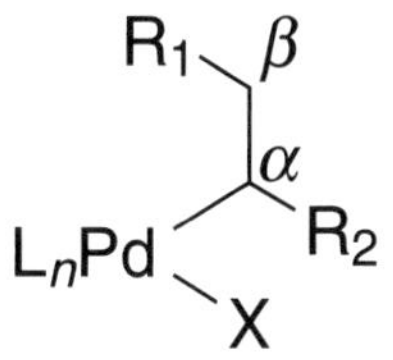

Fig. 6.11 The α- and β-bonds before beta-hydride elimination occurs are marked. The β-bond has an extra hydrogen atom that will return to the complex during the reaction step.

6.4 Synthesis of polyacetylene

As was mentioned as early as Chapter 1, polyacetylene is a generally insoluble polymer and so its synthesis is rather complicated. Having synthesized the polymer, its processing is no less easy. This is important, because it is necessary in part to process the polymer in order to characterize it. The Ziegler–Natta route described at the beginning of this chapter is a common route to the production of polyacetylene, but it is difficult to tailor the chemistry to reproduce the required molecular weight or microstructure. This is because so many factors come into play such as temperature, the concentration of titanium catalyst, or the concentration of triethyl aluminium, the pressure of the acetylene vapour used for the reaction, and even the storage conditions for the different chemicals used. In any case polyacetylene may be synthesized by a variety of different routes including the polymerization of the monomer acetylene ($HC{\equiv}CH$) with and without catalysis. The latter requires high pressures to compensate for the lack of a catalyst, and will not be dealt with here. Other methods include the polymerization of monomers other than acetylene to form a polyacetylene chain (cyclooctatetraene, which is included in Fig. 6.15, has been used), and a fourth possibility involves the final chain formed from the decomposition of a precursor polymer. We shall consider here some of the more successful methods for the synthesis of polyacetylene.

6.4.1 The Ziegler–Natta catalysis route to polyacetylene

It is possible to produce three forms of polyacetylene. In Figures 1.2a and 1.2c we show the *cis* and *trans* forms of polyacetylene. In fact there are two *cis* forms as illustrated in Fig. 6.12. These two forms are commonly known as *cis-transoid* and *trans-cisoid* forms. These are not identical; one can play the same game with the *trans* form and see that the results are chemically identical. The *cis* form has different energies for its own structural isomers. The *cis-transoid* form is the more stable and it may well be that the *trans-cisoid* form is wholly unstable as some calculations show, although it is often referred to as being metastable. The reasons for this are rather complicated, which is why computer simulations are required. However, we can at least address the relevant issues. Because the double bond is shorter than the single bond, for a given bond angle, the chain is thinner in the *trans-cisoid* form than the *cis-transoid* form. This would mean that the *cis-transoid* form has a greater density of electrons than the *trans-cisoid* form, because it is confined to a smaller volume. Because of these points, the bond angles are not the same for the two *cis* forms, and the energies of the different forms do not need to be the same either.

An important difficulty with the polymerization of acetylene with an organometallic catalyst is that the polymer is not necessarily the main reaction product. Given the similarity with benzene it is not perhaps a great surprise that many methods to produce polyacetylene end up with more benzene; after all, benzene is the first three monomers of polyacetylene (i.e. it is a trimer of acetylene), and it is inherently stable. Other common reaction products of the synthesis of polyacetylene are shown in Fig. 6.13.

It is not just the difficulty in obtaining the required product in enough quantity that is a major issue. Even with 100% conversion to polymer one needs to think about how to purify it; for example, catalyst will still be present in the polyacetylene; for this reason insoluble catalysts are to be discouraged given that polyacetylene itself is insoluble and so the catalysts cannot simply be rinsed out. A common catalyst is a mixture of triethyl aluminium and tetrabutoxy titanium, $Ti(O(CH_2)_3CH_3)_4$, which is perhaps now more commonly used than the titanium isopropoxide used by Natta and colleagues. Both triethyl aluminium and tetrabutoxy titanium are soluble in organic solvents and therefore can be easily washed away to purify the reaction product. Ziegler–Natta catalysis with these catalysts produces good quality polyacetylene crystalline films. The polyacetylene itself can be collected as a precipitate on either the surface of the catalyst solution or on the reaction vessel. Acetylene gas, which is used for the polymerization, will dissolve in the solution, which allows the polymer to precipitate on the walls of the vessel. Although the polyacetylene films created this way are of a good quality, they are highly disperse, i.e. they have a very broad distribution of polyacetylene molecular weights.

(a)

(b)

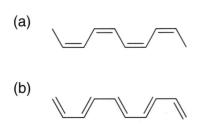

Fig. 6.12 The two different *cis* forms of polyacetylene. The *cis-transoid* form (a) is more stable than the *trans-cisoid* form (b).

(a)

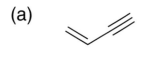

(b)

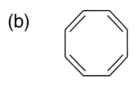

Fig. 6.13 A poor choice of reaction route to polyacetylene will often yield more benzene than polymer. Other common reaction products include (a) vinyl acetylene and (b) cyclooctatetraene. The structure of these molecules, like benzene, shows a clear link to acetylene.

Acetylene is polymerized just below atmospheric pressure ($\sim 9 \times 10^4$ Pa) at a temperature of $-78°C$, which means that it is cooled by dry ice (solid CO_2), usually mixed with acetone, ethanol, or methanol. The reaction must proceed under argon, and acetylene gas is passed through a mixture of the catalysts, which also allow initiation of the reaction. A few mmol/L of catalyst is required for polymerization to occur, and the ratio of aluminium to titanium is of the order of unity, but maybe with up to four or five times more aluminium. As polymerization proceeds the catalyst solution, initially a pale yellow becomes red. For more concentrated catalyst solution the polymer precipitates on the available surfaces virtually immediately, and the rate limiting part of the reaction is the speed at which acetylene gas is allowed to diffuse into the catalyst. Most syntheses require a high concentration of catalyst because the films thus produced are generally dense and of good crystalline quality. The choice of solvent, catalysts, and reaction conditions will change the *cis/trans* ratio, so these must be carefully considered. A high temperature ($150°C$) Ziegler–Natta polymerization in *n*-hexadecane can yield virtually 100% *trans*-polyacetylene.

The solvent used for a Ziegler–Natta catalyst route to polyacetylene must be completely dehydrated so organic hydrophobic solvents such as toluene have been used. Small quantities of water will attack the catalyst system, ruining the reaction. For this reason, other catalysts have been tried with some success. Luttinger catalysts are good examples, and these consist of a reducing agent such as sodium borohydride ($NaBH_4$) and a nickel complex or salt. With Luttinger catalysts it is possible for water to be a solvent, but more commonly polar solvents such as tetrahydrofuran (THF) or ethanol are used. The advantage of the Ziegler–Natta route is that the polyacetylene that is produced is of a better quality than that with other catalysts.

There are other catalytic routes to the polymerization of acetylene. Some of these require a co-catalyst such as the Ziegler–Natta and Luttinger routes, and some do not. Rhodium complexes are popular catalysts for the polymerization of acetylene due to their resistance to oxygen and moisture. It is not the purpose of this book to provide an exhaustive list of all methods capable of synthesizing polyacetylene, but one method, which requires a co-catalyst, is important: the metathesis route. It is possible to polymerize acetylene using metathesis catalysis in the presence of *n*-butyllithium and tungsten hexachloride (WCl_6) but a much more important metathesis route to polyacetylene is the Durham route, which is considered next.

6.4.2 Durham polyacetylene

Metathesis polymerization requires the metal-catalysed reaction between two dienes (hydrocarbons with two double bonds). We showed an example, butadiene, in Fig. 2.11. We also showed hexatriene in Fig. 2.11, but hexadiene, or more specifically 1,5-hexadiene, is an ideal candidate for such a polymerization reaction. (The 1,5 refers to the location of the

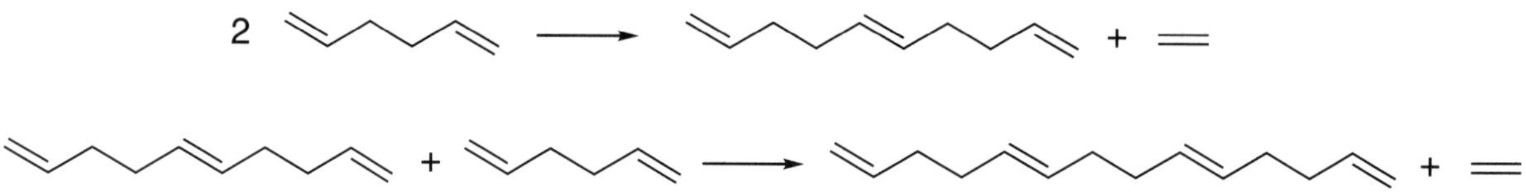

Fig. 6.14 Metathesis polymerization requires that at least two double bonds be present in the reactants to be polymerized. Here the polymerization of 1,5-hexadiene is shown; the end product is (non-conjugated) poly(propyl acetylene).

double bonds on the molecule. For the first and third carbon–carbon bonds to be double we should have 1,3-hexadiene.) The metathesis polymerization of 1,5-hexadiene is shown in Fig. 6.14. The polymerization of hexadiene actually results in poly(propyl acetylene), a polymer that is unsaturated but not conjugated. If we were to replace the 1,5-hexadiene by butadiene we should obtain a route to polyacetylene, but that reaction is not effective. The hexadiene example shown in Fig. 6.14 is an example of what is known as acyclic diene metathesis polymerization and is usually written as ADMET polymerization. A different form of metathesis polymerization is used to synthesize polyacetylene, and this is known as ring-opening metathesis polymerization (ROMP), which was pioneered in Durham, and is often known as 'Durham polyacetylene'.

The importance of Durham polyacetylene is that it is created from an intermediate *precursor* polymer which is soluble in, for example, isopropanol. This polymer thermally decomposes to polyacetylene which means that it can be processed to whatever form is required. This is important if one is to make devices, because the alternative to the Ziegler–Natta and Luttinger routes is to remove the polymer from the sides of a reaction flask and use it as is. Unfortunately, polyacetylene cannot be molten because it decomposes below its melting temperature. Durham polyacetylene was used in polymeric transistors, which was possible due to the ease at which films can be made with this synthetic route. The charge transport is not as good as in the Ziegler–Natta polymers, partly because it is more difficult to effectively dope Durham polyacetylene, but this can be improved, by, for example, stretching the film during thermal conversion to polyacetylene; the stretching process improves order and crystallinity—important properties for charge transport. It is also possible to end-label the polymer, which allows the possibility that these ends phase separate from the precursor polymer to create order. By phase separate we mean that the two components do not mix, not that the end-groups detach from the polymer. If the end groups form their own phase, the rest of the (still attached) polymer should be more ordered. This has been achieved and good charge mobilities reported.

The synthetic route to Durham polyacetylene is shown in Fig. 6.15. The monomer for this reaction is 7,8-bis(trifluoro methyl)tricyclodeca-3,7,9-triene, which, when polymerized, will thermally decompose to form polyacetylene. The first step is to form this monomer which is created in the reaction between cyclooctatetraene and hexafluoro-2-butyne. This reaction is straightforward with a high yield of 7,8-bis(trifluoro

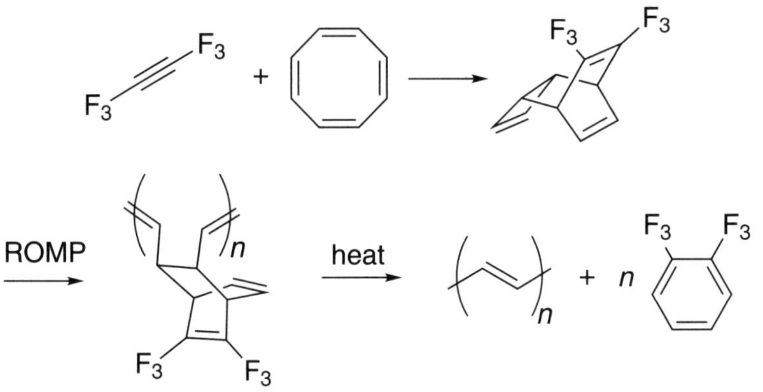

Fig. 6.15 The ROMP route to Durham polyacetylene. The first reaction creates the monomer 7,8-bis(trifluoromethyl)tricyclodeca-3,7,9-triene, which is ring-opened in the polymerization step. Finally, thermal decomposition of this precursor polymer creates polyacetylene.

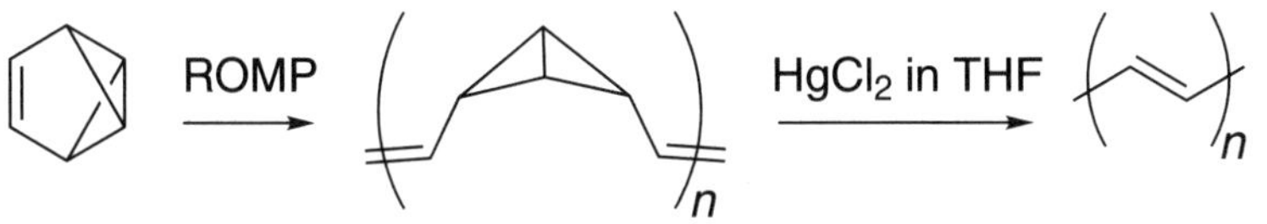

Fig. 6.16 There are other ROMP routes to polyacetylene. Here benzvalene is polymerized in a ROMP reaction. Polybenzvalene may be isomerized to polyacetylene in a solution of $HgBr_2$ or $HgCl_2$ in tetrahydrofuran.

methyl)tricyclodeca-3,7,9-triene. The ROMP reaction is catalysed by a mixture of tetrabutyl tin $(C_4H_9)_4Sn$ and tungsten hexachloride (WCl_6) in similar molar ratios. In ROMP reactions the catalyst binds to the growing chain by attacking the cyclobutene ring. (Cyclobutene, C_4H_6 is a four-sided ring made up of one double bond and three single bonds.)[3] Breaking the double bond in the cyclobutene part of the 7,8-bis(trifluoromethyl)tricyclodeca-3,7,9-triene allows the chain to grow (Fig. 6.15). This ring-opening method can also be performed with Ziegler–Natta titanium and aluminium catalysts. The Ziegler–Natta catalysts yield largely *trans*-polyacetylene, whereas the metathesis catalyst gives a mixture of the *cis* and *trans* forms. The final step of thermal decomposition can be performed at temperatures as low as 40°C or as high as 150°C; the reaction time is much longer for the lower temperatures—several days—whereas only a few minutes are required for the higher temperatures. Again a different ratio of *cis* and *trans* forms are obtained at different temperatures. At lower temperatures the *cis* form is predominantly produced, but at higher temperatures this isomerizes to the *trans* form.

ROMP can be used with other monomers to directly produce polyacetylene, a good example of which is the ROMP of cyclooctatetraene. Another example of ROMP to create a precursor polymer is in the polymerization of benzvalene as shown in Fig. 6.16. Here the monomer for the ROMP reaction is benzvalene, which is readily synthesized and highly reactive. The reactivity stems from the high strain energy in its bonds required to maintain its shape. Polybenzvalene will readily gel (i.e. the individual polymer chains will form physical crosslinks creating an elastic material that will not dissolve in, but rather absorb, solvent)

[3] In Fig. 6.15 there are only four hydrogen atoms in the cyclobutene ring because it is bonded to the rest of the precursor monomer.

and so it needs to be kept in non-concentrated solution to prevent this from happening. In fact the polyacetylene is formed by an isomerization process in a solution of tetrahydrofuran with mercury, zinc, or silver salts as catalysts. This process produces largely *trans*-polyacetylene with the advantage that there are few other reaction products to be removed during or after the synthesis. Otherwise the polymer produced is similar to Durham polyacetylene in that it has a relatively low crystallinity and similar conductivity.

6.5 Synthesis of poly(*para*-phenylene)

We first met poly(*p*-phenylene) in Section 3.1, where we mentioned its importance as a stable blue-emitting polymer. Like polyacetylene, PPP is insoluble in the usual organic solvents and there is at present no route to high-quality defect-free films. Nevertheless, correct preparation of the monomer to have alkyl side chains will not affect the conjugation because the side chains only involve one of the sp^2 electrons leaving the $2p_z$ electron responsible for semiconductor behaviour unaffected.

Synthetic routes to PPP involve the polycondensation (coupling) reactions described in Section 6.3. If pure PPP without side chains were required, Yamamoto coupling reactions would be unlikely to produce more than about fifteen monomers together, although other reactions can do a little better, but not by much. Aside from the Yamamoto polycondensation route a standard Stille reaction is possible, and is shown in Fig. 6.5.

Another possible route to PPP is with a Suzuki coupling reaction, an example of which is shown in Fig. 6.17. As we first mentioned in Section 6.3.2, boronic acid is of the form $R-B(OH)_2$, where R is often a derivative of a phenyl group. The reaction shown in Fig. 6.17 is therefore a classic example of Suzuki coupling.

The synthesis of PPP is now relatively mature, and very high molecular masses (in excess of 300 kg mol^{-1}) can be achieved. Variants of PPP are important for different reasons. A large problem with PPP, with and without the solubilizing side chains, is the possibility of backbone bending, which puts strain on the bonds and thus increases the possibility of defects and destroys the conjugation. We have discussed this problem in Section 3.7, and one solution is represented by the fluorenes, an example of which is shown in Fig. 6.7a. The alkyl side chains meet to strengthen the backbone covalent bond, without affecting the conjugation. A happy result of this chemistry is to shift the band gap to higher energies creating a blue-emissive polymer. Polyfluorene chemistry *via* Yamamoto coupling or Suzuki cross-coupling reactions are in principle very similar to those described for PPP above.

6.6   Synthesis of poly(phenylene vinylene)

Like polyacetylene, but unlike PPP, PPV is synthesized both directly, or by a precursor route. The use of precursors risks introducing chemical defects into the reaction products, but often has the advantage of greater flexibility, in the same way as for production of polyacetylene (Section 6.4).

6.6.1   Direct synthesis of PPV

A classic route to PPV is through the Heck coupling reaction. To achieve Heck coupling to PPV one reacts, for example, dibromobenzene with divinylbenzene (Fig. 6.18). If one refers to Fig. 6.10, the dibromobenzene is R_1 (X is Br) and divinylbenzene is R_2. The reader might like to stop here and consider for a moment why there are two bromine (other halogens should also work) or vinyl groups on the monomers. It is certainly true that the first two steps of Fig. 6.10 should be even quicker if there are two reactive sites, at either end of the ring, but the real reason concerns the product of this Heck reaction (Fig. 6.19). The reader will note that at one end there is a vinyl group, and at the other there is a halogen. Either of these groups may take the place of R_2 or R_1 respectively, and one would then be able to perform another Heck reaction to further elongate the chain; this is exactly the same argument that was presented when one was discussing metathesis polymerization (Fig. 6.14).

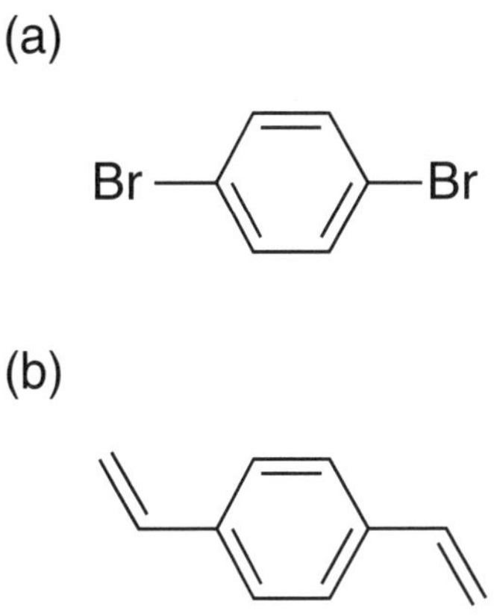

The Heck route allows rather pure PPV to be grown, and higher molecular weights can be achieved if side chains are added, e.g. replacing dibromobenzene (Fig. 6.18a) by a more soluble monomer such as that shown in Fig. 6.7b. The Heck reaction, like other routes such as Yamamoto coupling, allows alternating copolymers to be synthesized. The Heck reaction is not the only condensation route to PPV; there are others, such as the Wittig and Horner polycondensation reactions, but space precludes a detailed discussion of these.

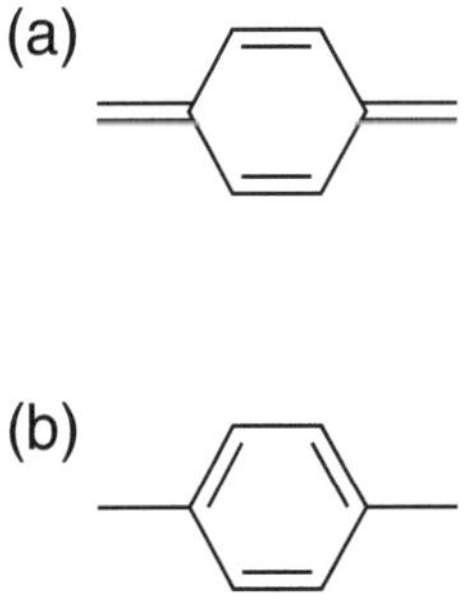

6.6.2   Precursor routes to PPV

The precursor routes to PPV involve the polymerization of a quinodimethane derivative, which can be converted into the appropriate PPV. Quinodimethane (Fig. 6.20a) refers to a quinoidal form, and can be thought of as the quinoid PPP monomer shown in Fig. 3.1b. In the Gilch route to PPV, which is the most popular of the precursor routes, a *p*-xylene (Fig. 6.20b) derivative is stripped of one of its halides when attacked by a strong base to obtain quinodimethane. The reaction with the base forces the formation of the quinodimethane derivative shown

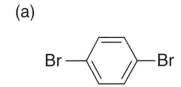

(a)

(b)

Fig. 6.18 Monomers used in the Heck coupling route to PPV include (a) dibromobenzene and (b) divinylbenzene.

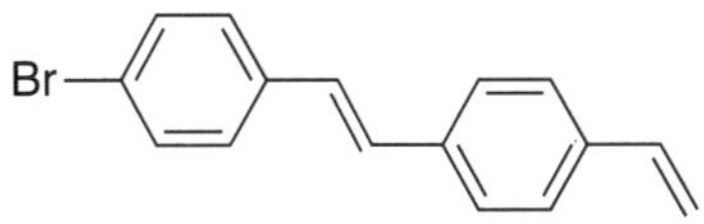

Fig. 6.19 The product of the Heck reaction between dibromobenzene and divinylbenzene.

(a)

(b)

Fig. 6.20 (a) *p*-quinodimethane and (b) *p*-xylene or 1,4-dimethylbenzene.

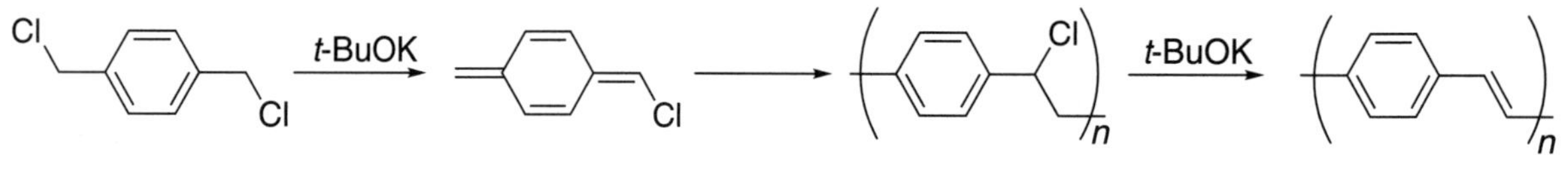

Fig. 6.21 The Gilch route to PPV. A derivative of *p*-xylene is reacted with a strong base (e.g. potassium *tert*-butoxide) to lose a halide (it is possible to lose both halides, but this is less likely at the low concentration of base used), creating a quinodimethane derivative, which can be polymerized. Heating in vacuum will strip the precursor polymer of the remaining halogen, forming PPV. Either chlorine or bromine can be used for the halogens.

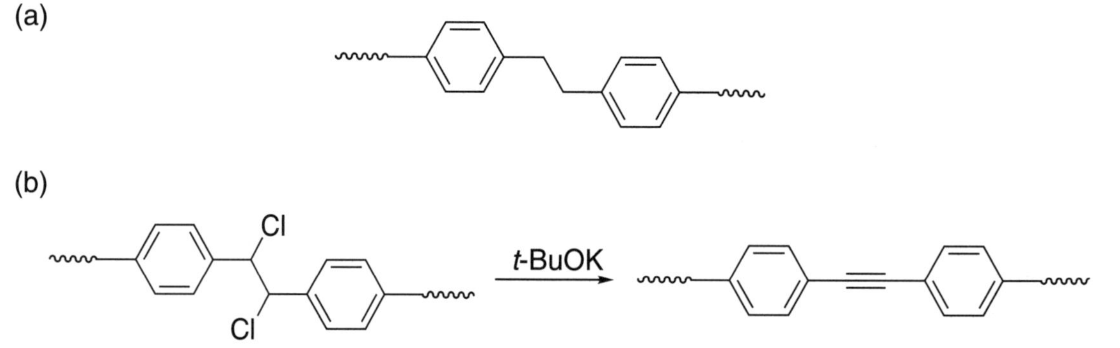

(a)

(b)

Fig. 6.22 The Gilch route to PPV may create chemical defects. If the coupling is incorrect, the conjugation may be destroyed (a) or a triple bond may be introduced (b).

in Fig. 6.21. This molecule can then polymerize though the process involves some rather complicated chemistry that is well beyond the scope of the book. On polymerization further base will remove the final halide. (It is possible to perform a 'one pot' synthesis, so that the same base as that that removed the initial halide, removes the final one, however, it is generally better to perform this polymerization sequentially.) The synthesis of PPV derivatives by this route is also possible, so much more soluble polymers can be synthesized with alkyl side chains attached to each monomer. Indeed, polymerizations to molecular weights in excess of 1000 kg mol^{-1} can be achieved.

Other mechanisms are possible for synthesizing PPV via the polymerization of a quinodimethane derivative. These include the Vanderzande and Wessling routes. The Wessling is perhaps the more popular of the two, but both require the final conversion of the precursor polymer to take place in vacuum at high temperature. The differences in the chemistry stem from the absence of halogens in the different reactions. If X is the first halogen stripped in the Gilch route (Fig. 6.21), and Y the halogen stripped during the conversion of the precursor polymer, then the Wessling route has both X and Y replaced by a SR, where R is often (CH$_3$)$_2$ (dimethylsulfonium). The Vanderzande route has X remaining as Cl, but Y contains a sulfinyl (SO) group, such as an *n*-butylsulfinyl group.

There are advantages and disadvantages to all three of these alternatives; all three are still used, but the primary disadvantage of all three is

(a)

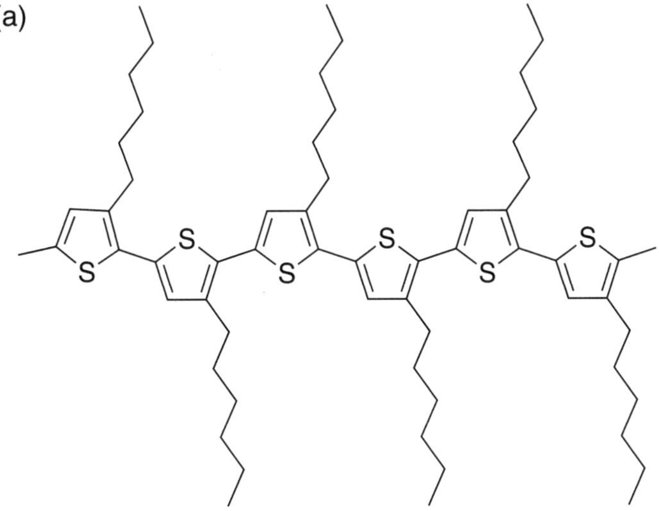

(b)

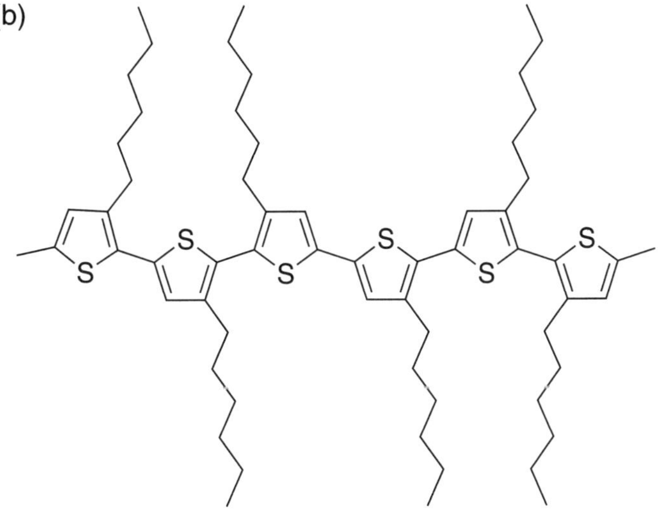

Fig. 6.23 The presence of side chains on polythiophene helps the polymer form better crystalline structures as well as aiding solubility. High-quality crystalline structures aid charge transport and so lead to large conductivities. The regio-regular poly(3-hexylthiophene), P3HT, (a) is such a polymer. However, if the alkyl chains are not ordered in this way (regio-irregular), charge transport mobilities are lower. In regio-regular P3HT, the side chains are coplanar, which is not the case for the irregular material. In the case of regio-irregular P3HT (b) the side chains will appear in different planes, affecting the stacking of the polymer and thus both intra- and inter-molecular charge transport.

that the polymer may contain many chemical defects. As an example, the chemistry shown in Fig. 6.21 requires *head-to-tail coupling* of the quinodimethanes. This means that the side that remains halogenated reacts with the non-halogenated side of another quinodimethane derivative. Indeed, this is the most likely outcome; the distribution of charge is such that the halogenated side is more negative, whereas the other side becomes nucleophilic (electron-donating), which means that the reaction is biased in favour of the preferred outcome. Unfortunately, there are other possibilities. A tail-to-tail coupling would destroy the conjugation, creating the product shown in Fig. 6.22a, whereas a head-to-head reaction would leave two halogens, which when stripped by the base, would leave an acetylene (triple) bond, as shown in Fig. 6.22b. The rest of the chain may be pure PPV, but such impurities will certainly affect the electronic and optical properties of the polymer. Since it is virtually impossible to control these chemical defects, they are not beneficial to the polymerization, and if pure polymers are necessary a direct route must be used.

6.7   Synthesis of polythiophenes

Polythiophene and related polymers are very popular materials to work with on account of their high conductivity and easy processability when doped; iodine is popular for p-type doping. There are several means to achieve polythiophene, so we shall be selective. Polythiophene can be synthesized by a variety of step-growth polymerization reactions including the Stille, Yamamoto, and Kumada coupling reactions. Another route is based upon a metathesis reaction, which is interesting in that it is a form of chain-growth polymerization. We shall discuss these below, but it is first worth noting the importance of alkyl solubilizing substituents in these polymers. Not only do they make the polymer more tractable, but they also can give it better conductivity. The conductivity clearly does not come from the side chains themselves, but rather from the ordering that the side chains make possible. A common polyalkylthiophene is P3HT or poly(3-hexylthiophene), and ordered and disordered P3HTs are shown in Fig. 6.23. When the side chains originate at the same point on the thiophene unit, the side chains are coplanar allowing more efficient stacking of the polymer; in Fig. 6.23b the third and sixth alkyl chains break coplanarity, and therefore disrupt crystallization. The ordered polymer is known as regio-regular P3HT, and is often written as rr-P3HT. The form described here is a head-to-tail regio-regular P3HT. There are other kinds of rr-P3HT, but the failure to realize rr-polythiophenes generally occurs when the chemical reaction introduces a mixture of different couplings such as head-to-head and tail-to-tail couplings, as discussed in the synthesis of PPVs (Fig. 6.22).

The reader may well be wondering about the "3" in P3HT. This refers to the point on the ring where the the alkyl chain is attached. The starting point is taken as the sulfur, and one moves around the ring as shown in Fig. 6.24. Convention decrees that the shorter route round the ring is taken, so we write P3HT rather than P4HT, which would otherwise be identical. Given 3-alkylthiophene monomers, the head-to-tail coupling is the desired synthesis, and this is a 2-5' coupling. This, and the undesired head-to-head (2-2') and tail-to-tail (5-5') couplings are shown in Fig. 6.24.

6.7.1   Condensation reaction routes to polythiophenes

Synthesis of polythiophene, including P3HT and similar substituted polythiophenes, can be performed by the standard Yamamoto route (Section 6.3.4), with dibromothiophene (with or without substituent) polymerized to give the final polymer. The Yamamoto route is not the best method to create polythiophenes, and the first truly successful poly-condensation is that known as the McCullough method, which enhances the regio-regularity of the final polymer. Other polycondensation methods have since been developed to provide regio-regular polythiophenes, but space constraints means we shall limit ourselves here to a description of the McCullough route.

To generate a regio-regular polymer, asymmetry in the molecules to be polymerized is required. If we were to use a dibromothiophene, either halogen would be attacked in the polymerization reaction so it is necessary to discriminate between the 2 and 5 positions. A molecule such as 2-bromo-3-hexylthiophene has the required property. The problem posed by using a molecule such as this is that it is hard to see how it can be polymerized because it only has one obvious reactive position on the molecule, which is the bromine atom. However, by cooling the reaction with carbon dioxide to $-78°C$ and reacting the dibromothiophene with lithium diisopropylamide ($LiN(C_3H_7)_2$), and then with magnesium bromide diethyl etherate ($MgBr_2O(C_2H_5)_2$) or zinc chloride dissolved in $O(C_2H_5)_2$, a molecule with the required asymmetry can be polymerized by an appropriate Kumada coupling reaction (Fig. 6.25).

Lithium diisopropylamide is a very strong base and so is a proton acceptor. The stronger the base, the more effective it is at accepting protons. In this case, however, it is performing a standard condensation reaction. These reactions are very sensitive to water and must be performed with great care. In water the lithium diisopropylamide or the organometallic reagents can easily be hydrolysed, which may be a problem given that the reaction is performed in THF, a solvent which readily absorbs water from the atmosphere. For this reason, the chemistry must be performed under well controlled conditions using apparatus known as a Schlenk line to prevent atmospheric water contamination. The second reaction creates the monomer required for the polymerization reaction, which takes place with a nickel catalyst.

6.7.2   Grignard metathesis

A metathesis reaction is also available for the synthesis of regio-regular polythiophenes, with the advantage of more amenable temperatures than the various cross-coupling reactions; the McCullough reaction is not unique in this respect. The Grignard metathesis (GRIM) reaction is for the functionalization of the monomers; the metathesis reaction is shown in Fig. 6.26a and will always take the form $AB + CD \rightarrow AC + BD$. In this case, one can equate 2,5-dibromo-3-alkylthiophene to A, RMgCl to B, the bromo-chloromagnesium-3-alkylthiophene to C, and RBr to D. The polymerization reaction involves the use of Grignard reagents for catalysis, and this reaction is shown in Fig. 6.26b.

The polymerization reaction shown in Fig. 6.26b is a good example of a Kumada coupling reaction, with limitations as briefly discussed in Section 6.3.3. It has recently been suggested that this GRIM synthesis of

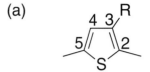

(a)

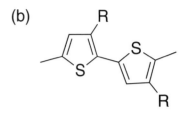

(b)

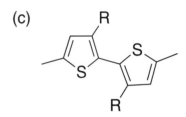

(c)

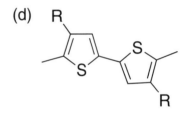

(d)

Fig. 6.24 (a) The alkyl side chain is located on the 3 position of the thiophene ring. There is no difference between the 3- and 4-positions since a simple rotation of the thiophene will take the alkyl chain from the 4-position to the 3-position. (b) The correct (for regio-regularity) head-to-tail coupling is a 2-5' coupling, whilst the head-to-head (c) and tail-to-tail (d) couplings are 2-2' and 5-5' respectively. Note that these couplings refer to the bond position and not the alkyl side chain, which is taken as being on the 3-position. The apostrophe refers to the second ring, which means that, for example, for 2-5' coupling the coupling joins the 2-position on the first ring to the 5-position on the second ring.

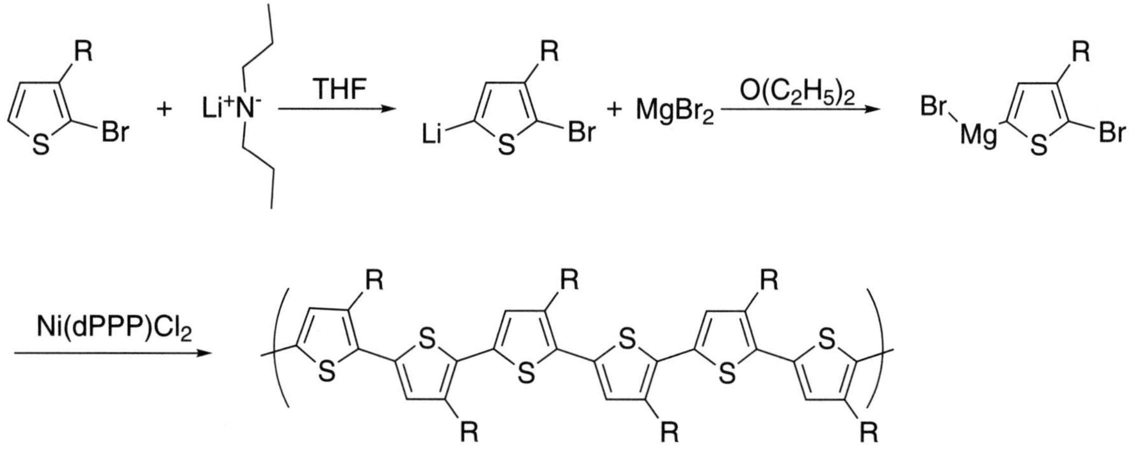

Fig. 6.25 The McCullough synthesis of regio-regular alkyl-substituted polythiophenes takes place through two steps before polymerization can occur. The first of these is the reaction of 2-bromo-3-alkylthiophene with lithium diisopropylamide to create 2-bromo-3-alkyl-5-lithium-thiophene. This is reacted with magnesium bromide to replace the lithium with the appropriate metal halide. A Kumada coupling reaction using dichloro(1,3-bis(diphenylphosphino)propane)nickel as the catalyst allows a reliable head-to-tail polymerization.

polythiophenes may be a form of living polymerization. In this case the reaction is not a polycondensation reaction, but rather a chain-growth polymerization. This is a very important result because chain-growth polymerization opens up new chemistry leading to 'tailor-made' semiconducting polymers, and also is a very powerful means of creating uniform polymers. The McCullough group at Carnegie Mellon University in Pittsburgh took a leading role in the development of the GRIM route to regio-regular polythiophenes, and it was they who demonstrated the likely chain-growth mechanism.

The mechanism for the GRIM reaction to synthesize regio-regular polyalkylthiophenes proposed by McCullough and co-workers is shown in Fig. 6.27. In general it is little different from the polycondensation coupling reactions that we introduced in Section 6.1, and it was initially expected that this reaction was a step-growth polymerization like the others discussed in this chapter. However, experiments showed that it had much in common with other chain-growth (addition) reactions, such as a rapid initial growth, which is indicative of a chain-growth mechanism. The low dispersity that can be obtained with this method is also indicative of a living polymerization, and so means that high-quality block copolymers can be synthesized, and indeed this has been done. Step-growth polymerizations can join to chains together, which means that the catalyst can be found anywhere on the chain after the transmetallation step. In the case of the GRIM reaction, the nickel reagent is located at the chain end during the polymerization (Fig. 6.27b). Indeed the McCullough group have argued that the nickel plays a more significant role as an initiator than as a catalyst, which means that it remains as part of the growing chain, even remaining associated with

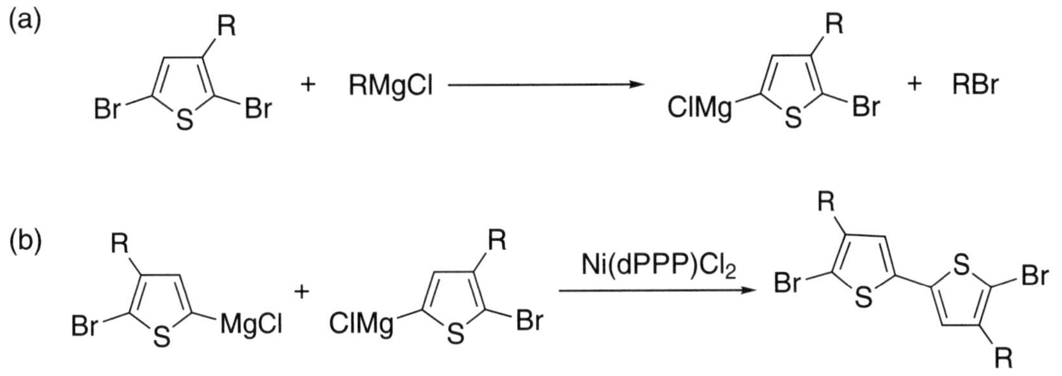

(a)

(b)

Fig. 6.26 The Grignard metathesis reaction requires a bromo-chloromagnesium-3-alkylthiophene to polymerize. The MgCl may be on either the 2- (20% yield) or 5- bond (80% yield) on the thiophene and is produced by the metathesis reaction shown in (a). The bromine atoms remaining on the product (b) can be reacted with a magnesium chloride Grignard reagent to achieve the necessary functionality so that the reaction can continue. This tail-to-tail coupling only occurs at the start of the reaction. Further (polymerization) steps are favoured thermodynamically and kinetically to produce head-to-tail rr-poly(3-alkylthiophene). The catalyst, Ni(dPPP)Cl$_2$, is versatile, being the same catalyst that we discussed for Kumada coupling reactions.

the chain after the reductive elimination step. In such a description the nickel reagent does not recover the chlorine to reform Ni(dPPP)Cl$_2$, so it cannot be considered a catalyst.

6.7.3   Synthesis of poly(3,4-ethylene dioxythiophene)

PEDOT is one of the more important polythiophenes because of its role in different technologies. PEDOT has an extremely large conductivity, which means that it can reasonably be referred to as a synthetic metal, and is also being used as an electrode in electrolytic capacitors. Its large hole mobility means that it can be used as a hole-injection layer at an anode, or even a synthetic anode itself. It helps greatly that PEDOT has good transparency to visible radiation, and so can be used in opto-electronic applications. PEDOT was originally designed by Bayer as an anti-static coating, for which it is still used. The most common form for PEDOT is a colloidal complex with polystyrene sulfonic acid, and was commercialized by Bayer under the trade name Baytron P, where the P merely means 'polymer'. In 2008 Baytron was renamed Clevios. The synthesis usually takes place with polystyrene sulfonic acid (PSS) to ensure solubility in water. The PEDOT itself is insoluble in water, so the complexation is important. Usually, PEDOT is synthesized with PSS in aqueous solution by oxidizing the EDOT monomer using Na$_2$S$_2$O$_8$ (sodium persulfate) and Fe$_2$(SO$_4$)$_3$ (iron (III) sulfate) as oxidising agent and initiator. The oxidized EDOT readily polymerizes under such conditions. The hydrophobic PEDOT molecules have a tendency to aggregate and so the surrounding polystyrene sulfonic acid stabilizes the dispersion and creates a *colloid*. Colloids may take many forms such as aerosols

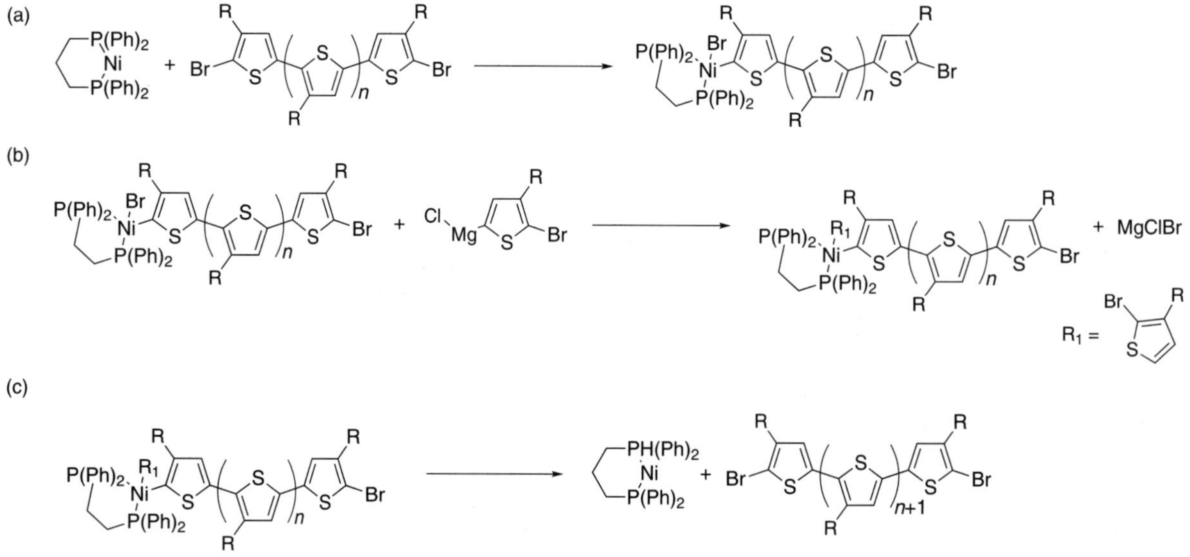

Fig. 6.27 Polymerization by the Grignard metathesis route also incorporates oxidative addition (a), transmetallation (b), and reductive elimination (c) steps.

(solids dispersed in gases), but in this case we have a solid polymer dispersed in water.

6.8   Polyaniline synthesis

Polyaniline is a relatively straightforward synthesis, similar to PEDOT, usually involving the oxidation of aniline hydrochloride, as indicated in Fig. 6.28. Aniline ($C_6H_5NH_2$) is a toxic liquid, but its salt with HCl is solid and therefore preferable from safety considerations. A strong oxidant such as ammonium peroxydisulfate is used for the reaction, which is strongly exothermic, meaning that relatively small quantities of the reactants are used to avoid explosions. The reaction can proceed at room temperature, although it may also be cooled to $\sim 0°C$, in which case larger mass polymers are obtained. The product shown in Fig. 6.28 is a salt, and, like the undoped fully conjugated form in Fig. 3.7b, has a strong green colour and conducts reasonably well, with a conductivity of ~ 500 S m^{-1}, which would not quite qualify it as a synthetic metal. The salt may be removed by washing in aqueous ammonia, but the product, coloured blue and no longer doped, does not conduct. If aniline is used rather than aniline hydrochloride, the same reaction will proceed but producing the emeraldine form of Fig. 3.7b rather than the salt shown in Fig. 6.28.

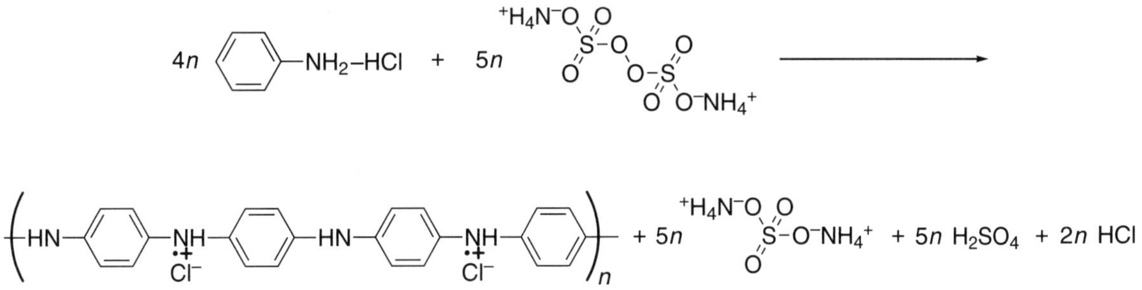

Fig. 6.28 The synthesis of polyaniline (emeraldine) hydrochloride. The reaction is exothermic and no catalyst is required. The two reactants (aniline and aluminium persulfate) are prepared separately as aqueous solutions with hydrochloric acid, and then mixed. The product is formed as a precipitate.

6.9 Electrochemical synthesis of polypyrrole

Many of the polymers discussed above can be synthesized with different techniques. Polycondensation reactions have dominated these because of their versatility in producing tractable materials. Electropolymerization results in a polymer film grown from, and generally attached to, an electrode. The growth of polypyrrole and polythiophene layers is popular with this method. Because of the restrictions on the synthesis, it is generally less preferred, although its protagonists struggle to understand why anyone would want to polymerize any other way. The advantages of electropolymerization include the high level of control of the films that are made with the technique, which means that devices based on these polymers are more reproducible. Because electropolymerized films are formed *in situ*, at point of use, there is no need to worry about solubilizing side chains, which confers another advantage on the technique.

The requirement for electropolymerization consists of monomer and salt (which is required for doping) in a suitable solvent contained within an electric cell. Most films are deposited on an anode, which means that electrons are drawn from the monomer at this anode creating radical cations, which can react. In Fig. 6.29 we show the reaction scheme for the electropolymerization of polypyrrole. The polymerization proceeds by the continual and alternate extraction of protons and electrons. The first step is to ionize the pyrrole, a step which is very often referred to as oxidation; we recall the description of oxidation in Section 2.4.5 as the removal of electrons. The cations formed during the polymerization are very reactive, but it may be something of a surprise that they react with each other; after all, like charges repel. In fact this is a rather rapid step. The black circle shown in the bond next to the N^+ (the α-position) in Fig. 6.29a, and 6.29b indicates an unsatisfied bond. The two single bonds meet, but there is only one hydrogen atom present, which leaves the molecule even more reactive than it would normally be as an ion.

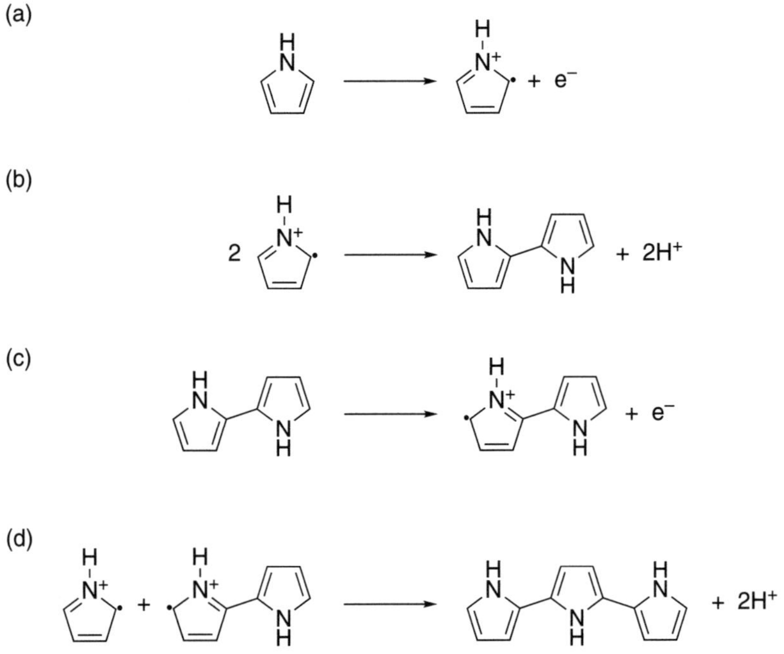

(a)

(b)

(c)

(d)

Fig. 6.29 The electropolymerization of pyrrole takes place in four stages. (a) ionization (oxidation) of pyrrole; (b) formation of bipyrrole; (c) ionization of birpyrrole; and (d) formation of ter-pyrrole. In each of these cases protons and electrons are extracted by the relevant electrodes due to the applied electric field. Polymerization continues with terpyrrole replacing bipyrrole, and so on. The solvent for these processes is water.

This is an example of a *radical cation*. The unpaired bond allows it to dimerize, as shown in Fig. 6.29b. Evidently, any Coulombic repulsion due to the charges is not enough to overcome the benefit of satisfying the bonds. This point demonstrates nicely that it is unwise to assume that molecules in a chemistry should be taken as point charges; in fact one should consider a molecule as a sea of charges, which is why, for example, they can easily form dipoles.

The polymerization of pyrrole described in Fig. 6.29 is known as the Diaz mechanism for the electropolymerization of pyrrole. This is the most commonly quoted mechanism, but the reader should be aware that there are other mechanisms described for electropolymerization reactions. The controversy is partly due to the difficulty in testing the mechanism of the reaction due to the inaccessibility of the reaction product, i.e. the polypyrrole film. Nevertheless, the Diaz mechanism probably has the most support. The speed of the reaction is one parameter that can differentiate between mechanisms, and this can be followed by observing the rate of increase of film thickness. In general, the reaction is retarded as longer oligomers are produced. (An oligomer is a small polymer of maybe fewer than twenty repeat units,[4] the material properties of which may, to some extent, depend on the molecular mass.) The reduction in rate coefficient reflects the difficulty that the larger molecules have in aligning. These oligomers fall out of solution, and are deposited at the anode. The polymer formed at the anode is not necessarily linear, but can be branched. A common cause of the side reactions that take place are β-coupling reactions, whereby the pyrrole does not react at the α-position, but rather at the second carbon

[4]For polymers that can be easily synthesized, chemists tend to view fewer than 100 units as oligomeric, but for more involved syntheses, often involving rigid molecules that limit is very often much lower, to perhaps as few as ten monomers. A physicist would prefer the limits to be the other way around; stiff, rod-like, polymers require more units to be considered polymers.

along from the nitrogen; the reader can easily verify that this destroys the conjugation. Nevertheless, by a careful choice of parameters, the electropolymerization of pyrrole can achieve high-quality metallic films, with conductivities of up to $\sim 5 \times 10^4$ S m^{-1}. The large values are due to the polymer deposited on the anode being doped (p-type) during the electropolymerization described here; The electropolymerization drives anions from the electrolyte (e.g. Cl$^-$) to the anode where they become part of the film. Here the level of doping is high, with typically $\sim 30\%$ of the film being electrolyte. Different concentrations and different salts affect the final quality of the film. The use of larger anions, such as polymeric sulfonates, provide very good conductivities, and also improve the film properties. Rigid molecules, such as polypyrrole, are rather brittle, and a large concentration of a flexible polymer as an anion improves the ductility of the final product.

The synthesis of polythiophene may be described in a completely analogous way to that of polypyrrole. One simply needs to replace NH in Fig. 6.29 by sulfur and the positive charge on the nitrogen by one at the α-position on the ring, and the synthesis of polythiophene is fully described. Thiophenes cannot be polymerized in water, and so acetonitrile or a similar solvent must be used.

6.10 Further reading

Polymer synthesis is well served by many books, but these do not normally concentrate on conjugated polymers, which have a tendency to be a little more awkward due to their solubility requirements and so on. Within this limitation, it is worth having a look at the books by Cowie (1991) or Hiemenz and Lodge (2007) for a good introduction to the basics of polymer chemistry. A text for those interested in polymerization techniques might be that by Odian (2004). The lack of books on conjugated polymer chemistry has fuelled the appearance of numerous review articles and book chapters on the subject; those by Babudri et al. (2004), Pron and Rannou (2004), Bolognesi and Pasini (2007), and Yamamoto (2010) are worthy of reading. Coupling reactions are highlighted, but other aspects (e.g. metathesis reactions) are also covered.

6.11 Exercises

6.1. A polymer (the monomer molecular mass of which is 252 g/mol) is synthesized with a number average molecular weight $\overline{M}_n = 403$ kg/mol. It is observed to have chain lengths that (approximately) follow a Poisson distribution. What is the weight average molecular weight of this polymer? Comment on the nature of the polymer thus synthesized. Another batch of the same polymer is observed to have a Gaussian distribution with $\overline{M}_n = 510 \pm 12$ kg/mol. What is its approximate weight average molecular weight?

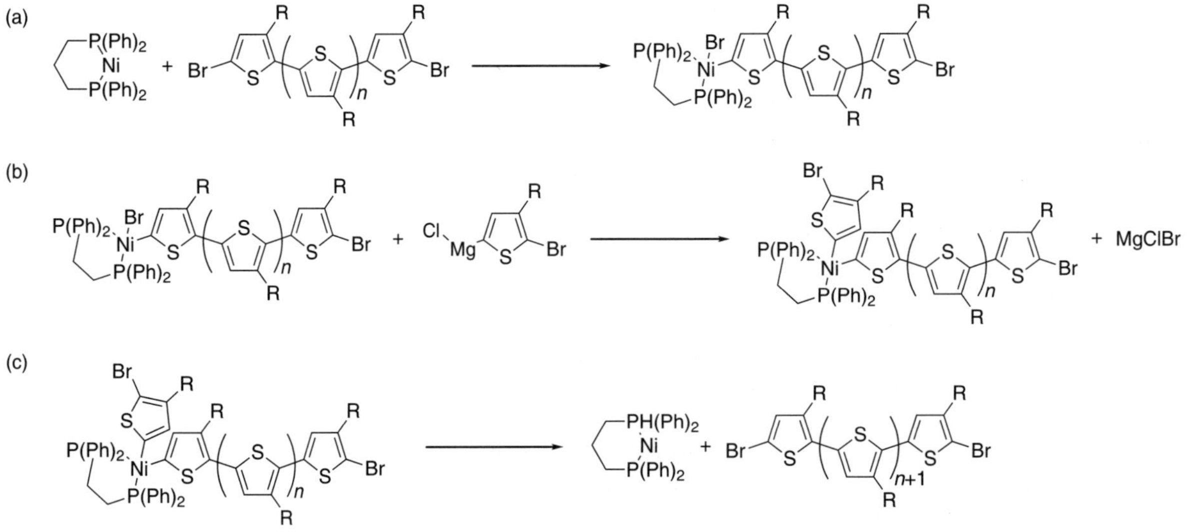

(a)

(b)

(c)

Fig. 6.30 This version of Fig. 6.27 is not quite correct.

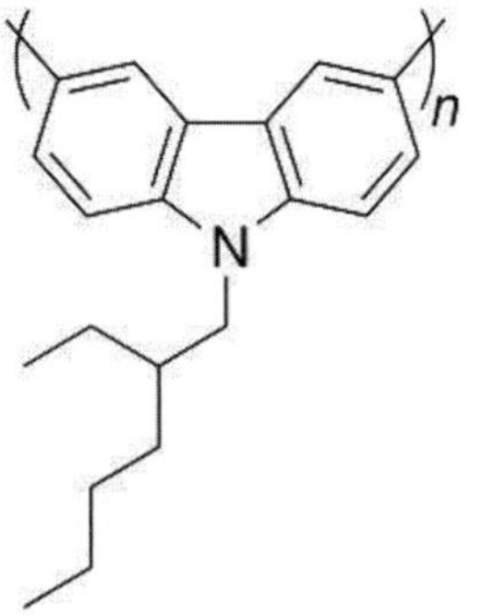

Fig. 6.31 The chemical structure of poly(9-(2-ethylhexyl)-9*H*-carbazole-3,6-diyl).

6.2. The Grignard metathesis route to creating P3HT (Fig. 6.27) was described by the group of Tsutomu Yokozawa of Kanagawa University. On visiting the University of Bordeaux, Yokozawa noticed this figure from an early draft of the book. He quickly spotted mistakes. What are they, and why is Fig. 6.30 incorrect? (Fig. 6.30 was the version of Fig. 6.27 that Professor Yokozawa had seen.) One mistake is chemical, i.e. the reaction scheme has a mistake in it, and the other is practical; a good-quality product would not be obtained.

6.3. Sketch the *cis-cisoid* form of polyacetylene. Do you think this form is likely to be stable?

6.4. Prove that the weight average polymerization index, X_w, is given by

$$X_w\left(t\right) = \frac{1 + p\left(t\right)}{1 - p\left(t\right)}. \tag{6.9}$$

6.5. A sample of a main-chain polycarbazole (Fig. 6.31) with an alkyl terminated solubilizing unit is synthesized by a cross-coupling route is observed to have a weight average molecular weight of 3.9 kg/mol. If 3.0 mol of monomer were used in its preparation, how many monomer units went unreacted?

The number average molecular weight was measured as 2.4 kg/mol. Explain any discrepancies that you may find, given the information above.

The physics of polymers

7

Polymer chains are an interesting subject for a physical treatment because they are not particularly well defined. Silicon is easily defined as having a diamond-like structure with each atom situated 0.5 nm from its nearest neighbours. We also know that each silicon atom has a mass of 28 g/mol, with the acceptance that one in twelve atoms might have a mass of 29 or 30 g/mol. The mass issue for conjugated polymers is much more complicated because most are *disperse* (Section 6.1.1) because they are synthesized by step-growth polymerization. This means that we cannot treat all of the molecular masses as the same. One could suppose that from a mathematical point of view dispersity is not a major problem, especially if we are working with *high* polymers, which are large enough so that a distribution of molecular masses means that the approximations to a uniform system can be made. However, many conjugated materials do not satisfy the 'high' polymer requirement, and dispersity can therefore significantly affect material properties such as surface composition.

Another example where polymers are less well defined by inorganic semiconductors is the copolymerization of two monomers to make a random copolymer, which means that no one polymer is the same as the next. We mentioned random copolymers in Section 6.3 when we pointed out that undesired monomers can find themselves on a polymer chain. Sometimes it is desirable to synthesize a polymer with a random mix of monomers, and their interactions would need to be calculated if we are to understand fully the physics of the final material. One way to understand how a polymer interacts with similar species (in a solid or molten phase) or with solvent molecules (in solution) is to calculate the strength of the interaction between each two individual monomers or solvent molecules as a function of their separation. Such a calculation is unnecessarily complicated given the number of molecules that this might involve. If we have four molecules this would involve six terms. The first molecule would interact with three others, and the second molecule would interact with another two (the interaction with the first having already been counted), and there would be one remaining interaction for the penultimate molecule. If we take 10^6 molecules one can see how quickly such calculations would get out of hand, because there would be $10^6!$ or 8.3×10^{5565708} interactions. To circumvent these problems we use *mean-field* theory, which is to treat individual polymer chains as if they were to experience a uniform environment wherever they might be in their particular medium. In the case of one polymer chain in a particular solvent, there are only three interactions that we need to consider: each polymer monomer with other polymer monomers, each monomer with solvent molecules, and each solvent molecule with other solvent molecules. This simplifies the problem greatly, and mean-field theories have had great success in explaining many aspects of polymers.

The physics of flexible polymers has a long and illustrious history. The application of mean-field theory to polymers played a large part in the award of the Nobel Prize for Chemistry in 1974 to Paul Flory for his work, both experimental and theoretical, in the physical chemistry of macromolecules. Flory's career began at DuPont under Wallace Carothers, but it was in academia in Cornell University and before that at Standard Oil in New Jersey that his most important work was published. At Standard Oil he developed mean-field theories for the mixing of polymers, for the size of polymer chains, and for the swelling of crosslinked polymer systems. These theories have been developed by many workers, but the ideas that underlie them remain as important today as when they were developed. Indeed Flory's 1953 book '*Principles of Polymer Chemistry*' remains a necessary text for many undergraduate and postgraduate students of polymers. Conjugated polymers are comparatively stiff molecules and so have many important differences with the flexible polymers described by Flory. Indeed they also can have similarities too; if the conjugated polymer is long enough, it is flexible on a larger length scale; i.e. comparatively stiff does not mean completely stiff. This means that all polymers are flexible on sufficiently long length scales, and similarly, all polymers show a certain degree of stiffness on short length scales, although in truly flexible polymers this is close to the size of one monomer. The length scale that describes the stiffness of a polymer molecule is known as the *persistence length*.

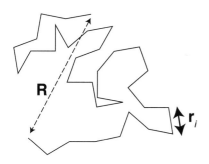

Fig. 7.1 A freely-jointed chain is one in which the monomer units can have any orientation, independently of the orientation of the adjacent monomers. The end-to-end distance **R** is the vector sum of individual monomer positions $\mathbf{r}_i$.

7.1   Persistence length

7.1.1   Freely-jointed chains

An *ideal* polymer chain is sometimes known as a freely-jointed chain. In this case the chain is treated as an assembly of monomers that are connected by bonds that are able to rotate freely. The analogy of a polymer *chain* is a good one. The end-to-end distance of a freely-jointed chain is given by the vector sum of the N individual monomer positions (Fig. 7.1),

$$\mathbf{R} = \sum_{i}^{N} \mathbf{r}_i. \tag{7.1}$$

The average size of this end-to-end distance cannot be computed from eqn (7.1) because $\langle \mathbf{R} \rangle = 0$. A more useful parameter is $\sqrt{\langle R^2 \rangle}$, which is non-zero and is calculated from

$$\langle R^2 \rangle = \langle \sum_{i,j}^{N} \mathbf{r}_i \cdot \mathbf{r}_j \rangle = Nl^2 + \langle \sum_{i \neq j}^{N} \mathbf{r}_i \cdot \mathbf{r}_j \rangle = Nl^2, \tag{7.2}$$

where l is the size of a monomer unit. The key to the simplification is the consideration of $i = j$ and $i \neq j$ separately. Because the monomer vectors are not correlated the terms for $i \neq j$ average to zero. For $i = j$, we have the simple result $\mathbf{r}_i \cdot \mathbf{r}_i = l^2$.

This result (eqn 7.2) is an important result in polymer physics. It states that the end-to-end distance of a polymer chain of N (monomer) steps is proportional to the square root of the number of steps. Of course, this is an average result; there is a distribution of lengths for the end-to-end distance of a polymer chain, but this defines the root-mean-square end-to-end distance. Given the generality and simplicity of the result, it is perhaps unsurprising that it is important in many aspects of statistical mechanics beyond polymer physics. The result in eqn (7.1) is a classical random walk and can be applied to subjects as diverse as diffusion in gases undergoing random collisions and stock prices in economics.

An important point to note concerning the ideal polymer chain is that it is valid only when there are no net interactions with its surrounding medium. A polymer in a good solvent will have a thermodynamic benefit from making monomer–solvent contacts relative to monomer–monomer contacts. This will cause the change to expand, because an expanded chain includes more solvent molecules in its volume than a collapsed chain. Such a chain is subject to a *self-avoiding* random walk. A chain in a poor solvent, by contrast, will lower its energy if there are more monomer–monomer contacts relative to monomer–solvent contacts; this will result in a smaller chain than the ideal result. Of course, a fully extended or completely collapsed chain will have an entropic penalty, and so the final chain size is a compromise between *enthalpic* interactions and entropic effects. We return to this in more detail in Section 7.5.2.

7.1.2 Worm-like chains

The freely-jointed chain model does not require a realistic view of the connectivity of monomers within a polymer. The *stereochemistry* of the monomer–monomer bonds requires fixed bond angles. A fixed bond angle does not mean that a polymer chain has no flexibility, because it is possible for a chain to mimic a freely-jointed chain over a period of very few monomers (Fig. 7.2). Some chains are more rigid than others, however. It may well be that polyethylene, as shown in Fig. 7.2, may well approximate rather well to a freely-jointed chain, but conjugated polymers have monomers that are correlated over much longer length scales than simple saturated polymers. These longer persistence lengths in conjugated polymers are due to any stereochemical property that inhibits bond rotation. Phenyl rings, pendant (solubilizing) groups, and sp^2-orbitals all increase the energy barrier to bond rotation. (Bond rotation does not simple mean that a bond in the middle of the chain will rotate, swinging the rest of the chain around it; there is a lot of co-operative motion involved in bond rotation which takes place in three dimensions, rather than the two that are represented on paper.)

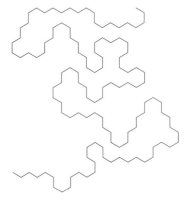

Fig. 7.2 The chain shown above has a fixed bond angle of 120°, but 'memory' of this is typically lost over only a few monomers. (By convention, chains are drawn with bond angles of 120°, but saturated chains have bond angles closer to the expected tetrahedral angle of 109.5°.) In three dimensions this 'memory loss' is even more apparent.

The persistence length, a is defined by

$$a = \frac{l}{1 - \cos \alpha} = \frac{l(1 + C_\infty)}{2}. \tag{7.3}$$

where α is the angle between bond vectors (Fig. 7.3) and C_∞ is known as the Flory characteristic ratio, which is defined by

$$C_\infty = \frac{\langle R^2 \rangle}{nl^2} \tag{7.4}$$

for a chain of n bonds. Note that we differentiate here between N monomers and n bonds simply because a monomer need not be restricted to one bond, as for example is the case for polyacetylene.

The calculation of the mean square end-to-end distance is not as simple as eqn (7.2) for the freely-jointed chain on account of the restrictions imposed on the local conformation by the bond angle. This means that $\langle \mathbf{r}_i \cdot \mathbf{r}_{i+1} \rangle = l^2 \cos \alpha$ and $\langle \mathbf{r}_i \cdot \mathbf{r}_{i+2} \rangle = l^2 \cos^2 \alpha$, or, more generally

$$\langle \mathbf{r}_i \cdot \mathbf{r}_j \rangle = l^2 \cos^{|j-i|} \alpha. \tag{7.5}$$

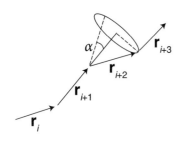

Fig. 7.3 The bond does not fully constrain the chain, which may also rotate azimuthally. The bond length may be written as $l = |\mathbf{r}_i|$. Note that $\alpha = \pi - \theta$, where θ is the bond angle and α the angle between bond vectors.

The mean is important here; if one considers, for example, $\alpha = \pi/4$, then we have two possibilities for adjacent bonds $\mathbf{r}_i \cdot \mathbf{r}_{i+2} = l^2$ and 0. The mean, however is $\langle \mathbf{r}_i \cdot \mathbf{r}_{i+1} \rangle = l^2 \cos(\pi/4) = l^2/2$. The reader can easily convince him or herself that this is true for all bond angles and also for three bonds. The torsional (azimuthal) angle plays no role in any of these calculations because the scalar product of two bond vectors is independent of this property, because a plane can always be found that contains these two bonds. For typical values of α, which might be considered for bonds that are close to tetrahedral, the correlation will quickly decay with increasing number of bonds. As an example, if we consider a tetrahedral bond over five monomers, $\cos^5 \alpha = 3^{-5} \approx 0.004$.

We can then modify eqn (7.2) for the more general case to give

$$\langle R^2 \rangle = Nl^2 + l^2 \sum_{i,j \neq i}^{N} \cos^{|j-i|} \alpha. \tag{7.6}$$

This exact equation has no useful analytical solution, so we need to make approximations. We therefore assume that there are many monomers either side of a given monomer j. Because $\cos^{|j-i|} \alpha$ decays rapidly it makes little difference if we assume that there are an infinite number of monomers either side of the jth monomer. Equation (7.6) can then be simplified to give

$$\langle R^2 \rangle = Nl^2 + 2l^2 \sum_{i}^{N} \sum_{j}^{\infty} \cos^j \alpha = Nl^2 + 2Nl^2 \sum_{j}^{\infty} \cos^j \alpha, \tag{7.7}$$

where the factor 2 indicates our assumption that there are an infinite number of monomers either side of the jth monomer. The final sum is a standard arithmetic series

$$\sum_{j}^{\infty} \cos^j \alpha = \frac{\cos \alpha}{1 - \cos \alpha}, \tag{7.8}$$

which can be easily shown if one expands its denominator as a binomial series

$$(1 - \cos \alpha)^{-1} = 1 + \cos \alpha + \cos^2 \alpha + \cos^3 \alpha + ..., \qquad (7.9)$$

We therefore have

$$\langle R^2 \rangle = Nl^2 \left(\frac{1 + \cos \alpha}{1 - \cos \alpha} \right), \qquad (7.10)$$

which gives us the definition of the Flory characteristic ratio

$$C_\infty = \frac{1 + \cos \alpha}{1 - \cos \alpha} \qquad (7.11)$$

which is in agreement with eqn (7.3).

The equation for $\langle R^2 \rangle$ in eqn (7.10) is general to all angles. For very stiff chains, however, α is small and the mean square end-to-end distance may be approximated to

$$\langle R^2 \rangle = \frac{4Nl^2}{\alpha^2}, \qquad (7.12)$$

by using the small angle approximation $\cos \alpha \approx 1 - \alpha^2/2$. Equation (7.12) does not provide obvious insight into the stiffness of the chains, for which we need to substitute $C_\infty = 4/\alpha^2$ into the persistence length from eqn (7.3) to obtain

$$a = \frac{2l}{\alpha^2}. \qquad (7.13)$$

So, if we take $\alpha = \pi/10$ as an example, the persistence length is more than six times longer than the bond length. The approximation that α is small describes polymers known as *worm-like chains*, and this worm-like chain model is often referred to as the Kratky–Porod model.

Clearly, if the chain is much longer than the persistence length it may be considered as a random (Gaussian) coil. This is because we can define an end-to-end distance based upon units that are the size of the persistence length rather than basing the step size on that of a monomer. However, if the polymer is not significantly longer than one persistence length we must assume that the chain is a rod. Unfortunately, although the statement that the chain behaves as a rod stands, the approximations inherent in the calculation of the persistence length fail. For example, the assumption of an infinite number of monomers either side of the jth monomer in eqn (7.7) may not be valid, because $\cos^j \alpha$ may not decay to zero fast enough with increasing j. Nevertheless, this treatment provides sufficient insight into what makes a stiff polymer behave like a rod.

7.1.3   Stiffness of conjugated polymers

Conjugated polymers are inherently stiff, but this property is not routinely measured because of the difficulty in obtaining such information. Persistence lengths are typically obtained by small-angle neutron scattering (SANS), which is an expensive technique because of the requirement to run a synchrotron or nuclear reactor to obtain the neutrons; a day of measurement time may cost around $30,000. (X-ray scattering is also possible, and generally much less expensive, but is much less useful because one requires heavy, electron dense elements to be present to provide a useful contrast with the solvent in which the polymer is dispersed.) Persistence lengths are not an important determining property of transport properties. It is expected that the larger the persistence length, the more likely that the material has better properties. However, it is far easier and cheaper just to measure the transport properties routinely rather than engage in an academic study of material structure–property relationships.

Although conjugated polymers are stiff molecules this is not necessarily reflected in their persistence lengths, possibly because of defects. Polyacetylene may well be synthesized in its *cis* form but may also contain *trans* impurities. In this case there will be slightly different bond angles in each of the stereoisomers, and these will further decrease the persistence length. For molecules with large solubilizing side chains, *cis-trans* isomers are likely to have an even greater effects on the persistence length compared to a simple molecule such as polyacetylene.

It is difficult to predict stiffnesses and persistence lengths in polymers; one simply has to measure them. There are some guidelines that can be followed to predict whether or not polymers will be stiff. In simple conjugated molecules such as polyacetylene, stiffness arises due to the conjugation. It is generally not possible to rotate around a double bond. To see why this is the case, one should return to Fig. 2.13 and consider the π_z bond. Rotation about this bond from the lowest energy state creates a bond of a higher energy; in some cases the rotation could even create an anti-bond. Bond rotation here is simply achieved by moving two adjacent p_z orbitals to the other side of the molecule. The inability to rotate about double bonds generally improves stiffness in the molecule. It is perhaps a little strange to discuss a physical property with arguments concerning electronic energy levels, but it is nevertheless not unreasonable, because σ bonds resulting from sp^3 orbitals have a pyramidal structure and so have a much greater flexibility than π bonds, arising from sp^2 orbitals, for which its forced planarity results in increased stiffness.

It is generally the case that chains with rings in them such as PPP (Section 3.1) are stiff, as an examination of Fig. 3.1 will reveal. Rotation of the ring around the single bonds does not change directionality, so the polymer retains a rod-like shape. Of course, the presence of substituents on the ring will deform it and so will change the shape of the molecule a little. Other molecules such as PPV (Section 3.2) also have ring rigidity built into them, but the vinylene group will not always have the same directionality and so, unlike for PPP, the polymer cannot be visualized as a rod. Other polymers containing a ring in the backbone, such as PANi (Section 3.5), are more flexible because of the angle of the C-N bond which also allows limited rotation.

Doping is also known to affect stiffness. For example, SANS measurements of poly(3-butylthiophene) have shown that its persistence length

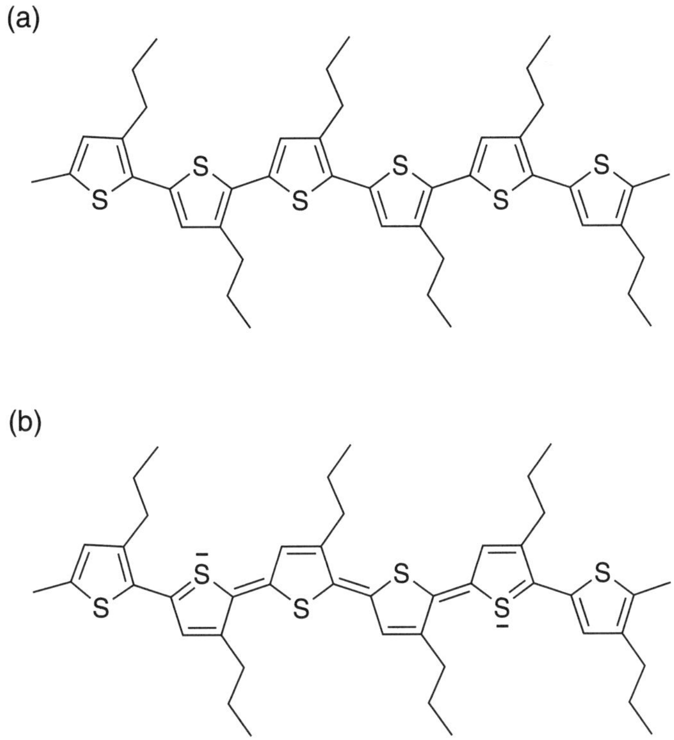

(a)

(b)

Fig. 7.4 Poly(3-butylthiophene), P3BT, (a) has a persistence length of 5.5 nm in its undoped form in solution. Doping, however, forces the quinoid conformation (b) with increased stiffness and a rod-like shape.

increases from 5.5 nm to a minimum of 85 nm when $NOSbF_6$ is added to the solution. (NO^+ acts as the dopant leaving SbF_6^- counter-ions free in solution.) The presence of the dopant creates a negative charge on the sulfur, introducing a double bond triggering the quinoid shape. In Fig. 7.4b we show the quinoid conformation, which is terminated at either side by the doping. It is also possible that these charges, and not just the conformation, is responsible for the enhanced rigidity, because of electrostatic repulsion between the NO^+ associating with the chain.

7.2 Crystallinity

The existence of crystalline phases is a phenomenon that exists throughout nature, causing many beautiful structures from the many precious metals to minerals such as the calcium carbonate structures that make up mother-of-pearl (nacre). The existence of crystals is perhaps surprising given that the second law of thermodynamics would seem to preclude such ordered structures with their high entropic cost. It is nevertheless energy arguments that are responsible for the existence of crystalline materials. If one considers the bonding orbital shown in Fig. 2.10, we see that there is an internuclear separation at which the potential energy is a minimum. This potential energy curve is quite generic, and so if one considers the potential energy of an amorphous (i.e. non-crystalline)

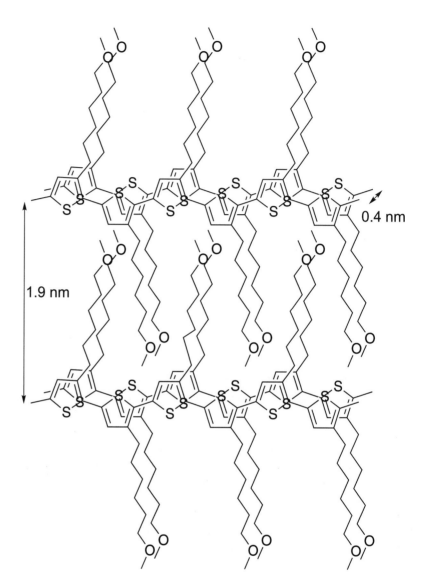

Fig. 7.5 Regio-regular poly[(3-hexylmethoxy)thiophene] exhibits different crystalline polymorphs. This structure, in which the alkane chains intermesh, is particularly common. The overlapping aryl units improve transport properties by facilitating hopping.

material one can understand the energetic driving force for crystallinity. An amorphous material by definition has atoms located with a distribution of nearest-neighbour distances, which will necessarily raise the free energy with respect to the crystalline lattice. The question is simply the balance between the energy gained by having all atoms on a crystal lattice being at their equilibrium positions, and the energy cost of having a beautifully ordered structure. This balance generally favours crystallinity, although a perfect crystal at equilibrium is not possible to achieve because lattice defects caused by missing atoms (vacancies) or the addition of extra atoms (interstitials) must exist for energetic reasons.

Polymers can form crystalline structures for similar reasons. Crys-

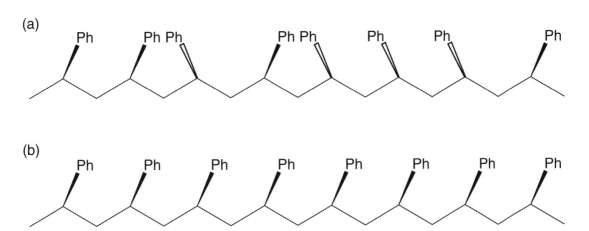

Fig. 7.6 (a) Atactic polystyrene is the most common form of polystyrene. The phenyl groups appear on different sides inhibiting crystallization. (b) Isotactic polystyrene can crystallize, because the phenyl groups appear in the same position on each monomer.

talline packing is not always due to the internuclear separation, but can be explained by polymer architecture. Solubilizing alkyl chains can define crystalline phases, as exemplified in Fig. 7.5 for poly[(3-hexylmethoxy)thiophene], although solubilizing chains can also render crystallinity more difficult by disrupting the packing of the main chain. Such disruption is particularly likely when the side chains are such that the polymer is not regio-regular. Other effects promoting crystallization include π-stacking, where aromatic groups (rings) interact in a rather complex mechanism through their delocalized electrons to stabilize an ordered layered structure of rings,

There are different forms of defect in crystalline materials. Apart from the point defects mentioned above, line defects (dislocations) can also exist. These are the insertion or removal of a plane of atoms into part of the crystalline lattice. Grain boundaries are another form of defect, which are formed when crystals nucleated (seeded) at different sites meet. The existence of vacancies and interstitials notwithstanding, it is possible to produce perfect crystals for many purposes. Turbine blades are made from pure (single) crystals because defects would significantly increase the risk of failure during operation. Closer to home, change transport properties in devices are affected by line defects and grain boundaries (i.e. the boundaries between different crystalline regions), which form particularly effective traps. The silicon-based semiconductor industry requires the fabrication of single crystals for its transistors. However, polymers do not form such single crystals. Although it is possible to design routes to create single crystal organic materials, these are expensive and so provide no real cost benefit over inorganic materials. If one wishes to make an organic crystal, one would generally use epitaxial methods. (Epitaxy is the growth of a crystal on a crystalline substrate. The crystalline order of the substrate is a template for the growth of single crystals.) Growing polymers by such methods is generally not

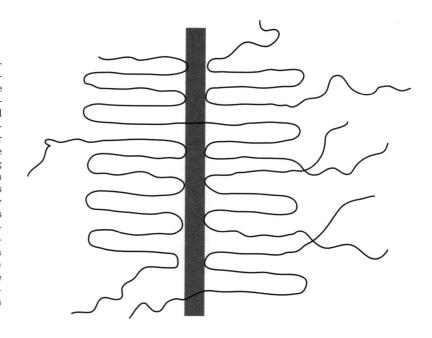

Fig. 7.7 Chains diffuse from amorphous regions to join crystalline regions. These crystalline regions are a mixture of chains forming lamellae. These lamellae may be stacked upon each other to form bigger structures known as *spherulites*. Polymer chains may contribute to more than one lamella; the part of the chain crossing from one lamella to another is known as a *tie fibril*. Typically, the thickness of these lamellae is of the order of a few (typically between 10 and 20) nm. This lamellar thickness is a compromise between the driving force for crystallization (i.e. the speed at which chains can straighten to join the growing crystal) and the surface energy of the folds. The shaded region delineates the two lamellae; the region between two lamellae is very often amorphous.

possible, although it is possible to produce oligomeric crystals.

Not all polymers crystallize. Amorphous polymers generally consist of polymers whose side groups appear on different sides of the polymer backbone. An example for polystyrene is shown in Fig. 7.6. Sometimes, branching is an explanation behind an amorphous structure. Anything random in a structure will inhibit crystallinity.

The properties of conjugated polymers are linked to their crystal structure in such a way that any changes in the structure will change their electronic and optoelectronic properties. Crystallization is heavily dependent on the processing route and different forms of crystallinity may occur depending on the sample history. In general, crystallization is seen as a 'good thing' because it encourages better charge transport with both band transport and hopping benefiting from an ordered crystalline structure. These advantages come at the price of defects. Polymers do not form the purest crystals, and as noted above, defects and grain boundaries are common. Although crystallinity is usually understood to take place in a temperature range between the glassy and molten (liquid) states, the use of sample heating (annealing) to create crystalline structures generally results in poor quality crystallinity because the kinetics of polymer diffusion to form crystalline structures is very slow. Solvent 'annealing' is generally a more appealing route to crystalline structures because the presence of solvent allows the polymer more space in which to move, because it is easier for solvent molecules to vacate space for the polymer than other polymer molecules. By having a concentrated solution and slowly removing the solvent good quality structures can be formed which are manifested by their electrical properties. It should go without saying that such solution-based crystallization works only for polymers that can be dissolved in a solvent! The optoelectronic properties are affected in different ways. Inhomogeneous broadening (Section. 4.2) is determined by the purity and crystalline quality of the materials, so better crystalline structures lead to sharper emission and absorption peaks. Grain boundaries will also scatter light and therefore affect the quality of the emitted radiation.

Polymers that crystallize are generally referred to as semi-crystalline because they usually co-exist with an amorphous phase. Chains from the amorphous phase diffuse towards and become part of a growing crystalline region. The crystalline region itself is formed from lamellae. Chains fold back upon themselves, as well as upon other chains in order to form these lamellae (Fig. 7.7). As lamellae grow, spherulites are formed (Fig. 7.8). These spherulites form a characteristic 'Maltese cross' shape when viewed in an optical microscopy under crossed polarizers (Fig. 7.9). Such images demonstrate that semi-crystalline polymers are birefringent. Polarized light will not be transmitted if the direction of the crystalline polymer backbone (i.e. the *director*) is perpendicular to the direction of polarization. (In order to extinguish light in one direction, it is necessary that the length scale associated with the separation distance of chain backbones be much less than the wavelength of light. This is readily realized in molecular systems such as semi-crystalline polymers.) The crossed polarizer (or *analyser*) eliminates transmission from light due to polymer backbones perpendicular to those rejected by the initial polarizer. The polarizer and analyser are, of course, located either side of the sample. Should they both be located before the sample, no light would be incident on the sample.

Microscopy is a ready means of detecting semi-crystalline behaviour, as is neutron or X-ray scattering, but other methods exist too. Heating or cooling a sample so that it crystallizes is associated with a latent heat, which can be detected by calorimetry. When a coil joins the growing lamella it must straighten at a certain entropic cost. The enthalpic energy gain must be greater than this entropic cost otherwise crystallization will not take place because there is a net energy barrier. Obviously the entropic cost is only too large when the lamellae are too big, and so this consideration limits the lamellae size. The melting temperature (T_m) of the lamella also depends on its size. If we denote the latent heat of melting (fusion) to be ΔH_m, then the energy gain related to its melting is

$$\Delta G_m = 2a^2 \sigma_f - \frac{\Delta H_m \left(T_m \left(\infty \right) - T_m \left(l \right) \right)}{T_m \left(\infty \right)} \rho l a^2, \qquad (7.14)$$

where ρ is the density of the lamella and la^2 is its volume. The first term in this equation $2a^2 \sigma_f$ is the fold energy, where σ_f is the surface energy of fold. This fold energy term should not be neglected, because it helps put a minimum size on the lamellae. If the lamellae are two small, then the exposed folds (Fig. 7.7) would present a significant energy cost. By setting $\Delta G_m = 0$, lamellae (of length l_0) are in equilibrium with the surrounding molten polymer. When $\Delta G_m = 0$ the crystallization

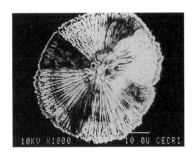

Fig. 7.8 A spherulite of PPP as imaged by scanning electron microscopy. The scale bar is 10μm. Taken from Mani et al. *J. Solid State Electrochem.* **2** 242 (1998), with kind permission from Springer Science and Business Media.

Fig. 7.9 An image of poly(9,9-dioctyl fluorene-*alt*-benzothiadiazole) taken through with an optical microscope using crossed polarizers, giving rise to the Maltese cross pattern, which indicates the presence of spherulites. Reprinted with permission from Banach et al. *Macromolecules* **36** 2838 (2003). Copyright (2003) American Chemical Society.

temperature for a given size of lamella can be determined using

$$T_m\left(l_0\right) = T_m\left(\infty\right)\left(1 - \frac{2\sigma_f}{l_0 \rho \Delta H_m}\right). \tag{7.15}$$

For smaller lamellae a lower temperature is required than for larger values of l. Although it is easy to understand that the melting point of a finite lamella must be different from the bulk (infinite) value—the fold energy contribution cannot be ignored and its effect can only be on the melting temperature—this does not explain clear why this temperature must be lower than the bulk value. In fact the folds act as a nucleating site for melting, which must decrease the melting temperature. If the temperature of a molten polymer is lowered, crystallinity will not become apparent at the bulk melting temperature, and so a degree of undercooling is observed. This undercooling (i.e. $T_m\left(\infty\right) - T_m\left(l_0\right)$) is not a physical property of the system and depends on sample history. Changing the size of the lamellae will change $T_m\left(l\right)$, and so any value of $T_m\left(l\right)$ or undercooling is physically possible so long as $l_0 \rho \Delta H_m > 2\sigma_f$ and $T_m\left(l_0\right) > T_g$, the glass transition temperature. indexglass transition—(It is generally accepted that crystallization is nucleated, and, if nucleation sites are not initially present, this is a thermally activated process. At the melting temperature, $T_m\left(\infty\right)$, no thermal activation is possible, because any fluctuations in temperature that might create a nucleation site will raise the local temperature into a point whereby the molten polymer is the thermodynamically favoured state.

Although there are different values of lamellar size that can exist at any given temperature, the value that satisfies eqn (7.15) at $T_m\left(l_0\right)$ must be a minimum size. Any smaller, and the fold energy dominates and $\Delta G_m > 0$. However, longer lamellae will minimize the overall free energy because they reduce the amount of folds present. Of course, longer lamellae will be slow to form because of the amount of molecular organization required to create such structures. Given that lamellae of length l_0 will not grow because they are in equilibrium with the melt, further growth will not reduce the overall energy of the polymer, and furthermore, given that large values of l are likely to grow too slowly to be observed, a dominant value of l will emerge.

Although the lamellar length is obtained from thermodynamic considerations, the growth of spherulites is generally considered on a rather more empirical basis. Here the fractional volume, ϕ_c, of a material that is crystallized has a time dependence given by the *Avrami equation*,

$$\phi_c = 1 - \exp\left(-K(T)\,t^n\right), \tag{7.16}$$

where $K(T)$ is a temperature-dependent constant and the exponent n determines the nature of the crystallization. Historically n was an integer between zero and 4, but more recently it is simply considered a fitting parameter. Larger values of n indicate a stronger dimensionality to the crystallization; $n = 3$ indicates crystals spontaneously initiated

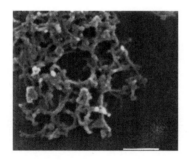

Fig. 7.10 A scanning electron micrograph of a polyacetylene fibril network. The scale bar is 1 μm. Taken from Kim et al. *Synth. Met.* **105** 207 (1999), with permission from Elsevier.

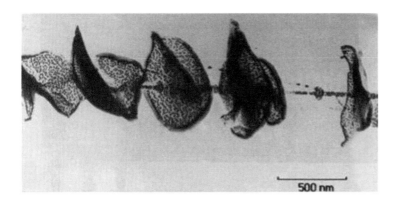

Fig. 7.11 Transmission electron micrograph of a polyethylene shish kebab. Reproduced from A. Keller *Faraday Discuss. Chem. Soc.* **68** 145 (1979) with permission from The Royal Society of Chemistry.

growing in three dimensions, whereas crystallization confined to a surface (but again spontaneously initiated) would have $n = 2$. If nucleation sites continually appear then n would be greater; in the original form this would be by unity, i.e. for three-dimensional growth one would then have $n = 4$. This form is rather too rigid but does contain insight; one should expect a greater exponent for growth in more dimensions. Generally, however, eqn (7.16) applies only to the early stages of crystallization.

7.2.1 Fibrils

Spherulites form from molten crystals on cooling, or indeed when a polymer glass is heated. Different processing routes allow for the formation of different crystals. A common processing method is to draw a polymer from solution or from its molten state. Drawing involves stretching the polymer either in the molten state or in a highly concentrated solution. Extensional flow first causes stretching, but this causes thinning (to conserve volume), though eventually the drawn polymer forms fibrils, which can be considered as very thin (a few nm) fibres. Although fibrils can be considered as small fibres, there are differences. Fibres are not necessarily crystalline, so a diffraction experiment (X-ray or neutron) would be required to determine the crystalline nature of the sample. An example of a fibrillar structure of polyacetylene is shown in Fig. 7.10, although in this particular case it has not been confirmed that the structures are crystalline. Not only is the structure of the fibrillar crystal very different to a spherulitic crystal, but the chains are more likely to be elongated than lamellar. The flow properties of the drawing process are responsible for the fibrillar shape, and allow for the entropic tendency to form lamellar crystals to be overcome. Fibrils therefore do not generally represent a lowest energy state.

It should not be surprising that these structures are not necessarily unique in a given material. A drawn polymer may well contain spherulites as well as fibrils; such a structure will depend on the preparation method. Indeed, crystalline fibrils can be used as a nucleating point for lamellae, creating a *shish kebab* morphology (Fig. 7.11), schematized in Fig. 7.12. Shish kebabs of P3HT are shown in Fig. 7.13, including

Fig. 7.12 Fibrils can be used to nucleate the growth of lamellae, creating a shish kebab morphology.

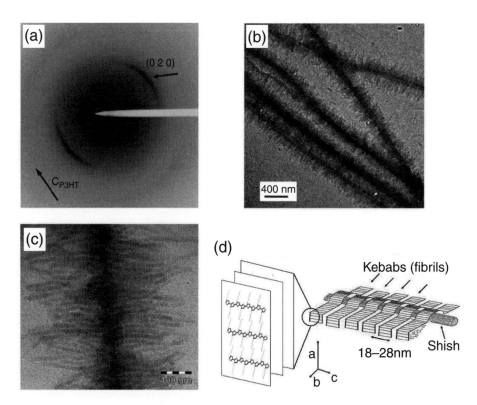

Fig. 7.13 Image data for P3HT shish kebab fibres. (a) An electron diffraction pattern shows clear diffraction peaks. The peaks are picked out by a length scale in the sample, which is that of π-stacking in the chains. (b) and (c) Transmission electron microscopy images of shish kebabs. Note the difference in structure between the fibrils here and those shown for polyethylene in Fig. 7.11. (d) Schematic diagram showing the P3HT semi-crystalline structure. The packing is similar to that shown in Fig. 7.5. Taken from M. Brinkmann et al. *Adv. Funct. Mater.* **19** 2759 (2009).

a diffraction pattern demonstrating that a good level of order exists in such fibres.

7.2.2   Crystalline structure

Crystallography is a large area of condensed matter science involving characterizing crystalline structures. Each crystalline structure can be reduced to a *unit cell*, which, when repeated throughout space forms the crystal. The unit cell fits on a lattice, known as a *Bravais lattice*. The simplest kind of lattice is a simple cubic in which all the atoms or molecules sit at the edges of a cube. A body-centred cubic lattice is another form of Bravais lattice on a cubic structure, in which another atom or molecule sits in the centre of the simple cubic structure. It is not necessary that all crystalline structures be based on cubic formalisms; the length of the sides need not all be the same, and the angles between them need not be 90°. There are seven kinds of crystalline structure in total, housing fourteen Bravais lattices, and three of the crystalline structures are shown in Fig. 7.14.

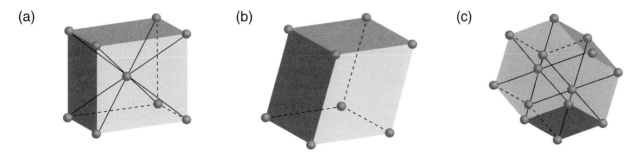

Fig. 7.14 Some examples of crystal systems: (a) An orthorhombic crystal, which has different lengths of the three sides but defining angles between these sides are all 90°. (b) A monoclinic crystal, which also has sides of different lengths, but only two of the defining angles between the sides are 90°. (c) A hexagonal structure. In (a) a body-centred orthorhombic crystal is shown. This is an example of one of the four Bravais lattices for the orthorhombic system. The other kinds of Bravais lattices have slightly different structures: Without the molecule or atom at the body-centre, it is known as a simple orthorhombic lattice. An atom or molecule centred on two opposing faces is known as base-centred, and the fourth lattice consists of an atom or molecule centred on all six faces (face-centred orthorhombic lattice).

Polymers form all these crystalline structures. An example of a common crystal structure structure is shown in Fig. 7.15 for PPP. Such 'herringbone' structures are formally known as *orthorhombic*, which means that the lengths of each of the three dimensions of the lengths of the unit cell are unequal, but that the angles between them are all 90°. Many conjugated polymers form crystals of this type, including polyacetylene, and PPV. Other well studied polymers include polyaniline, which in its pure form is monoclinic, which is the same as orthorhombic except that one of the three defining angles is not 90°.

The presence of dopants in a conjugated polymer must affect its crystalline structure because the dopant ion (required for overall charge neutrality) will in general not be of a size commensurate with the crystal lattice, which often means that the form of the crystal has to change to accommodate the dopant impurity. In many cases doping a polymer will change the nature of the crystalline structure. For low levels of doping the overall effect is usually the introduction of defects into the lattice.

7.3 Liquid crystals

The fourth state of matter is a statement much abused by researchers wishing to promote their own discipline. It describes a form of matter not easily characterized by solid, liquid, or gas. Plasmas, gels, and colloidal systems are examples of areas where researchers have designs on this title for their own field. The plasma community seem to be the most successful at this claim, pointing out that gases cannot conduct electricity, but plasmas interact strongly with electric and magnetic fields. Plasmas are often prefixed by 'gaseous', which is meant to differentiate them from solution plasmas or electrolytes. In fact, it merely tells you that a plasma is a gas. A gel might be considered a liquid, because it is largely made up of solvent even though it does not flow. It therefore has

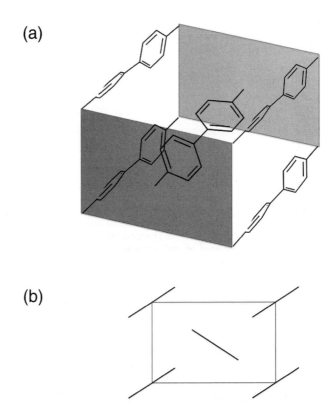

(a)

(b)

Fig. 7.15 (a) PPP forms a herringbone pattern-type crystalline structure. Note the splay between phenyl rings, due to inter-ring torsion. (b) A schematized unit cell for polymers such as PPP. Polyacetylene also displays a similar crystalline structure.

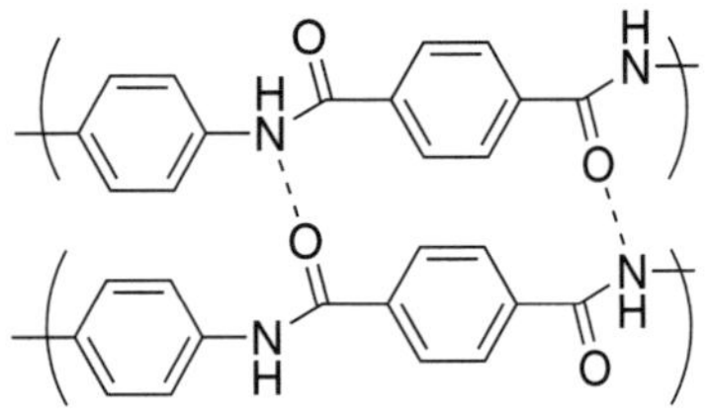

Fig. 7.16 The chemical structure of Kevlar[TM], which forms fibres stabilized with hydrogen bonds, which give the structure its impressive rigidity.

solid-like properties, so perhaps it is more convenient to call it a solid. This colloidal system is a dispersions of solid particles in a liquid, and the ubiquity of these leaves a very strong claim that colloids may be a fourth state of matter. Milk is a colloidal system, and it looks and behaves like a liquid, so that claim must also be discarded. Liquids are generally considered a disordered state of matter, so an ordered substance that flows can therefore be considered a fourth state of matter. Liquid crystals have this unusual property. The claim of liquid crystalline materials to be a fourth state of matter is rather undone by the word 'liquid' in the title, so one should accept that there is no fourth state of matter and simply enjoy the physics of unusual materials. The four examples here are not the only 'fourth states of matter' out there. Bose–Einstein condensates are another, and there are more. But for now, we shall simply consider the rather unusual properties of liquid crystals and discuss their considerable importance in polymer electronics.

Liquid crystals have the encompassing property of long-range order in one or two dimensions. (Three-dimensional long-range order cannot be a property of a fluid, as it would represent a solid crystalline material. Despite this, some liquid crystals get remarkably close to three-dimensional behaviour, for example the *smectic B* phase has a very high degree of order.) Of course, long-range is a relative term, and here it would be rare for liquid crystalline materials to show order at lengths greater

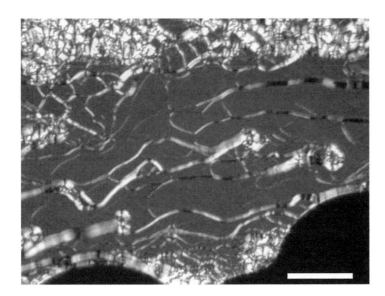

Fig. 7.17 As well as forming a semi-crystalline state, PPP exhibits several liquid crystalline phases. One of these is the chiral nematic (cholesteric) phase, which is shown here. This phase can exist only with molecular asymmetry, which is given to the molecule by substituent solubilizing side chains. The scale bar (bottom) is 50 μm. Taken from K. Akagi *J. Polym. Sci. A: Polym. Chem.* **47** 2463 (2009).

than a few microns, so liquid crystals exhibit mesoscopic phenomena. Polymers may exhibit both semi-crystalline and liquid crystalline morphologies, with the liquid crystal phase lying between the molten phase at higher temperatures and the semi-crystalline phase below. Indeed, there may be more than one liquid crystalline phase observable, either as a function of temperature, or indeed in different parts of the material at the same temperature. Materials which undergo a temperature-induced liquid crystalline phase are known as thermotropic. Another class of materials, which undergo a liquid crystalline transition at a given solvent concentration, are known as lyotropic materials. Some materials can exhibit both lyotropic and thermotropic behaviour.

Liquid crystals are prevalent in consumer goods. The most common is the liquid crystalline display, but this is being replaced in many areas, such as computer displays, by LED-based systems, and represents a new opportunity for polymer electronics! A famous example of a liquid crystalline polymer is Kevlar (Fig. 7.16), which is a trademark of DuPont, and is less commonly known as poly(*p*-phenylene terephthalamide). Kevlar is an example of a lyotropic polymer, dissolving in sulfuric acid (H_2SO_4), in which it must be processed. Efficient packing in Kevlar is due to hydrogen bonding between the carbonyl and amide groups, and this packing is what allows the Kevlar fibres to be used in various applications where its strength is necessary, such as in bullet-proof vests and tennis rackets, amongst others.

Like semi-crystalline polymers, liquid crystals are birefringent, as revealed by polarizing microscopy. An example of a polarizing microscope image of PPP is shown in Fig. 7.17. This (cholesteric) and other liquid-crystalline structures (exhibited by both polymers and small molecules) are schematized in Fig. 7.18, and a list of various forms of liquid crystals is included in Table 7.1. Shape is very important in the formation of liq-

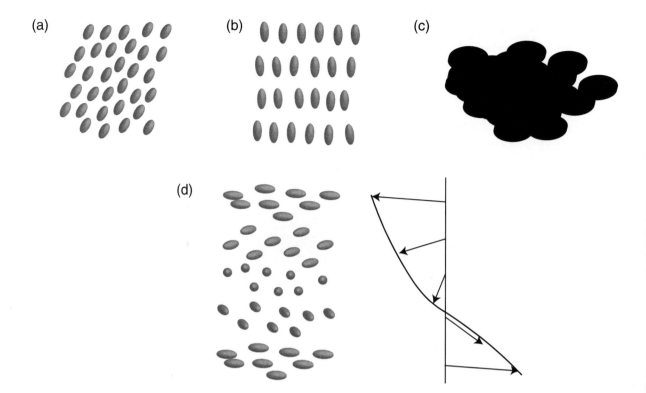

Fig. 7.18 Different forms of liquid crystals. (a) Nematic structure. Here only positional order is exhibited. The molecules all point in the same direction subject to the limitations of thermal motion. The direction is defined by the *director*. (b) Smectic A liquid crystals have one-dimensional order, with the molecules confined to layers, but there is no lattice position within these layers for the molecules to reside. These molecules are again pointed in the same direction given by the director. (c) Discotic liquid crystals. These disc-shaped molecules are defined by their director and only have positional order. (d) Cholesteric liquid crystals, also known as chiral nematic liquid crystals, have nematic-like ordering, which is twisted with each layer. The diagram on the right-hand side shows how the director changes with each layer. Molecules forming such a structure must be asymmetric.

uid crystalline and semi-crystalline structures,. In the case of cholesteric molecules, asymmetry must be built into the structure in order that there is a 'twist' in each layer. The origin of the twist is clear if one considers the original cholesteric molecule, cholesterol (Fig. 7.19), where the angle of the alkyl units gives the molecule its twisted alignment.

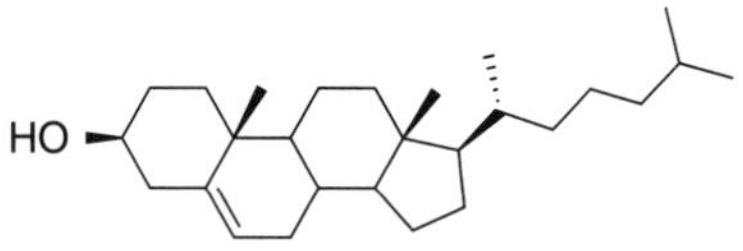

Fig. 7.19 The chemical structure of cholesterol.

7.4   Amorphous polymers

Semi-crystalline and liquid crystalline states are not the only states that a polymer can have. Indeed, it is only under special circumstances that a polymer exhibits some form of crystalline phase; at low temperatures a glass is normally formed, and at high temperatures a molten phase. If the molten phase were not to exist because, for example, the polymer were to decompose at high temperatures, then one might be limited to crystalline structures, although even then amorphous glassy phases could be instigated, perhaps through solvent treatment. Some polymers

Table 7.1 Properties of some liquid crystalline phases.

Phase	Positional order	Orientational order
Liquid	None	None
Nematic	None	Yes
Discotic	None	Yes
Cholesteric	None	Yes
Smectic	1-D	Yes
Columnar	2-D	Yes
Crystalline	3-D	Yes

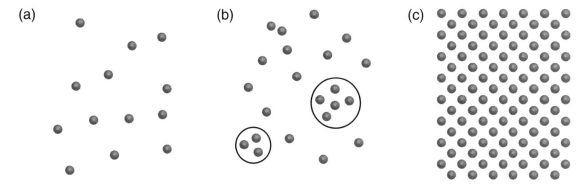

Fig. 7.20 Gases, liquids, and solids can be viewed through the interplay between thermal motion and kinetic energy: (a) Molecules and atoms in the gaseous state have a large thermal energy (with kinetic, vibrational, and rotational energy components) compared to the binding energy, and so they do not bond. (b) In the liquid state thermal energy is comparable to interatomic and intermolecular binding energies so atoms and molecules will either move freely or form short-lived clusters (ringed). (c) In the solid state thermal energy is not enough to escape the energy that binds atoms and molecules together. This effect is strongest on a crystalline lattice (as schematized here) because the attractive part of the interatomic or intermolecular potential is maximized, but it is also the case for amorphous solids.

exhibit only amorphous phases; as we discussed in Section 7.2, atactic polymers will not generally crystallize.

7.4.1 The glass transition

We have made reference to the glassy state in various parts of the book. There it sufficed to assume the reader's understanding of glasses as being amorphous and the components non-mobile. Glasses are generally misunderstood; they are not liquids with incredibly high viscosity and they do not flow, at least not on any time scales that can be readily identified with liquid behaviour. The fact that stained glass windows in medieval churches are thicker at the bottom than at the top is a result of the way the glass was prepared. Glasses are resistant to strain, and this gives them an important quality of solids. If a material forms crystalline phases, then glasses are generally metastable, with the crystalline phase being of lower energy.

When one tries to understand the glass transition, it is perhaps instructive to first consider the differences between gases, liquids, and solids as an energetic problem. Gases consist of molecules with enough space in which to move with little impediment from each other. The kinetic energy of the molecules and atoms that make up the gas is considerably greater than the interatomic or intermolecular energies that would want to bind them together. In the case of solids, the kinetic energy of the molecules is only enough to allow atoms to vibrate about a fixed point, usually, but not necessarily, on a crystalline lattice. The binding energies are much greater than any thermal energy that would want to force them apart. Liquids are somewhere between these two extremes. The intermolecular and interatomic forces do succeed in withstanding the motion of the molecules and atoms, but not for very long. In liquids, small, solid-like, temporary clusters of molecules and atoms are formed (Fig. 7.20), but thermal energy will be enough to quickly break these up. In a glass, thermal energy is not sufficient to allow the atoms and molecules forming the glass to escape their immediate environment, whereas in a molten state above the glass transition this is not the case. The glass transition is actually a kinetic transition because the glass transition temperature depends on the rate of cooling; it is not a thermodynamic phase transition, even though it shares some features with thermodynamic (second-order) phase transitions such as a discontinuity in the volume expansivity of the material. If cooling were infinitesimal, there would be no glass transition and so glasses represent a higher energy state than the corresponding crystalline structure.

The rigid structure of conjugated polymers generally means that they are not liquid, but glassy or crystalline, at room temperatures. This is a consequence of their rigidity, and not of the covalent bonds. Polybutadiene, for example, has double bonds contained within its backbone, but is liquid at room temperature. Of course, rigidity is a consequence of many double bonds, and the polybutadiene backbone has three single bonds for every double bond. Above the glass transition temperature the polymer will exhibit some crystallinity, although the crystalline state is usually in competition with the glassy state and so the temperature window where semi-crystalline structure can be observed is quite small. Above the melting temperature liquid crystallinity is often observed; poly(9,9-dioctylfluorene), a thermotropic conjugated polymer, exhibits smectic A and nematic liquid crystalline phases (Fig. 7.18) above the melting point. The nematic phase corresponds to higher temperatures. Above these liquid crystalline phases a molten polymer would exist were it not for the fact that it decomposes at these high temperatures.

7.4.2   Polymer melts

Conjugated polymer melts will generally exist only at high temperatures, which will often be in excess of 200°C. Under these circumstances, the polymer chains are ideal with a mean end-to-end distance given by eqn (7.2), which is in contrast to polymers in a good solvent, whose end-to-end distance is somewhat larger. Molten polymers are in one sense simply liquids, because they flow, and will fill available space. However, there is a rich amount of physics hidden in how polymers flow because of their *viscoelastic* nature. Viscoelasticity is, as its name suggests, a property whereby liquids exhibit both viscous and elastic behaviour. On short time scales polymers exhibit elastic properties; if one was to pull at a polymer melt, there would be an elastic restoring force, but at longer time scales the polymers still flow, i.e. they are viscous.

The unusual *rheological* (flow) properties of polymer melts are due to the polymers being entangled. One can imagine polymers behaving rather like spaghetti; if one were to take a handful of spaghetti from a plate, one would in most likelihood remove it all because the pasta would be entangled. These *entanglements* mean that polymers do not move from side-to-side, but rather on a path defined by the shape of the polymers along their length (Fig. 7.21). This is a process known as *reptation*, the theory of which was partly responsible for Pierre-Gilles de Gennes winning the 1991 Nobel Prize for Physics. Pulling at the polymer material would cause an opposite reaction from the entanglements pulling the polymer back, which gives rise to the elasticity of polymer melts. However, after a long enough period, depending on the size of the polymer, the chain would have been able to move out of its 'tube' and so the elasticity would be lost, i.e. the polymer would flow.

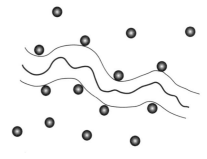

Fig. 7.21 A polymer moves along a path (known as a tube) defined by entanglements caused by the presence of other chains. This motion is known as *reptation*.

7.5 Polymer solutions

As has been pointed out elsewhere in this book, the great benefit of polymers for electronics is that they can be solution cast, which means that their processing is inexpensive. Polymer solutions can be categorized into dilute, semi-dilute, and concentrated solutions, each of which is treated differently. Concentrated solutions are those in which there are numerous monomer–monomer contacts, and dilute solutions are where there are few monomer–monomer contacts, at least from different polymers. In the case of dilute solutions, each chain is assumed to be isolated from other chains in a sea of solvent. From a theoretical viewpoint both of these situations are similar because they can be analysed in a mean-field approach. Mean-field theory involves a rejection of the idea that it is necessary to count individual monomer–solvent, monomer–monomer, and solvent–solvent interactions to understand the energetics of the system. The individual monomer is imagined to be in an average field containing other monomers and solvent molecules. The average field is sufficient to predict the behaviour of the polymer. Semi-dilute solutions, by contrast, lie somewhere between these two cases, and a mean-field theory is unable to predict average behaviour because the properties of the polymer environment may change. Formally, semi-dilute solutions are dominated by fluctuations, which cannot be accounted for in mean-field theory. (The reader with a grounding in other areas of physics where fluctuations are important will recall that they are associated with criti-

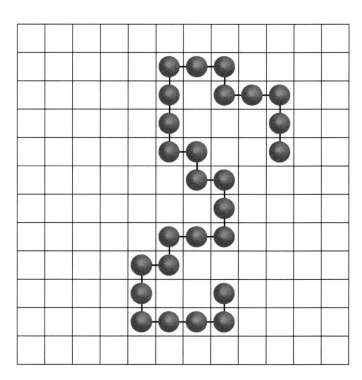

Fig. 7.22 A polymer can be treated as a series of connected monomers on a square lattice (cubic in three dimensions) where the 'empty' points contain solvent molecules.

cal points. Formally, the critical point referred to in this case is the onset of chain overlap; the concentration at which dilute chains start to 'feel' monomers of neighbouring polymers.) To account for the behaviour of semi-dilute solutions, *scaling theory* is required. (We give an example of scaling theory in Section 7.5.2 when we discuss the conformation of a polymer in a good solvent.) Here, however, we shall introduce the methods of mean-field theory.

7.5.1   Lattice theory

To understand the behaviour of a polymer in a solvent one needs to describe the energy of the system. This energy will allow us to predict whether or not the polymer will be soluble in the solution, or will phase separate out of solution. We can picture the polymer as being a series of connected monomers on a lattice, as shown in Fig. 7.22. Each of these monomers will feel an interaction with either other monomers of the same kind, or with solvent molecules. It does not matter in this instance whether or not monomer–monomer interactions are between two monomers from the same polymer. This enthalpic term will either discourage mixing (poor solvent) or favour mixing, for the case of a good solvent. Of course, entropy always favours mixing, and we start with a consideration of the entropy of mixing of a polymer in a solvent. (The related problem of the entropy of mixing of two solvents is presented in Appendix C. We can recover the result for two solvents here rather than a polymer in a solvent by setting the polymer chain length $N = 1$.) The entropy is a function of the size of the lattice, so it is useful to consider the entropy per lattice site, which has the advantage of being an intrinsic property of the system. If we consider a lattice of n sites, then there are n configurations for a solvent molecule, so the entropy for that solvent molecule is $k_B \ln n$. If, however, we reduce the concentration of solvent molecules by a volume fraction of $1 - \phi$, the entropy (per molecule) becomes $k_B \ln (n (1 - \phi))$. The entropy of mixing of the solvent molecules is the difference in entropy between the unmixed state and that of the mixed state.

$$\Delta S_s = k_B \left(\ln (n (1 - \phi)) - \ln n \right) = -k_B \ln (1 - \phi). \quad (7.17)$$

The same logic applies to a polymer of volume fraction ϕ, and we need not account for the polymer having N monomers, because we are considering the entropy per lattice site. We therefore write, for the entropy per lattice site of the polymer,

$$\Delta S_s = k_B \left(\ln (n \phi) - \ln n \right) = -k_B \ln \phi. \quad (7.18)$$

The entropy of mixing, ΔS, per lattice site is the sum of these two terms relative to their respective concentrations,

$$n \Delta S = -n k_B \left(\frac{\phi}{N} \ln \phi + (1 - \phi) \ln (1 - \phi) \right). \quad (7.19)$$

We have multiplied the equation by n on both sides because it helps explain the multiplicative factors ϕ/N and $1-\phi$. Since ΔS is the entropy per lattice site, $n\Delta S$ is the total mixing entropy (an extrinsic property and thus less useful). The contributions of the polymer and solvent in eqns (7.17) and (7.18) are per unit lattice site and therefore need to be multiplied by the number of molecules occupying these lattice sites, which is $n(1 - \phi)$ and $n\phi/N$ respectively. Only at this point is the size of the polymer important, because a longer polymer requires that more lattice sites be occupied by that polymer, and therefore there are fewer polymers, for a given ϕ, that can sit on the lattice.

Of course entropy is not the only thermodynamic property controlling solubility; if it were, all liquids would be solvents. In a closed system (where the are no molecules entering or leaving) the (Gibbs) free energy consists of an entropic term and an enthalpic term. Using a lattice, one can consider energetic interactions with nearest neighbours. One need not consider interactions with next-nearest neighbours or those further away, on the grounds that these are small corrections and are in any case important only if we wish to calculate bond energies accurately; this is done for the *Madelung energy* of ionic crystals such as NaCl. We therefore require an en enthalpy of mixing to account for the interactions of monomers and solvent molecules with their environment. Let us consider a polymer with an interaction energy of ϵ_{mm} between monomers of the same species. The solvent molecules will also have an interaction energy, which we denote ϵ_{ss}. If we again consider a polymer volume fraction of ϕ such that there are $\phi n/N$ polymer molecules, and we allow for z nearest neighbours, the energy of the polymer and solvent *in the unmixed state* will be

$$H_{\text{pure}} = \frac{nz}{2}\left(\phi\epsilon_{\text{mm}} + (1-\phi)\,\epsilon_{\text{ss}}\right). \tag{7.20}$$

This energy is the sum of the interaction energies for z contacts for each monomer and solvent molecule, and the factor $1/2$ is to prevent double counting of each energy. In order to consider the energy of mixing, it is necessary to introduce the energy of interaction between unlike species, ϵ_{ms}. This energy is given by

$$H_{\text{mix}} = \frac{nz}{2}\left(\phi^2\epsilon_{\text{mm}} + (1-\phi)^2\,\epsilon_{\text{ss}}\right) + nz\phi\,(1-\phi)\,\epsilon_{\text{ms}}, \tag{7.21}$$

and one immediately notes the ϕ^2 and $(1-\phi)^2$ terms, which indicate the mixed system. To understand these squared terms, one needs to consider that the energy of a lattice point not only depends upon which molecule is situated at that point, but also its neighbours. The probability that a monomer is located at that point is thus given by ϕ, and the probability that a given neighbouring point is located by the same species is ϕ^2. Of course, there are z positions for a neighbouring position to be occupied by a monomer of the same species so we need to multiply that by z. Over the whole lattice one needs to divide by 2 in order to eliminate double counting of monomer–monomer interactions. The same logic applies to the solvent molecules. We neglect the effect of monomers being connected in this calculation; i.e. the probability of two monomers being next to each other is greater than $z\phi^2$. Finally, the term $z\phi\,(1-\phi)\,\epsilon_{\text{ms}}$ represents the energy of interaction of the different species with each other, and here there is no possibility of double counting.

One must not confuse the energy of the mixed state with the energy of mixing. The energy of mixing is the difference in energy of the mixed state and the pure state, i.e.

$$\Delta H = H_{\text{mix}} - H_{\text{pure}} = k_{\text{B}}Tn\chi\phi\,(1-\phi), \tag{7.22}$$

where

$$\chi = \frac{z}{2k_{\text{B}}T}\left(2\epsilon_{\text{ms}} - \epsilon_{\text{mm}} - \epsilon_{\text{ss}}\right). \tag{7.23}$$

χ is known as the monomer-solvent interaction parameter, or often the *Flory–Huggins* interaction parameter, after the pioneers of the subject. χ is temperature-dependent, often with the form $\chi = A + B/T$, where A and B are constants. However, the form of χ is not so fixed, and may even include a composition (ϕ) dependent term. We can then add the two components (eqns 7.19 and 7.22) to obtain the total free energy of mixing

$$\begin{aligned}\frac{\Delta G}{nk_{\text{B}}T} &= \frac{1}{nk_{\text{B}}T}\left(T\Delta S + \Delta H\right)\\ &= \frac{\phi}{N}\ln\phi + (1-\phi)\ln(1-\phi) + \chi\phi\,(1-\phi).\end{aligned} \tag{7.24}$$

The free energy of mixing in eqn (7.24) is divided by nk_BT for simplicity; it is common to write energy units in terms of k_BT because these represent useful units in which to work, with values of order unity common. The division by n is rather important, because it means that the right-hand side of eqn (7.24) is *intrinsic* to the polymer solution in question, because it does not depend on the quantity of solution, although it does depend on the composition.

Equilibrium between polymer and solvent will be reached when the amount of solvent swelling a polymer is cancelled by its outflow. This is simply Gibbs phase equilibrium and can be considered by differentiating the Gibbs free energy (eqn 7.24) with respect to ϕ to obtain the (exchange) chemical potential, $\Delta\mu$,

$$\frac{\Delta\mu}{nk_BT} = \frac{1}{N}\ln\phi - \ln(1-\phi) + \chi(1-2\phi) + \frac{1}{N} - 1. \qquad (7.25)$$

We note that the chemical potential is formally defined as being the derivative of the Gibbs free energy with respect to the number of molecules n_i of each species, but since, in a closed system, this is proportional to the composition ϕ, we can safely replace n_i by ϕ. At equilibrium the chemical potential does not necessarily equal zero, but simply that the chemical potential of the two phases be equal. This means that the outflow of any one species from one phase is cancelled by the same species going back into that phase from the other phase.

Gibbs phase equilibrium between the two components does not necessarily mark the point at which a solution becomes unstable. There are concentrations which are metastable, and these are located between the limit of stability and the phase equilibrium. At the limit of stability, the change in chemical potential with respect to the number of molecules entering (or leaving) the system must equal zero, i.e. $\mathrm{d}\Delta\mu/\mathrm{d}\phi = 0$. We therefore differentiate eqn (7.25) again,

$$\frac{1}{nk_BT}\frac{\mathrm{d}^2\Delta G}{\mathrm{d}\phi^2} = \frac{1}{N\phi} + \frac{1}{(1-\phi)} - 2\chi = 0. \qquad (7.26)$$

This second derivative of the free energy (eqn 7.26) predicts the limit of stability. The free energy for an arbitrary system is shown in Fig. 7.23, and the limit of stability and phase equilibrium are indicated. If we wish to know the value of χ at which the this boundary is just reached, we need to know the values of ϕ and χ (ϕ_c and χ_c) at the critical point. In other words, we need the *third* derivative of the free energy to equal zero,

$$\frac{1}{nk_BT}\frac{\mathrm{d}^3\Delta G}{\mathrm{d}\phi^3} = \frac{-1}{N\phi^2} + \frac{1}{(1-\phi)^2} = 0, \qquad (7.27)$$

which is achieved when

$$\phi_c = \frac{1}{1+\sqrt{N}} \qquad (7.28)$$

and

$$\chi_c = \frac{1}{2}\left(\frac{\sqrt{N}+1}{\sqrt{N}}\right)^2. \qquad (7.29)$$

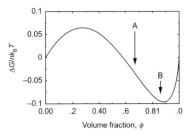

Fig. 7.23 $\Delta G/nk_BT$ as a function of ϕ for $\chi = 1.5$ and $N = 300$. The limit of stability and the point at which polymer and solvent are in phase equilibrium are marked by arrows (A and B respectively). There are in fact two limits of stability and points of phase equilibrium, but in this case, as is generally true of polymer solutions, the other points are too close to $\phi = 0$ to be discernible. Concentrations between A and B are metastable, and concentrations greater than the phase equilibrium value are stable (miscible).

We conclude from eqn (7.29) that the longer the polymer, the smaller the χ parameter is at the onset of phase separation, and therefore the less likely the polymer is to be in the solvent. The two limits to this equation are $\chi = 2$ for small molecules and $\chi = 1/2—)$ for high polymers. The larger the polymer, the smaller the value of ϕ_c, which means that the mixture will be more miscible for larger solvent concentrations, which is not especially surprising.

We return to these ideas in Section 7.6.2, where we shall present calculations for mixtures of two polymers illustrating the importance of the concepts first addressed in this section. The miscibility of polymer mixtures has implications in the structure and properties of optoelectronic materials, whereas polymer solutions are important for processing of polymers, which is perhaps (debatably) less interesting. For this reason, we leave the detailed discussion until we address polymer blends. The reader can skip Section 7.5.2 (returning later!) and go directly to Section 7.6.2 if she or he wishes to continue with this theme.

7.5.2   Good, Θ, and poor solvents

The quality of solvent determines the conformation of the polymer in the solution. A polymer in a good solvent is extended, whereas a polymer in a so-called Θ solvent has a random walk conformation, and a polymer in a poor solvent is collapsed. We discuss each of these in turn.

Good solvent conditions

In a good solvent, monomer–solvent interactions are energetically more favourable than monomer–monomer interactions. In order to encourage monomer–solvent contacts, the chain has to swell in that solvent, which gives it a larger end-to-end distance than the random walk result discussed in Section 7.1.

To understand the swelling of a polymer chain in a good solvent we need to consider two competing effects. The chain wants to maximize the number of monomer–solvent contacts on the one hand, and on the other it wishes to maximize its *conformational entropy*, which means that a rod-like conformation is to be avoided. To maximize the number of monomer–solvent contacts we take the approach of minimizing the number of monomer–monomer contacts. We achieve this by introducing a concept of *excluded volume*, by which we mean a region of space surrounding a monomer in which another monomer cannot encroach. The repulsive energy of a polymer containing N monomers is thus

$$\frac{F_{\text{rep}}}{k_{\text{B}}T} = v_{\text{ex}}\frac{N^2}{2R^3} \tag{7.30}$$

in a volume whose lateral dimensions are given by R and v_{ex} is a constant of proportionality known as the *excluded volume parameter*. Again, one needs to consider the probability of a monomer interacting with another monomer as $\phi^2/2$. If R is the radius of the volume enclosing the polymer, then there are N monomers in that space of volume of the order of R^3, so that ϕ^2 can be rewritten as N^2/R^6. With this logic the repulsive energy per unit volume, F'_{rep}, can be written as

$$\frac{F'_{\text{rep}}}{k_{\text{B}}T} = v_{\text{ex}}\frac{N^2}{2R^6}, \tag{7.31}$$

which means that eqn (7.30) is recovered for a volume of R^3. Note that in eqns (7.30) and (7.31) the factor $1/2$ is explicitly shown, and again refers to the need to avoid double-counting monomers.

Whilst F_{rep} causes the chain to stretch, we must accept that it cannot do so completely, because there is an entropic cost that needs to be paid. The final conformation of the chain will be a compromise between this entropic effect, which suggests that the chain should have a random walk conformation, and the enthalpic (repulsive) term described in eqn (7.30). For small departures from a random walk confirmation, it is sufficient to describe the polymer as a spring which obeys Hooke's law; i.e. one that, when stretched, experiences a restoring force proportional to the distance stretched. The associated energy for such a chain is

$$\frac{F_{\text{el}}}{k_{\text{B}}T} \propto \frac{R^2}{Na^2}, \tag{7.32}$$

where the symbol F_{el} denotes an elastic energy term and a is a monomer size. We have not included a numerical pre-factor in this equation, because it contains no interesting physics.

The equilibrium conformation of the chain will be when the total energy ($F_{\text{chain}} = F_{\text{rep}} + F_{\text{el}}$) is minimized, i.e. when

$$\frac{\text{d}F_{\text{chain}}}{\text{d}R} = 0, \tag{7.33}$$

which has its solution when

$$R \propto aN^{\frac{3}{5}}. \tag{7.34}$$

This is an example of a *scaling theory*. In such theories we do not worry about individual pre-factors, only the relevant power laws. Clearly, scaling theories are not always usable, but when they are they give real physical insight without the distraction of unimportant and cumbersome multiplicative factors. This $N^{3/5}$ dependence is greater than the $N^{1/2}$ random walk result, as expected. In fact, this calculation, due to Paul Flory, has been experimentally tested, with results showing that $R \propto N^{0.58}$. This remarkable agreement is rather fortuitous (i.e. coincidental) because of the cancellation of approximations in both eqns (7.30) and (7.32) above. This is known as a *self-avoiding* random walk.

Θ solvents

It is possible to have a random walk conformation in a solvent. Such a solvent is known as a neutral or Θ solvent. Under these conditions the polymer and solvent have an equal affinity for each other. In essence, at the Θ condition, the polymer cannot tell the difference between itself and its solvent. A polymer in its own melt is a classic example of the Θ condition, and so such polymers have a random walk conformation. We can calculate the Flory–Huggins interaction parameter—(for the Θ condition by seeing at what point excluded volume effects are cancelled out by attractive monomer–solvent interactions. The energetics of interactions are considered in a mean-field formalism, and we start by considering the total number of contacts between similar monomers,

$$N_{\mathrm{mm}} = \frac{znNv_{\mathrm{ex}}\phi}{2V} \tag{7.35}$$

where V is the volume of a lattice, so that $n\phi/V$ is the polymer concentration. The presence of the excluded volume means that we cannot work in terms of only the volume fraction because we need to normalize v_{ex} by another volume to retain a dimensionless number. Again, the factor $1/2$ is required to prevent double counting of interactions. Similarly, the number of monomer–solvent contacts is given by

$$N_{\mathrm{ms}} = \frac{znNv_{\mathrm{ex}}(1-\phi)}{V}. \tag{7.36}$$

We write the number of solvent–solvent contacts as the number of contacts that would exist should no polymer be present, less the number of monomer–monomer and monomer–solvent contacts.

$$N_{\mathrm{ss}} = \frac{zn}{2} - \frac{znNv_{\mathrm{ex}}\phi}{2V} - \frac{znNv_{\mathrm{ex}}(1-\phi)}{V}. \tag{7.37}$$

Clearly, the total energy of the interactions described in eqns (7.35), (7.36), and (7.37) can be written as $N_{\mathrm{mm}}\epsilon_{\mathrm{mm}}$, $N_{\mathrm{ms}}\epsilon_{\mathrm{ms}}$, and $N_{\mathrm{s}}\epsilon_{\mathrm{ss}}$ respectively. The total interaction energy is thus

$$\begin{aligned} H &= N_{\mathrm{mm}}\epsilon_{\mathrm{mm}} + N_{\mathrm{ms}}\epsilon_{\mathrm{ms}} + N_{\mathrm{ss}}\epsilon_{\mathrm{ss}} \\ &= \frac{znNv_{\mathrm{ex}}\phi}{2V}\left(\epsilon_{\mathrm{mm}} + \epsilon_{\mathrm{ss}} - 2\epsilon_{\mathrm{ms}}\right) + zN\left(\epsilon_{\mathrm{mm}} + \epsilon_{\mathrm{ss}}\right) + \frac{zn}{2}\epsilon_{\mathrm{ss}}. \end{aligned} \tag{7.38}$$

We can simplify eqn (7.38) by substituting into it the χ-parameter from eqn (7.23) to give

$$H = -\frac{nNv_{\mathrm{ex}}\chi\phi}{V}k_{\mathrm{B}}T + zN\left(\epsilon_{\mathrm{mm}} + \epsilon_{\mathrm{ss}}\right) + \frac{zN}{2}\epsilon_{\mathrm{ss}}, \tag{7.39}$$

which is useful because the first term is the only one that depends on composition and is therefore the only term of interest to us. We use eqn (7.30) to obtain the total energy of the polymer,

$$\begin{aligned} H_{\mathrm{tot}} = F_{\mathrm{rep}} + H &= v_{\mathrm{ex}}\frac{N^2}{2R^3}k_{\mathrm{B}}T - \frac{nNv_{\mathrm{ex}}\chi\phi}{V}k_{\mathrm{B}}T \\ &= v_{\mathrm{ex}}\left(1 - 2\chi\right)\frac{N^2}{2R^3}k_{\mathrm{B}}T, \end{aligned} \tag{7.40}$$

where we have replaced $n\phi/V$ by N/R^3 and have removed the uninteresting constant terms. We should note that the assumption $n\phi/V \approx N/R^3$

is only valid in semi-dilute solutions, but we can use it more generally here, because we consider only the chain and the solvent contained within its volume. We note that if $\chi = 1/2$ this energy $H_{\text{tot}} = 0$, which is the requirement for the Θ condition. For $\chi < 1/2$ we have good solvent conditions, with the appropriate ($N^{0.6}$) scaling behaviour of the chain size. For $\chi = 1/2$ the repulsive effect of excluded volume is cancelled out by the energetic balance of the polymer in its solvent. At this point the polymer may as well be immersed in identical polymers rather than the solvent; the energies are the same. Under such circumstances, a polymer that does not 'know' the difference between itself and its solvent must undertake a true random walk conformation. When the Θ condition is satisfied, there is no excluded volume.

Poor solvents

Poor solvent conditions are defined through $\chi > \frac{1}{2} \left(\frac{\sqrt{N}+1}{\sqrt{N}} \right)^2$, (from eqn 7.29), under which condition the polymer will fall out of solution. In such circumstances, the chain will exclude solvent and the scaling relation for the chain size will scale as $N^{1/3}$, which is the limit for a collapsed sphere. This is not a difficult condition to achieve as most polymers will be insoluble in many solvents. It is not interesting though, because it is valid only in dilute solution and there are few applications for such solutions. (In semi-dilute or concentrated solutions the polymer will phase separate, forming an aggregate, which will precipitate out of solution, either at the top, on the walls of its container, or at the bottom.)

7.6 Polymer blends

Mixing two or more polymers is an important concept in polymer physics. These mixtures, known as *blends*, are advantageous because they allow the modification of material properties through the addition of a component whilst retaining the original advantages of the host polymer. If we consider polystyrene, then we have a versatile material that can be used in many applications, such as insulation and packaging. However, it remains a rather brittle polymer and so for added toughness polybutadiene can be added. The polybutadiene is rubbery at room temperature, because it has a glass transition temperature which, at $\sim$ 185 K, is considerably below room temperature. (Polystyrene has a glass transition of $\sim$ 373 K.) Polybutadiene is immiscible with polystyrene and phase separates, forming polybutadiene-rich domains in a polystyrene-rich matrix. This is known as high-impact polystyrene (HIPS) and is a very common polymer blend, whose properties depend on the amount of polybutadiene added. Car tyres also use mixtures of polystyrene and polybutadiene, but in this case, more butadiene is necessary. HIPS is itself blended to poly(2,6-dimethylphenylene oxide) (PPO) to reduce its glass transition temperature to facilitate its processing. (The high glass transition temperature of PPO at ~ 485 K makes it useful as a plastic resistant to high temperatures.) Another example concerns mixtures of poly(vinyl alcohol) (PVA) and poly(ethylene terephthalate) (PET). The PET provides the structural rigidity that plastic bottles require, and the PVA is impermeable to gases, which means that carbonated drinks can be kept in the bottle.

Polymer blends can play a useful role in optoelectronics. Both light-emitting diodes and photovoltaic devices require electron and hole carrying components, and one way to achieve this is to use a blend of polymers. The structures that these blends form will play a crucial role in their device properties. The use of polymer blends in optoelectronic devices is discussed in a later chapter. However, blends can also be useful in other situations. Modifying the properties of the active layer of a transistor so as to use a less (expensive) charge carrier by adding (cheaper) polymers is one way to achieve cheaper and perhaps more easily processable but just as effective devices, or similarly, tailoring transistor properties by blending two semiconductors has also been used. We discuss transistor blends in Section 9.5.1.

Most blends are immiscible, although not all. It is therefore important to understand phase behaviour of polymer blends, which is achieved through mean-field (lattice) theory. As we have seen for PPO and HIPS, the glass transition also has an important role to play, and so we discuss these in turn.

7.6.1   Glass transition

The glass transition temperature, T_g of a miscible polymer blend is obtained from the glass transitions, T_{g1} and T_{g2}, of the individual constituents by

$$\frac{1}{T_g} = \frac{\phi}{T_{g1}} + \frac{1-\phi}{T_{g2}}, \tag{7.41}$$

where ϕ is the volume fraction of the component with a glass transition temperature given by T_{g1}. Eqn (7.41) is known as the Flory–Fox equation and is largely empirical. There are other equations that describe the glass transition temperature of miscible blends, but, for our (and most other) purposes, the Flory–Fox equation is good enough. It certainly is very widely used, even if it is not fully correct.

7.6.2   Miscibility of polymer blends

It readily follows from the discussion in Section 7.5.1 that the total free energy of mixing of a polymer blend of chain lengths N_1 and N_2 is given by

$$\frac{\Delta G}{n k_B T} = \frac{\phi}{N_1} \ln \phi + \frac{1-\phi}{N_2} \ln (1-\phi) + \chi \phi (1-\phi). \tag{7.42}$$

Similarly, using the same methodology described in Section 7.5.1, we have, for ϕ_c and χ_c

$$\phi_c = \frac{\sqrt{N_2}}{\sqrt{N_1} + \sqrt{N_2}} \tag{7.43}$$

and

$$\chi_c = \frac{1}{2} \left(\frac{\sqrt{N_1} + \sqrt{N_2}}{\sqrt{N_1 N_2}} \right)^2. \tag{7.44}$$

A cursory check with $N_2 = 1$ will recover eqns (7.28) and (7.29).

In Fig. 7.24 we show the free energy for an arbitrary blend with equal chain lengths, $N_1 = N_2 = 100$. Such a blend is often called symmetric. Here we see that for the largest χ, there is a large hump in the free energy. It is clear here that a concentration of $\phi = 0.5$ is unstable. Any fluctuation in concentration due to Brownian motion will cause some regions to have $\phi > 0.5$ and others $\phi < 0.5$. Normally such fluctuations in concentration are transient, but here they lower the global free energy, which means that the mixture is unstable. Some concentrations will remain stable, however, and these are found at either side of what is known as the coexisting compositions. For symmetric blends only, these coexisting compositions are when $\partial \Delta G / \partial \phi = 0$. For $\chi = 0.035$ these are close to $\phi = 0.04$ and 0.96. The coexisting compositions are much more visible in Fig. 7.24 for $\chi = 0.025$, where they can be found at $\phi = 0.14$ and 0.86. Outside the range $0.14 < \phi < 0.86$, for $\chi = 0.025$ the mixture is stable because any energy gain in a fluctuation in concentration is less than the corresponding energy cost. At $\phi = 0.5$ there are only energy gains in fluctuations taking the concentration away from ϕ_c. One does not need $\phi = \phi_c$ for the mixture to be unstable. At concentrations just less than $\phi = 0.5$ the energy gain from a fluctuation δ to a concentration $\phi - \delta$ is greater than the corresponding energy cost of the fluctuation to a concentration of $\phi + \delta$, and so the mixture remains unstable. There are also regions of the phase diagram that are metastable. Regions of metastability, instability, and miscibility are shown for the $\chi = 0.025$ blend in Fig. 7.25.

Metastable states exist when an activation energy must be overcome in order for phase separation to occur. For mixtures with volume fractions in this region, an overall energy gain is possible, but this energy gain is smaller than the height of the 'hump' in the free energy. This is indicative of a first-order phase transition, whereby an activation energy must be overcome for the transition to occur. The boundary between the compositions which require an activation energy and those which do not is known as the *spinodal*. The boundary between the miscible region and the metastable region is known as the *binodal* or *coexistence*. The *phase diagram* of these polymer mixtures is obtained by plotting χ as a function of concentration, ϕ, and an example is shown in Fig. 7.26.

The derivation of χ leading to eqn (7.23) suggests that χ is inversely proportional to absolute temperature, but this is not experimentally the case. χ is not wholly enthalpic, and may have an entropic component. In this case it may be of the form, $\chi = A + B/T$, where A and B are constants. This is the most common form, and is the one used in Fig. 7.26. Phase diagrams of the kind shown in Fig. 7.26 are known as *upper critical solution temperature* (UCST) phase diagrams, because, for $T > \phi_c$, the mixture is always miscible. Lower critical solution temperature (LCST) phase diagrams, when the mixtures are miscible

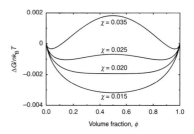

Fig. 7.24 $\Delta G / nk_B T$ as a function of ϕ for $\chi = 0.015$, 0.02, 0.025, and 0.035, where $N_1 = N_2 = 100$. This blend has $\chi_c = 0.02$ (eqn 7.44) and at $\phi_c = 0.5$ the free energy flattens. For $\chi > 0.02$ a hump appears in the free energy curve, indicating immiscibility and for $\chi < 0.02$ the mixture is completely miscible.

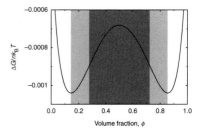

Fig. 7.25 $\Delta G / nk_B T$ as a function of ϕ for $\chi = 0.025$ for $N_1 = N_2 = 100$. This is the same calculation as shown in Fig. 7.24, but here shading marks unstable (darker) and metastable (lighter) regions of the phase diagram. Volume fractions for the miscible (stable) blends are shown unshaded.

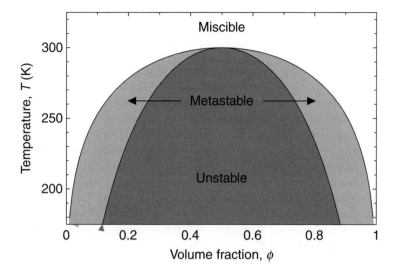

Fig. 7.26 A plot of absolute temperature, T as a function of ϕ allows miscible, metastable, and unstable polymer mixtures to be delimited. This plot is known as the *phase diagram* for the given polymer blend. In this case the blend is chosen to have $N_1 = N_2 = 100$ and $\chi = -0.02 + 12/T$.

for $T < T(\phi_c)$ are also common. In LCST mixtures, if χ has the form $A + B/T$, then B must be negative. There are also blends that are a mixture of LCST and UCST. These may occur when the different components have different volume expansivities, for example, but in any case are well beyond the scope of this book.

Nucleation

For metastable blends, the activation energy to phase separation is generally not prohibitive. For mixtures that are in the metastable region of the phase diagram and close to the spinodal, the activation energy is not particularly large, and *thermal nucleation* may take place. Here fluctuations in temperature or composition are enough to overcome the activation energy and trigger phase separation. Such phase separation is, however, rare, and in any case competes with nucleation driven by impurities. Impurities drive nucleation because they present the mixture with an alien surface which will be preferentially coated by one of the two components of the mixture. This occurs because there is an interfacial tension associated with the impurity and each of the components of the mixture. Whichever interfacial tension is the smallest will dominate, and wetting of the nucleation site will cause phase separation to proceed at that interface. At this point there is no activation energy and a wetting layer will form, and this could be a very strong effect. It is therefore necessary to have ultrapure materials in order to avoid such effects. Of course, the surface or walls of a container will also initiate demixing, so in practice it is virtually impossible to prevent metastable liquid mixtures from demixing.

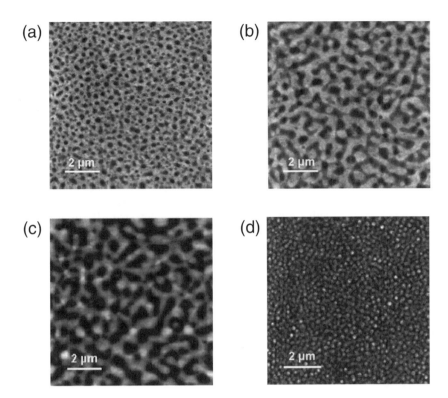

Fig. 7.27 Scanning force microscopy images of phase separated blends of poly(9,9-dioctyl fluorene) (F8) and poly(9,9-dioctyl fluorene-*alt*-benzothiadiazole) (F8BT). The amount of F8 by fractional volume, ϕ is 0.15, 0.24, 0.50, and 0.77 for (a), (b), (c), and (d) respectively. The phase inversion is clear as one goes from low concentrations, where the circular structures are the darker (lower) component to the largest volume fraction, where the circular structures correspond to the lighter (higher) component. It is not strictly accurate to link darker regions to being rich in F8, because these are topography images with the maximum contrast corresponding to a height difference of between 20 and 40 nm, depending on the image. The images in (b) and (c) show more continuous morphologies where it would be difficult to say that one structure is dissolved in the other. It is clear from these images that a uniform length scale of a few hundred nm is evident in each of the images, although it is different for different volume fractions. Image taken with permission from Higgins et al. *J. Phys.: Condens. Matter* **17** 1319 (2005) and is copyright IOP Publishing 2005.

Common length scale

The spontaneous phase separation of unstable mixtures (also known as *spinodal decomposition*) is an interesting and important route to pattern formation. The competition between enthalpy (favouring demixing) and entropy (favouring mixing) results in a common length scale. To understand this, we can consider a mixture separating into different phases of different sizes. These different sized phases will result in different energies per unit volume at different points in the material. Although it is only true for symmetric mixtures that the free energies of the different phases are the same, the average free energy covering the mixture should not change at different points in the medium. To keep the overall energy per unit volume constant, each phase-separated domain (and their distance from each other) will be of equal size, at least in the early stages of phase separation. Uniformity in domain size and homogeneity in the energy per unit volume results in a common length scale of phase separation.

Very small length scales, meaning lots of domains, are thermodynamically prohibited because of the large amount of interface that this involves which means that phase separation will raise the overall energy of the system rather than lowering it. Very large lengths (large domains) will not have this problem, but they do require significant mass transport. These take too long to form, and shorter length scales will dominate. In fact, the dominant length scale is often referred to as the fastest-growing length scale. Nevertheless, the dominant length scale arising out of spinodal decomposition depends on the relative competition between entropy and enthalpy. If enthalpy is dominant, there will be more domains, of smaller size. If the entropic component is a significant part of the free energy, there will be fewer domains, which consequently involves less interface per unit volume.

A uniform length scale does not restrict the morphologies that can form. An unstable mixture in which one component is at a much smaller concentration will have domains of the minor component dissolved in a matrix of the major component.[1] At concentrations close to ϕ_c, symmetry does not allow one component to be dissolved in the other, because there is no 'minor' component. In these circumstances a bicontinuous morphology results. These structures can be seen in Fig. 7.27.

7.7   Block copolymers

Phase separation of polymer blends will result in structures with a degree of ordering on micron length scales. Unfortunately, the mechanism of the phase separation means that it is difficult to obtain beautifully ordered structures. In the case of spinodal decomposition, complete ordering requires that the whole material is at a uniform temperature and 'senses' the onset of phase separation at the same time. Because spinodal decomposition is extremely sensitive to these parameters, ordering is usually not great over length scales of a few microns. Impurities also spoil the phase separation by introducing different mechanisms of structure growth to the material. Once spinodal decomposition is initiated, coarsening starts to occur, whereby the domains further grow. Structures are very often visible to the naked eye, but are certainly commonly of more than 10 μm in size. This large size is of little use in many organic electronic devices such as photovoltaics, because the exciton transport lengths are of the order of nm. Block copolymers provide a means of circumventing both these problems. It is now possible to have block copolymers which provide nm-sized structures that are uniform over many microns, with perfect fidelity.

There are different forms of copolymer, including graft copolymers, random copolymers, alternating copolymers, and block copolymers. Two of these examples are sketched in Fig. 7.28. Block copolymers are usually immiscible. It is possible to synthesize block copolymers of miscible

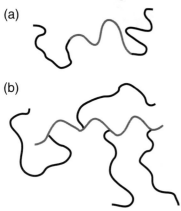

Fig. 7.28 (a) A block copolymer consists of two or more polymers joined by a covalent bond to form a longer linear chain. This image schematizes an ABA triblock copolymer. If the three blocks were made of different homopolymers they would form an ABC triblock copolymer; two polymers would be a diblock copolymer. (b) A graft copolymer, sometimes known as a comb polymer. Here linear chains extend from a chain backbone.

[1] An exception is when the minor component is much more viscoelastic than the major; then the minor component can form the continuous phase.

Fig. 7.29 Block copolymers can take a variety of different structures depending on the ratio of the block lengths and, to a lesser extent, parameters such as temperature. If one block is extremely small, then a spherical morphology will result, with the spheres organized in a body-centred cubic structure (a). If the blocks are of similar length, then a lamellar structure results (c). The cylindrical morphology (b) lies between these two. Other morphologies exist, but are harder to realize experimentally,

blocks, but it is difficult to understand why this would be done, when the synthesis of a random copolymer of the same components is usually much easier to perform and would generally result in similar material properties. Immiscible block copolymers are constrained by the covalent bonds linking the blocks from forming the larger structures of spinodal decomposition. Nevertheless, different structures can be formed with block copolymers, and these are shown in Fig. 7.29.

Block copolymers, like polymer blends, have a phase diagram. The transition from the two-phase region to the one phase region is known as the *order–disorder transition* (ODT), and the two-phase region is referred to as being *microphase separated*. The phase separation is on the nanoscale, rather than the microscale and one supposes that the nomenclature has its origin in the era of microtechnology rather than nanotechnology. The main block copolymer morphologies are spherical, cylindrical, and lamellar. Prediction of the different morphologies is theoretically rather straightforward, but nevertheless, beyond the scope of this book. Suffice to say that the physics involves limiting the interfacial energy contribution and the stretching of the polymer chains. There is some compromise between these different factors, and the use of self-consistent mean-field theory allows a full prediction of the possible structures. The two important parameters governing the structure are the interaction parameter between the two components and the ratio of the chain lengths. If the different blocks are highly asymmetric, then the morphology will be spherical; in block copolymers a spherical structure allows the minimization of the surface area for the smaller block, with the bigger block forming a matrix (Fig. 7.29).

7.8   Further reading

There are several books on the physics of polymers. *Polymer Physics* by Rubinstein and Colby (2003) and '*The Physics of Polymers*' by Strobl (2007) are worth reading. Both are comprehensive but readable accounts. The former is particularly good for mathematical derivations, whereas the latter is very good in areas such as crystallization, Liquid crystalline polymers are well-catered for in the book by Donald, Windle, and Hanna (2006). The level of this latter book is slightly more advanced than that of the two books on the physics of polymers, which are equivalent in level to this book. Nevertheless, the reader should find much of use in the book on liquid crystalline polymers. An excellent pedagogical account of much of this chapter is contained within the text by Jones (2002). All of these books are relatively recent, but there are two timeless texts that are worth considering from the Nobel Laureates Paul Flory (1955) and Pierre-Gilles de Gennes (1979).

7.9   Exercises

7.1. A sample of P3HT (Fig. 6.23) has a molecular mass of 41.5 kg/mol. (a) Given its density is 1.1 g/cc, calculate the end-to-end distance of a P3HT chain in the melt. (b) A persistence length of 2.4 nm has been determined for this molecule. What is the bond angle if the monomer length is 0.37 nm? (c) Calculate the end-to-end distance for P3HT as a worm-like chain? (d) Calculate the end-to-end distance for P3HT as a freely-jointed chain using a monomer length of 0.37 nm, rather than the density given in (a). Comment on the difference between the results.

7.2. The rheological properties of polymers is of considerable interest. In their liquid form (i.e. above their glass transition temperature T_g, assuming no crystalline behaviour), the temperature dependence of their viscosity, η is often written using the *Williams–Landel–Ferry* (WLF) equation,

$$\eta = \eta_{0W} \exp\left(\frac{b}{f_g + \alpha\left(T - T_g\right)}\right), \qquad (7.45)$$

where b is a constant for that particular polymer and η_{0W} is a molecular weight-dependent parameter, which also varies between polymers. $\alpha = 4.35 \times 10^{-4}$ K^{-1} and $f_g = 0.022$ are *universal* parameters, independent of the choice of polymer. Another equation, often used for the viscosity is the Vogel–Fulcher–Tammann (VFT) equation,

$$\eta = \eta_{0V} \exp\left(\frac{B}{T - T_0}\right), \qquad (7.46)$$

where, again, B is a constant for that particular polymer and η_{0V} is a molecular weight-dependent term. T_0 is a temperature, which is approximately 50 K below the glass transition temperature of the polymer.

Prove that the WLF equation reduces to the VFT form and derive the relationships between the different constants.

7.3. In order for crystalline lamellae to grow, their rate of melting must be less than the rate at which polymers join the growing crystal. The melting and growth rates can be written in an Arrhenius form as

$$\tau_{\mathrm{m}} = \tau_0^{-1} \exp\left(\frac{-\Delta S}{k_{\mathrm{B}}}\right) \qquad (7.47)$$

and

$$\tau_{\mathrm{g}} = \tau_0^{-1} \exp\left(\frac{-(T\Delta S - \Delta G_{\mathrm{m}})}{k_{\mathrm{B}}T}\right) \qquad (7.48)$$

respectively, where τ_0 is a constant, ΔS is the entropy cost on joining the lamellae, and ΔG_{m} the corresponding change in free energy. Using these equations for the melting and growth of a crystal to find an overall growth speed, show, stating any assumptions that you may make, that the length of crystalline lamellae is given by

$$l^* = 2a + \frac{2\sigma_{\mathrm{f}} T_{\mathrm{m}}(\infty)}{\Delta H_{\mathrm{m}} \rho\left(T_{\mathrm{m}}(\infty) - T\right)}. \qquad (7.49)$$

You will need the result that the configurational entropy of a polymer of N monomers of size a is

$$S = S_0 - \frac{3k_{\mathrm{B}} R^2}{2Na^2}, \qquad (7.50)$$

where R is the end-to-end distance of the polymer chain and S_0 is a constant.

A given polymer of density $\rho = 0.88$ g/cc is observed to have crystalline lamellae the size of which increase with temperature (i.e. dl/dT) by a factor of three as the temperature increases from 406 K to 412 K. If the fold surface energy is 5.2 mJ/m^2, the monomer size, $a = 0.5$ nm, and the lamellar size at 412 K is 5 nm, what is the latent heat of fusion of the polymer? Is the fold surface energy a parameter given to easy measurement? If so, how would you measure it? If not, justify your answer.

8 Surfaces and interfaces

Surfaces and interfaces play a critical role in many aspects of science and technology because different physics exists at surfaces. For example, we have seen in Section 5.5.1 how Fermi levels are equalized and bands bend to accommodate the Fermi levels when a metal–semiconductor interface is formed, or that excitons may find the activation energy barrier leading to their dissociation overcome at an interface (see Section 4.3), so we can ascertain the importance in understanding and controlling surfaces and interfaces.

Surfaces and interfaces are important because their properties are different from the bulk of a mixture, and therefore the presence of a surface may change the material properties. This is particularly the case at the nanoscale when confinement comes into play. Surface layers are generally of the order of the size of the molecules contained in the material. A silicon-based device has a surface layer commensurate with the size of the silicon atom, which is 0.58 nm. Polymers are of course much larger and so will have a more significant surface layer. Some polymers, such as many proteins, can be more than a micron long when stretched out, but conjugated polymers will not approach these lengths, and in normal mixtures one can suppose that the surface layer is more like several nm. Nanoscale materials are important because many optoelectronic properties take advantage of such length scales. The diffusion length of an exciton before it decays radiatively is ~ 10 nm, and so the creation of materials which take advantage of such length scales is important. However, a 10 nm polymeric material is essentially all surface, because each molecule would be expected to have at least some monomers in contact with the surface of that material.

8.1 Interfacial energy

The creation of a surface costs energy, so the inclusion of many surfaces in a system raises the overall free energy of that system. The simplest model of surface energy is to consider a solid sliced into two pieces, thereby creating two surfaces. If we break one bond of energy ϵ per atom in creating these surfaces, the energy per unit area is

$$\gamma = \frac{\epsilon}{2V_0^{2/3}} \tag{8.1}$$

where V_0 is a volume associated with the unit whose bond has broken. The factor of 2 in the denominator is due to there being two surfaces created for every bond broken. γ is then the *surface energy* or *surface tension*; the two terms are identical, and their respective usage depends on whether one prefers units of J/m^2 or N/m. (Formally surface tension, is a vector, and we shall see how direction can be applied in Section 8.1.1.) Using the logic described here, one can also relate the surface energy to the work required to separate the material into two components, W, by

$$\gamma = \frac{W}{2}.$$ (8.2)

A feel for the order of magnitudes that one may obtain in surface energies may be considered by using this equation with water. The bond energy of water may be replaced by the latent heat of vaporization rather than the energies of breaking chemical bonds, which would give $\epsilon = 7 \times 10^{-20}$ J. The density of water at 5°C is 1 g/cc, so one molecule of water has a volume $V_0 = 3.0 \times 10^{-29}$ m^3, from which we obtain $\gamma = 0.4$ J/m^2. The experimentally measured value for the surface energy of water is $\gamma = 73$ mJ/m^2, a factor of five smaller than our value. Given the simplicity of the model we have used, the discrepancy is trivial. Water has a relatively large surface tension, in comparison with polymers, whose surface energy may be typically a factor of three or four smaller than that of water.

A more sophisticated calculation of surface energies would stem from considering molecules as having attractive interactions due to long-range intermolecular forces. These forces are usually taken as van der Waals forces with an intermolecular potential scaling with separation r^{-6}. The first step in considering the surface energy is to start with the interaction between two molecules and to generalize that to the interaction between a molecule and a surface. The pair potential between two identical molecules separated by a distance r may be taken as $-C\rho_1/r^n$, where $n = 6$ for van der Waals interactions, and C is a constant specific to the molecule with units of Jm^n/kg. The negative sign indicates that the energy decreases as the molecules approach each other. To find the energy of interaction between a molecule and a semi-infinite wall made of the same species separated by a distance D, we integrate the pair potential using the interaction between the molecule and an annulus (ring) as the starting point (Fig. 8.1),

$$W(D) = -2\pi C\rho_1 \int_D^\infty dz \int_0^\infty \frac{x\,dx}{(x^2 + z^2)^{n/2}}$$

$$= \frac{-2\pi C\rho_1}{(n-2)(n-3)D^{n-3}},$$ (8.3)

where z is the distance between the molecule and the depth at which the annulus is located, and x is the radius of the annulus. The distance between the annulus and the molecule is given by r, where $r^2 = x^2 + z^2$.

If we now replace the individual molecule (point) with another semi-infinite wall (density ρ_2) parallel to the first wall, the interaction poten-

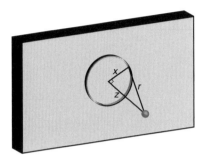

Fig. 8.1 Geometry for the calculation of the potential between an object (here, a small sphere), located at the apex of the triangle, and a semi-infinite wall at distance D from the molecule. The interaction between the sphere and an annulus of thickness dx and depth dz located distance z from the sphere ($z > D$) and of radius x is calculated for all z and x.

tial becomes

$$W_a(D) = \frac{-2\pi C_a \rho_1 \rho_2}{(n-2)(n-3)z^{n-3}} \int_D^\infty \frac{dz}{z^{n-3}}$$

$$= \frac{-2\pi C_a \rho_1 \rho_2}{(n-2)(n-3)(n-4)D^{n-4}}, \tag{8.4}$$

where the subscript 'a' indicates that the parameter W_a is an energy per unit area (units J/m^2), and C_n has appropriate units, Jm^n/kg^2. The choice of units of J/m^2, i.e. the per unit area qualification, is of course necessary because otherwise the interaction between two parallel semi-infinite blocks would be infinite. The usual situation is that $n = 6$, so we can write

$$W_a(D) = \frac{-\pi C_a \rho_1 \rho_2}{12D^2}. \tag{8.5}$$

If we set $A = \pi^2 C_a \rho_1 \rho_2$, then we can rewrite eqn (8.5) in such a way as to find the surface energy,

$$2\gamma = \frac{A}{12\pi D_0^2}. \tag{8.6}$$

A is an important parameter known as the Hamaker constant, and it describes the strength of *dispersive* forces between two parallel surfaces. It is dependent on the medium in which these surfaces are placed, as well as the nature of the surfaces themselves, but in air its values are typically $10^{-20} - 10^{-19}$ J for polymeric materials. The Hamaker constant can be either positive or negative, depending on whether or not the surfaces are attracted to each other. D_0 is a distance of the order of the interatomic spacing.

8.1.1   Young equation and spreading coefficient

We can also consider the interfacial energy between two liquids, γ_{12}, which is given by the Dupré equation,

$$\gamma_{12} = \gamma_1 + \gamma_2 - W_{12}, \tag{8.7}$$

where W_{12} is the work required to separate the two interfaces. This equation is easy to understand if we rewrite it as $\gamma_{12} + W_{12} = \gamma_1 + \gamma_2$, because by considering the left-hand side we start with two materials (1) and (2) joined together, so that the interfacial energy is γ_{12}, and apply W_{12} to separate them. This leaves us with two surfaces (1) and (2) of energy γ_1 and γ_2 respectively.

A liquid droplet on a solid surface will experience different forces which define its equilibrium shape. The energy of the solid surface (substrate), γ_{SV}, will counter the interfacial energy between the liquid and solid, γ_{SL}. The former will pull the liquid to wet the surface, whilst the latter will try to cause the droplet to dewet, as shown schematically in Fig. 8.2. The energy of the liquid, γ_{LV}, will act along the line of contact with the substrate. Of course, energy is a scalar quantity, so it is wiser to consider

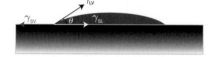

Fig. 8.2 The shape of a droplet is a balance between surface tensions where γ_{SL} is the interfacial tension and γ_{LV} and γ_{SV} are the liquid- and solid-vapour surface tensions respectively. The arrows show the directions in which these tensions act. The angle between the γ_{SL} and γ_{LV} is known as the contact angle, θ.

these parameters as tensions, and this allows us to equate tensions along the solid–vapour interface to give the *Young equation*,

$$\gamma_{SV} = \gamma_{SL} + \gamma_{LV} \cos\theta. \tag{8.8}$$

In the vertical direction $\gamma_{LV}\sin\theta$ is simply part of the reaction at the surface to gravity, from which we should learn nothing of interest, except in the case of liquid–liquid interfaces where there are vertical components both up and down that can be equated, because the liquid substrate allows deformation of the interface (see Question 8.1 in the exercises at the end of this chapter). Gravity causes spreading (thinning) of the droplet until the equilibrium contact angle is reached, and so is usually of little interest in such equations. The contact angle θ can vary between zero (complete wetting) and $180°$ (completely non-wetting). When $\theta < 90°$ the surface is hydrophilic and for $90° < \theta < 150°$ we simply consider the surface as hydrophobic, for the usual case of θ referring to the water contact angle. A surface with $\theta > 150°$ is known as superhydrophobic. The spreading coefficient is another important parameter commonly used and is defined by

$$S = \gamma_{SV} - (\gamma_{SL} + \gamma_{LV}). \tag{8.9}$$

The utility of defining S is that for $S > 0$ the liquid spreads the substrate and for $S < 0$ dewetting occurs.

8.2 Polymers at surfaces

The surface energy of polymers is generally something that is best measured rather than referred to in reference texts. Polymers generally have similar surface energies which are dependent on their molecular mass, and few studies have been performed to the required detail. In the case of conjugated polymers, this is especially the case. If one mixed two conjugated polymers, or a conjugated polymer with another polymer, it is not obvious which component would wet the surface or substrate. Generally, there is no great motivation to measuring surface energy for conjugated polymers. If polymers are mixed one can deduce which molecule is at the surface from appropriate diagnostic experiments. The importance of surface in polymer electronics has now been appreciated for a number of years with many research groups working in this area, not least those of the two authors of this book.

So why are surfaces important? We mentioned at the beginning of this chapter that surfaces are in general important in polymers, because the molecules are so large. We also hinted at the importance that surfaces and interfaces have with regard to excitons, which we introduced in Section 4.2. We know that an exciton may dissociate at an interface, so if we are creating a device such as a solar cell that requires the dissociation of excitons we need to take advantage of the presence of interfaces, because it is here that we can create charges to be collected at electrodes. However, if we are creating an LED the requirements are different; when electrons and holes meet they create an exciton, and this

(a)

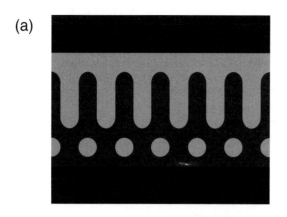

(b)

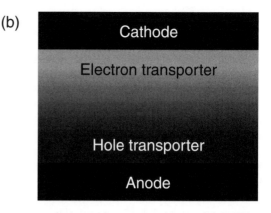

Fig. 8.3 (a) The photovoltaic device is expected to perform best with sharp interfaces which join phases with a direct path to electrodes. The circular phases close to the anode (bottom) are therefore not favoured. (b) A light-emitting diode requires the injection of holes at the anode and electrons at the cathode. These meet within the device and excitons are formed that can decay radiatively. It is understood that sharp interfaces impede the performance of the device and shallow gradients work best. In both devices an anode of indium tin oxide would be appropriate due to its optical transparency and high work function. The cathode is a low work function metal, of which calcium is a typical example; see Section 4.11 for further details.

must not dissociate back into the electron and hole, so interfaces are best avoided.

The different structures required for photovoltaic devices or solar cells are shown in Fig. 8.3. For the photovoltaic, light must arrive close to an interface, and when the exciton dissociates at this interface there must be a clear route for the charge carriers to reach the electrodes. If one considers a region of, say, the hole transporter dissolved within a matrix of electron transporter, then holes cannot make it easily to the anode and this would reduce device efficiency. A device consisting solely of the interdigitating structures shown in Fig. 8.3a would be ideal, if these structures were about 20 nm wide and the ends were between 10 and 20 nm from each electrode. In such a case most excitons would reach an interface and dissociate with a clear route to the electrodes. If this device were several microns thick it would be even better, because the collection efficiency of photons would be impressive. (We can think about this in terms of an internal quantum efficiency such as that defined by eqn (4.10) when we considered electroluminescence in Section 4.12. Photovoltaics have an internal quantum efficiency too, and thicker photovoltaics capture more photons, although not all photons captured in a thick solar cell will escape.)

For the case of LEDs, excitons need to be formed, so holes and electrons must be carried, preferably with equal mobility, into the device to meet. Experiments show that graded interfaces work best to create efficient charge injection. One can therefore envisage a device morphology something like that shown in Fig. 8.3b. It is clear that sharp interfaces are unwanted in LEDs because these create traps for excitons, where they can radiatively decay. LEDs, unlike photovoltaics, will benefit from miscible blends, or blends where wetting dominates structure formation. Isotropic phase separation in the bulk of the film creates interfaces that

(a)   (b)   (c)

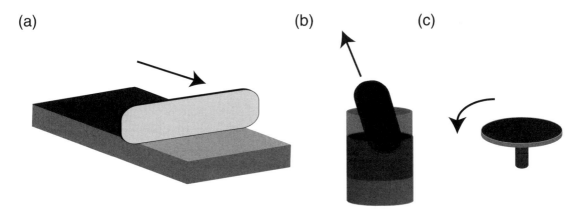

Fig. 8.4 (a) In doctor-blading, a blade moves a small distance above the surface to remove excess solution. The thin layer behind the blade dries rapidly, forming a good quality film. (b) Dip-coating requires the substrate be pulled out at constant speed from the solution. There is little remaining solution on the substrate, and this dries quickly. (c) Spin-coating is the rapid rotation of the substrate to initially remove excess solution, and then to accelerate drying.

one would wish to have suppressed.

Blends are ideal because the alternative is to work with single-layer or bilayer devices. Bilayers, where one film of a polymer with a large hole mobility is deposited on a layer of good electron mobility, are not as effective for LEDs because of the traps they create. (In Section 10.1.3 we shall see that bilayers can nevertheless still be used.) There has been some success with photovoltaics, but the problem is that these devices must be thin. Early attempts at photovoltaics worked with single layers, which suffer from an imbalance in the charge collection efficiency, as well as the inefficiency of having thin layers. It is therefore desirable to try to tailor devices through control of structure, or through design, such as using block copolymers.

8.3 Film formation

Virtually all devices involving conjugated polymers are made up of films of material, although not necessarily on the nanoscale. It is therefore useful to discuss briefly a few of the many different methods of film formation. It should be clear that there are others that we do not discuss below. Some, such as thermal evaporation are popular, but are limited to oligomers and small molecules (polymers have essentially no vapour pressure, even at elevated temperatures), and so are not discussed.

8.3.1 Doctor-blading

Most of the techniques to create a film involve the evaporation of solvent, leaving a deposited polymer layer. The trick is to find a method that provides a high-quality film. Simply leaving a droplet on a substrate and allowing it to evaporate has significant disadvantages: it is slow and creates a film of rather dubious quality, especially since smooth films are generally a prerequisite. Doctor-blading provides reasonable quality films by the use of dragging a blade across the substrate to spread the droplet (Fig. 8.4a). This removes excess solution, allowing quick drying. The technique is generally useful for films with thickness greater than ~ 100 nm. Doctor-blading is (ideally) performed with a motorized blade travelling at constant speed, and can be used on substrates of any size. It often finds use in colloidal suspensions of polymers such as PEDOT:PSS or even nanoparticles such as TiO_2 that have use as electrodes.

8.3.2   Dip-coating

Dip-coating involves immersing the substrate in a solution of the polymer. It is then withdrawn at constant speed by a motorized device (Fig. 8.4b). In these respects dip-coating is rather like doctor-blading, because excess solution is immediately removed, this time by gravity, leaving a thin layer that can rapidly dry. Dip-coating requires solutions in a container large enough to immerse the substrate, and so is less useful for large area surfaces. Viscous solutions are also not suitable for dip-coating because these tend to result in very non-uniform films.

8.3.3   Spin-coating

Spin-coating is intuitively a strange method of casting films. The solution is deposited as a drop on the substrate, which is held in place by a vacuum chuck. The substrate and solution is accelerated to between typically 1000 and 5000 r.p.m. over 3 or 4 s. In this period excess solution is thrown off, and the solution can then be allowed to dry. Simple hydrodynamic equations do not contain the complex non-equilibrium phenomena that occur when the film is being formed, with the thinning of the film (of thickness h) given by

$$\frac{\mathrm{d}h}{\mathrm{d}t} = \frac{-2\rho\omega^2 h^3}{3\eta} - e_\mathrm{e}, \tag{8.10}$$

where e_e is the areal evaporation rate (the dimensions can be understood by writing e_e as having units of $\mathrm{m^3 s^{-1} m^{-2}}$), ρ and η are respectively the solution viscosity and density, and ω is the angular spinning speed. This simple equation states that the evaporation rate due to the vapour pressure of the solvent is supplemented by a flow term, dependent on the film thickness during thinning, viscosity, and angular speed; density is not much of a control parameter because it will be ~ 1 g/cc for most solutions.

Spin-coating provides remarkably uniform films with roughness that can be less than 1 nm, and an overall film thickness that varies little over the entire substrate. It is, however, restricted to thickness ranges of (typically) between 10 and 500 nm. Thinner films are likely to dewet (also a problem for dip-coating and doctor-blading), whilst thicker films tend to be inhomogeneous. Certain solvents are not appropriate either; chloroform is too volatile, and water often not volatile enough. In the former case, e in eqn (8.10) begins to dominate and in the latter, residual solvent tends to remain in the final film. An inappropriate solvent limits the quality of the film rather than renders the technique unworkable, for example, PEDOT:PSS may be spin cast directly from its aqueous dispersion.

A difficulty inherent in spin coating, as well as in dip-coating and doctor blading is the difficulty of making multilayers. Devices generally rely on stacks of layers and spin coating one atop another is likely to dissolve part of the initial layer. To a degree, one can circumvent this problem by using *orthogonal solvents* for the casting process. The first layer may be cast from any useful solvent, and the second will be dissolved in a solvent that does not dissolve the first layer. Although simple in principle, the use of orthogonal solvents is not widespread, because they are not perfect, and some smearing of the interface occurs. In any case, it is difficult to find good solvents for conjugated polymers without restricting the choice of solvent used to the second layer. One can cast the upper layer onto a different substrate (usually glass) and insert this into water at a glancing angle. Water will often wet the glass better than the polymer, and so the polymer layer will float on the water, whereupon it may be picked up on the second substrate. This otherwise appealing method is not suitable for commercial devices because it is not suitable for large-scale production. We therefore need to consider other methods of creating multilayers.

8.3.4   Layer-by-layer deposition

The layer-by-layer (LbL) technique is restricted to polyelectrolytes, but its simplicity is remarkable. Most substrates contain some level of charge in water. Silicon substrates contain a native oxide layer and so many hydroxyl groups will be present. These will liberate their proton in water, leaving O^- and consequently a negatively charged substrate. Other surfaces, such as sapphire, will be positively charged. In Fig. 8.5 we schematize the process for a cationic (positively charged) substrate. Dipping this into an aqueous solution of polyanions (i.e. a polyacid solution) will result in the adsorption of a monolayer to the substrate. Only a monolayer can form because the negative charges on the anion layer will repel any further negative charges, inhibiting the adsorption of more layers. This may be easily washed; the electrostatic interactions between the polymer on the surface will resist desorption during a rinse. An oppositely charged polymer may then be adsorbed in a further step. This process can be repeated as often as necessary. The interfaces between these layers will not be perfectly defined, and roughness on the scale of a few nm will exist, but nevertheless a well defined multilayer will be formed.

The utility of the LbL technique is limited in polymer electronics, but it has, for example, been used to created graded interfaces for improved LED performance (Section 10.1.4). The disadvantages of the technique rest in its reliance on alternating monolayers of oppositely charged poly-

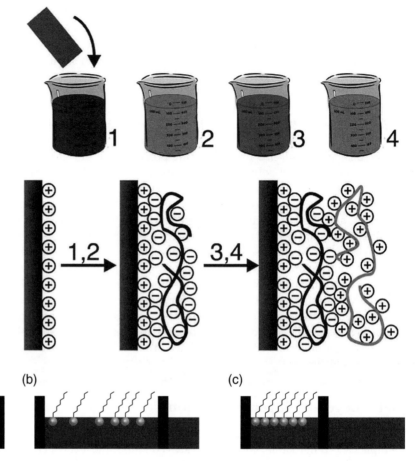

Fig. 8.5 Layer-by-layer deposition of charged polymer films. In the first stage a positively charged substrate is dipped into a polyanion solution (1) and then washed (2). Only one negatively charged layer is deposited on the substrate because of the Coulombic repulsion between like charges. This process can then be repeated with a polycation solution (3) followed by further washing (4). The process can be repeated until numerous layers have been deposited. In the figure, counterions are not shown. The opposite process beginning with a negatively charged substrate is also possible. Redrawn from G. Decher *Science* **277** 1232 (1997) © 1997 AAAS. Used with permission from AAAS.

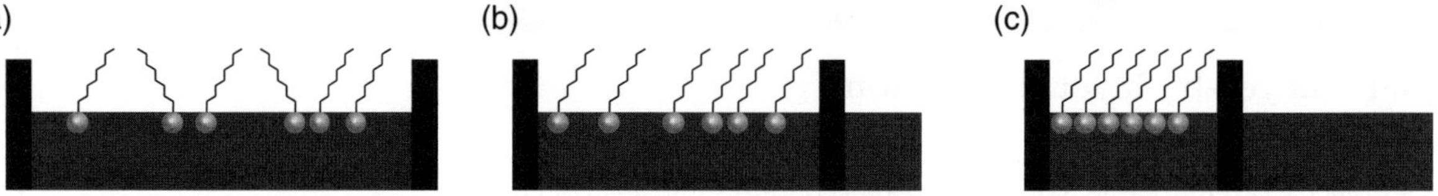

Fig. 8.6 A Langmuir film begins with an amphiphile spread at the air-water interface. (a) Shortly after deposition there is no ordering; individual molecules are scattered randomly across the surface. (b) The walls of the trough increase the surface pressure and eventually molecules begin to interact. (c) In some cases very good order can be achieved at high surface pressures, leaving a two-dimensional crystalline structure on the surface. Deposition is also possible with molecules that are not amphiphiles, but are hydrophobic. These molecules will sit on the surface of the water, but good crystalline structures are difficult to obtain.

mers. Thicker individual layers cannot be easily achieved by using higher molecular mass polymers, because these polymers tend to spread on the surface rather than extend to form thicker films.

8.3.5 Langmuir–Blodgett deposition

If sharp interfaces at the monolayer scale are required, then Langmuir–Blodgett techniques can be helpful. Here the requirements are for an amphiphile, which is deposited onto water from a solvent (chloroform is commonly used because its immiscibility with water and its volatility ensures it quickly evaporates) where it disperses at the surfaces. The hydrophobic part of the amphiphile ensures that it remains at the surface; if the amphiphile is too hydrophilic then Langmuir–Blodgett techniques are inappropriate. (Strictly the hydrophobic part is lipophilic, which refers to an affinity for fatty and other non-polar hydrocarbons.) The hydrophilic part will immerse in the water (Fig. 8.6a). The walls of the container holding the water (known as a *Langmuir trough*) can then be brought together to compress the amphiphile sitting on the surface of the water (Fig. 8.6b). In some cases astonishingly impressive order can be achieved (Fig. 8.6c), but this is unlikely to be the case for conjugated polymers, which are usually hydrophobic and will thus sit on the surface of the water. Nevertheless, LB deposition of conjugated molecules has been performed and rather precise structures have been obtained; the use of monolayers means that spacing between layers of different molecules can be controlled, which means that precise control of optoelectronic properties can be realized.

8.3.6   Spray-coating

For large-area deposition, the use of spray-coating technology is appealing. The key benefits of spray-coating are that it represents a relatively easy means of depositing material over a large area, and also because there is no wastage, unlike in, say, spin-coating where very little of the polymer becomes part of the film. Spray-coating is also relatively insensitive to the nature of the surface, so even if the surface is not particularly wettable, good coverage can still be achieved.

Spray-coating is a relatively undemanding technology; the material viscosity should be low enough to be processed by the device. Here, a well dissolved polymer (or other) solution is required in order that there be no aggregation or material build-up at the nozzles; typically this results in concentrations in solvent that are less than 1% by weight. A compressor is used to provide the necessary pressure to drive the solution up to the nozzles, at which point it mixes with an airflow which causes the formation of droplets (Fig. 8.7).

The solvent evaporates on contact with the surface, although some will have evaporated during the spraying, and this leaves a film, the quality of which depends varying on the solvent, the distance between the spray and the substrate, and the quality of the spray-coater.

8.3.7   Chemical grafting

The chemical synthesis of polymer films at surfaces has already been discussed in the case of electropolymerization in Section 6.9. Solution based synthetic methods can be applied to polymer film synthesis, if an initiator layer is attached to the surface. (It is also possible to synthesize the polymer *ex situ* with a functional group at an end that can attach to the surface.)

The synthesis of polymers at surfaces is useful when one requires thin polymer films that are robustly attached to surfaces. LbL and LB methods are restricted to depositing one monolayer at a time and are consequently of limited versatility. Coating methods will often not work for

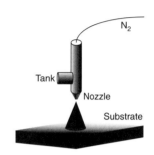

Fig. 8.7 A spray-coater consists of a gas flow (usually nitrogen) that mixes with the fluid held in a tank. The fluid is driven through the nozzle under pressure whereupon it is atomized. The resultant droplets provide an even coating on the surface.

Fig. 8.8 End-grafted polymers extend from a surface rather like the bristles of a brush.

thin films which very often dewet the surface. The chemical attachment of polymer layers to the surface avoids these problems. The most popular form of grafting is the end-grafting of polymers to form *brush* layers (Fig. 8.8). These brushes can have thicknesses from a few nm up to ~ 100 nm. Brush thickness, h, is related to its grafting density, σ_g, by

$$\frac{h}{\sigma_g} = Na^3, \tag{8.11}$$

where N is the polymerization index (number of monomers in the chain) and a^3 is the volume of one monomer. A grafting density of σ_g means that one grafted polymer molecule is associated with a surface area of $1/\sigma_g$, which means that h/σ_g is the volume of a polymer. The density, ρ, of the polymer can be written as

$$\rho = \frac{M_{w,m}}{N_A a^3}, \tag{8.12}$$

where $M_{w,m}$ is the monomer molecular weight and N_A is the Avogadro number. Combining eqns (8.11) and (8.12) we obtain

$$h = \frac{N \sigma_g M_{w,m}}{\rho N_A}. \tag{8.13}$$

Thicker brushes therefore require either larger molecular weights or grafting densities. Chemically grafting brushes *in situ* from a surface is the method of choice for thicker brushes. It is possible to synthesize functionalized polymers *ex situ* and then allow them to react with an appropriate functional group on the surface, but polymers which attach first will form *mushrooms*, i.e. self-avoiding random walk structures that do not stretch away from the substrate. These form an entropic barrier to later arrivals, kinetically limiting the grafting density. These limitations do not exist for the growth of polymers from the surface where, typically, a self-assembled initiator layer is deposited on the substrate and immersed in a solution of monomer and catalyst. The synthesis of conjugated polymer brushes has been achieved by some research groups (an example is shown in Fig. 8.9), but it has not been significantly used to practical effect. However, attaching conjugated polymer layers to small components or patterned surfaces will mean that this technique of film formation is likely to be increasingly important in future years.

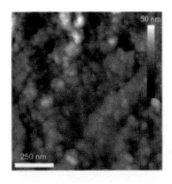

Fig. 8.9 Scanning force microscopy image of a polythiophene brush. This material was synthesized by a Kumada coupling reaction (Section 6.3.3). Taken from Sontag et al. *Chem. Commun.* 3354 (2009) with permission of The Royal Society of Chemistry.

8.4   Surface analysis

Most polymer electronic devices, be they field-effect transistors, light-emitting diodes, or photovoltaic cells, operate with the active component in the form of a thin film. Therefore the characterization of thin films and surfaces in general is profoundly important in developing our understanding of these devices. The approach to the analysis of surfaces depends on one's requirements. It may be important to obtain chemical structure at the surface, or information about the surface topography.

Some relevant techniques are summarized in Table 8.1. Briefly, they can be split into those techniques that provide information about the chemical components of the near surface region, typically the first 2 or 3 nm; those that provide information about the structure of the surface from detailed nm-resolution topographical information of the scanning probe techniques to simple information from an optical or fluorescence microscope; and those that provide information about the location of species as a function of depth in the sample. The most versatile of the latter depth profiling techniques is the suite of ion beam techniques in which an MeV beam of ions scattered on passing through the sample are detected (Rutherford backscattering); fragments emitted during nuclear collisions are detected (forward recoil spectrometry); and products of nuclear reactions within the film are collected (nuclear reaction analysis). The different possibilities means that there is a technique to analyse virtually all films.

The techniques of X-ray and ultraviolet photoelectron spectroscopy (XPS and UPS respectively) and scanning probe microscopy are perhaps worth studying in a little more detail as examples of techniques particularly useful for conjugated polymers. Not only can both of these provide surface information, but they can provide the means to obtain more fundamental information about the polymers themselves—the former because it can provide information on band structure (i.e. not only the chemical composition of the surface), and the latter because one variant (scanning tunnelling microscopy) can provide electrical information on single molecules and another (scanning near-field optical microscopy) can be used to interact directly with the optical properties of conjugated polymers.

8.4.1   Photoelectron spectroscopy

XPS, and its longer wavelength sibling, UPS, are techniques whereby an incident energy photon has a large enough energy to emit core electrons from a surface. (Photons can also be responsible for Compton scattering, and, although this process takes place, it occurs over a different energy range not relevant to the present discussion.) The kinetic energy of these photoelectrons is measured and depends upon the energy of the incident photon and the bond energy by

$$E = h\nu - E_{\mathrm{b}}, \tag{8.14}$$

where $h\nu$ is the energy of the incident photon and E_{b} is the binding energy of the ejected electron. The technique is then rather simple; in the standard XPS and UPS experiment (schematized in Fig. 8.10), the photon energy is constant so a scan of photoelectron energies reveals information about the different bonds present at the surface. The technique is very sensitive and can provide information beyond the presence of individual atoms. One might assume that an electron emitted from oxygen would tell us only that oxygen is present, but a carbonyl bond (C=0) has a different energy to a hydroxyl bond (O–H), and these

Table 8.1 Methods of surface analysis.

Information	Technique	Notes
Chemical analysis	X-ray photoelectron spectroscopy	Detailed bond information and atomic constituents
	Secondary ion mass spectrometry	Molecular constituents
Surface structure	Scanning force microscopy	Topographical information
	Scanning tunnelling microscopy	Topographical information
	Scanning electron microscopy	Topographical information
	Scanning near-field optical microscopy	Location of different optoelectronic materials
	Optical microscopy	
	Infrared and Raman spectroscopy	Identification of molecular species
Depth profiling	Ion beam analysis	Several techniques capable of locating molecules with resolution of several nm (decreasing with depth) up to a depth of microns
	Neutron reflectometry	Sub-nm resolution technique capable of studying buried interfaces. Usually requires selective deuteration for contrast
	X-ray reflectometry	Sub-nm resolution technique capable of studying buried interfaces. Requires the selective presence of heavy elements for contrast
	Dynamic secondary ion mass spectrometry	Secondary ion mass spectrometry with a sputtering capability
	Ellipsometry	Optical technique capable of sub-nm depth resolution

different lengths alter, only slightly, the energies of the different photoelectrons in a manner that can be resolved by appropriate analysis software. (This is expanded upon in Question 8.2 in the exercises at the end of this chapter.) The difference in the binding energy of the electrons when they are isolated atoms and when the atom is involved in a chemical bond is known as the *chemical shift*, and is not to be confused with that measured by nuclear magnetic resonance.

Photoelectron spectroscopy requires short wavelengths because, as can be seen from eqn (8.14), no photoelectrons can be emitted if we do not have an illuminating energy large enough to overcome the binding energy. Once this has been achieved, the intensity of the photoelectron emission is proportional to the intensity of incident radiation. In terms of the spectrum of X-rays, these are rather low-energy in comparison to those hard X-rays that can be produced by synchrotron sources. In fact, ultraviolet radiation is also appropriate for surface analysis. The related technique of UPS therefore follows identical physics.

The emitted electron results in a vacancy in the core electrons, which is immediately filled by an electron dropping down from a higher-energy orbital. The energy freed by that electron may be in the form of X-

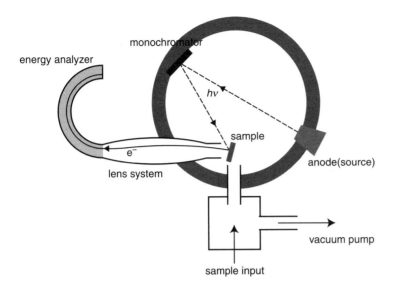

Fig. 8.10 Schematic diagram showing the essential components of a photoelectron spectrometer. X-rays are generated in a source and a monochromatic beam can be selected using a single crystal grating. Different manufacturers, however, have different methods of producing monochromatic X-rays or ultraviolet photons. Photoelectrons are ejected from the sample and are collimated by a lens system. A hemispherical analyser uses the cyclotron principle to determine the energy of the electrons. This information is sent to a computer so that the data can be analysed.

ray fluorescence or even in the emission of another electron, an *Auger* electron. The energy of these Auger electrons can also be considered in the electron spectrum.

The binding energy of the core electron is typically of the order of a few hundred eV, which might be considered large when compared to the ionization energy of various atoms and molecules, which rarely exceed 10 eV. However, electronic binding energies increase the closer they are to the nucleus. One could say that the binding energy of the core electron is equal and opposite to its orbital energy, but this would neglect the effect of removing the electron on neighbouring electrons. These electrons will change their state in order to minimize (shield) the effect of the hole created by the photoemission. These are known as relaxation effects and do not include the further relaxation of the ion by having outer electrons replace the core electron.

A difficulty in measuring material properties using photoelectron spectroscopy is the effect of charging on the surface. When the photoelectron and any Auger electrons exit the sample, it becomes positively charged, thereby inhibiting the loss of further electrons by the addition of a Coulombic potential barrier that must be overcome. For metal samples this is not a problem, because free electrons are able to replace the ejected electrons and minimize charging effects. Semiconducting polymers do not have that luxury because the band gap energy needs to be overcome before these electrons can be replaced. In order to avoid charging at the surface of the sample, electron flood guns are used to pour electrons onto the film to retain neutrality throughout the experiment.

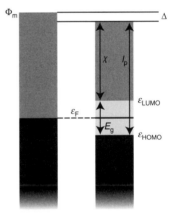

Fig. 8.11 The energy levels of a conjugated polymer in contact with a metal are shifted so that their Fermi levels equate. This schematic diagram ignores the band bending shown in Fig. 5.9, and the reader is directed towards the shift in the vacuum levels between the metal and polymer.

Measurement of the band gap

The spectrum of photoelectron energies obtained from XPS and UPS measurements can be used to determine density of states and electronic structure in a given material. Although it is not trivial to obtain detailed information, this is a very useful complement to optical techniques. We return to the physics described in Section 2.3 and consider the simple energy diagram for a semiconducting polymer, and compare this to a simple metal surface. These are shown in Fig. 8.11, which contains essentially the same information as Fig. 5.9. The band bending discussed in Section 5.5.1 is related to the equalization of the Fermi levels. However, this is not the only effect that occurs at an interface. If one considers a free metal surface, the electron density will tail off at the surface, and a non-constant electron density profile will be responsible for a surface dipole.

It is wise here to take a step back and consider the parameters shown in Fig. 8.11. The most important with respect to XPS and UPS measurements is the Fermi energy; all energies measured by these techniques are with respect to ϵ_F. For $h\nu < \Phi$ there is minimal photoelectron emission, at least at low temperature. However, for a metal sample, at $h\nu = \Phi$ photoelectrons will be ejected, and these will correspond to the Fermi level. For semiconducting polymers the Fermi level is in the middle of the band gap, so no electrons will be emitted until $h\nu = \Phi + \epsilon_{HOMO}$ has been achieved. Of course, we are not increasing $h\nu$ until electrons start to appear, but rather measuring the kinetic energy of photoelectrons after illuminating with radiation of energy $h\nu$. This is not a problem, because from eqn (8.14) we know $h\nu$ and we measure E, so the binding energy E_b is easily obtained. In Fig. 8.12a we show UPS data for a gold substrate, and the onset of electron emission can be seen at the Fermi energy. This gives us an immediate measurement of Φ. As one increases the binding energy (i.e. the electron kinetic energy decreases) one can see more structure in the spectrum. Essentially, but with some caveats, this is the density of states for gold. The structure bears little resemblance to that in Fig. 2.3, and this is because in the simple free electron model the number of electrons in different orbitals is not considered, merely 'unbound' or 'free' electrons. In gold, the 3d orbitals, 2.3 eV below the Fermi level contains many more electrons and consequently explains the sharp increase in intensity at this energy in Fig. 8.12a. (As an aside, $\lambda = hc/2.3e = 540$ nm, and the absorption of this and longer wavelengths when re-emitted explains the yellow colour of gold.) An important caveat that we must acknowledge is that the photoelectron spectrum is also mixed with secondary electrons.

We now turn to the polyfluorene film in Fig. 8.12b. This film is on a gold substrate and so its Fermi energy must necessarily be that of its substrate. However, we note the energy offset Δ. The offset is not obvious from the position of ϵ_{HOMO} because this is the sort of parameter that we should want to measure, so we do not know it to begin with. However, the ionization potential of the polyfluorene is a material property that can easily be measured, and in experiments such as this, the ionization energy of polyfluorene, $I_p \neq E_g/2 + \Phi_{Au}$, but rather $I_p = E_g/2 + \Phi_{Au} - \Delta$. (Generally, the Fermi level of semiconductors is situated at the midway point of the band gap, but the discussion at

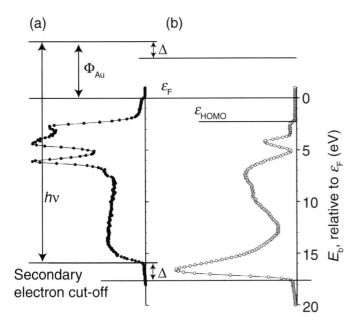

Fig. 8.12 UPS data for (a) a gold substrate and (b) a polyfluorene film on a gold substrate. In both cases the surfaces are irradiated with a monochromatic beam of $h\nu = 21.2$ eV The secondary electron cut-off (i.e. the energy of Auger electrons) is shifted in the polymer with respect to the gold by an amount Δ, but remains $h\nu$ below the vacuum level of each sample. The uppermost horizontal lines (again separated by Δ) indicate the vacuum level. Taken from Salaneck et al. *Mater. Sci. Eng. R* **34** 121 (2001), with permission from Elsevier.

the end of Section 2.3.2 concerning the applicability of this assumption should be noted.) A corollary of this phenomenon is that there is a shift in the energy at which secondary electrons (i.e. low energy electrons) appear relative to the gold substrate. The shift in the energy is caused by the surface dipoles mentioned above.

8.4.2 Scanning probe microscopy

Scanning probe microscopies (SPM) in general involve a tip, be it a tapered optical fibre or a sharp silicon nitride tip used in the most popular format, scanning (atomic) force microscopy. SPM requires that a tip be moved across the sample (or the sample across the tip) with the tip–surface interaction monitored. Changes in topography, optical, frictional, magnetic, or chemical properties can all be monitored to give information about the respective properties of the surface. A relatively simple feedback system controls the tip so that data can be continually obtained under the same operating conditions, without, for example, the tip crashing into the sample. The tip is supported on a cantilever, the back of which is used to reflect a laser onto a quadrant photodiode. As the tip interacts with the sample (Fig. 8.13), it exerts forces on the cantilever, bending it. This bending causes the laser to be reflected onto different parts of the photodiode, which not only may trigger the feedback mechanism, but also supplies information on the tip–sample interaction. The classic mode of operation is atomic force microscopy (AFM), which is best referred to as scanning force microscopy (SFM) because it is rarely used with a resolution whereby individual atomic forces are revealed. (Elaborate tip preparation methods can yield impressive results however, such as that for pentacene, shown in Fig. 3.13, where a

Fig. 8.13 Scanning probe microscopes generally involve a tip of some kind interacting with the surface. The nature of the interaction may vary (long-range forces, electrostatic, etc.) but the interaction will affect the cantilever upon which the tip is mounted. Any deflection of the cantilever due to this interaction will determine where a laser is reflected (on the top of the cantilever onto a quadrant photodiode), which will trigger a feedback mechanism determining how the tip is to interact with the sample.

truly atomic resolution is shown.) SFM is used to reveal surface topography. Although SFM is routinely used in many areas of science, including polymer electronics (see, for example, Fig. 7.27), we shall concentrate here on more specific techniques: scanning tunnelling microscopy reveals information about electronic structure, scanning near-field optical microscopy reveals information on optical properties, and Kelvin probe microscopy reveals the work function of the surface.

Scanning tunnelling microscopy

The first scanning probe technique was the scanning tunnelling microscope (STM), which was invented by Drs Gerd Binnig and Heinrich Rohrer of the IBM research laboratory in Zurich, earning both of them the Nobel Prize for Physics in 1986. With the STM a tip is scanned across the sample, as in the other scanning probe techniques. A small potential difference is held between the tip and the sample, and the tip is kept to around 1 nm above the sample. The space represents a potential barrier for charge carriers, but because the tip is held so close to the surface, electrons from the surface of the sample can tunnel through the 1 nm space to the tip (or in the other direction). The technique is therefore very sensitive to electron density distributions and can map the location of single atoms. With a very sharp tip (of the order of single atoms) images with height resolution of ~ 0.01 nm and a lateral resolution of ~ 0.2 nm is possible.

It is possible to confuse the principle of an STM with field emission, because in field emission a potential difference is placed between a sharp tip and a surface, causing electron emission. This is what happens in an STM, but the origin of the effects are different. Let us summarize field emission first. For an electron to leave a surface, it must have an energy of at least its work function greater than the Fermi energy of the material. It is possible to lower this energy by applying an electric field (field emission). This phenomenon is not dissimilar to the discussion in Section 5.5.2. An applied electric field can be used to reduce the potential barrier until at some point (a large enough field) the electron can tunnel through that barrier (Fig. 8.14a). This is the field effect, but if the material surface is brought very close to an STM tip, then there is no need for a field.

In principle, two adjacent materials, i.e. the tip and the sample, will have their Fermi levels aligned. If a positive bias is applied with respect to the sample, the Fermi level of the tip is raised in comparison with that of the sample, which means that electrons can travel from the tip to the sample, as schematized in Fig. 8.14b. If the sample is a conjugated polymer or other organic semiconductor, these electrons will occupy the LUMO levels. If a negative voltage is applied, the Fermi level of the sample is raised with respect to the tip, and electrons can flow from sample HOMO levels to the tip. The bias voltage and the current flow will thus give information about the sample density of states, in both occupied and unoccupied molecular orbitals. An example is shown in

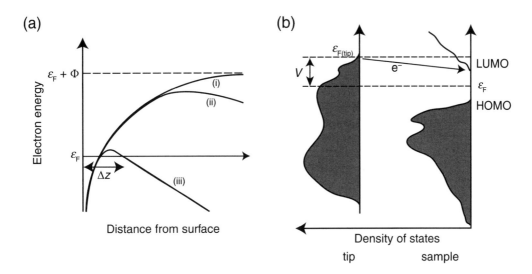

Fig. 8.14 (a) An electron can be emitted from a surface of a material of work function, Φ if it has an energy $E > \epsilon_F + \Phi$. Any energy below that will result in a surface image charge being able to pull the electron back to the surface. The potential energy curve for this situation is shown in (i). If a uniform electric field, E is applied, an additional energy $-eEz$ will lower the potential energy shown in (i). A small field (ii) will have little effect, but a large field (iii) will allow electrons with energy close to the Fermi level to reduce their overall energy by tunnelling through the potential barrier, Δz. (b) In the absence of an electric field, a small bias voltage between the tip and the neighbouring sample will allow electrons to tunnel through the vacuum barrier (assuming close proximity). In the case of a positive voltage bias, the Fermi level of the tip will be raised with respect to the sample. In this schematic image, electrons are able to tunnel through the tip into the LUMO of a conjugated polymer.

Fig. 8.15 for pentacene.

A means by which STM can be used to measure the conductance of individual polymers is to physically attach one to its tip and to pull that polymer from the surface. Many experiments in scanning probe microscopy involve the location of molecules on a tip. Tips, for example, can be readily coated with gold, and a thiol (SH) terminated polymer (the thiol readily attaches to the gold) can be attached to the tip. (The molecule can be attached to the tip before or after adsorption on the surface.) Detaching the molecule from the surface reveals the polymer-surface interactions as a function of distance. This is known as force spectroscopy. When a conjugated polymer is attached to an STM tip, the current through it for a given bias voltage will change. The rate of change of current or bias voltage with distance is monitored. This has been achieved for a polyfluorene oligomer (oligofluorene), and the concept shows promise for future experiments. In particular, single-molecule conductance measurements should reveal information regarding the relative importance of the conjugation length (intrachain transport) and hopping (interchain transport).

Scanning near-field optical microscopy

Scanning near-field optical microscopy (SNOM) is another component of the suite of SPM techniques. SNOM itself is a technique whereby

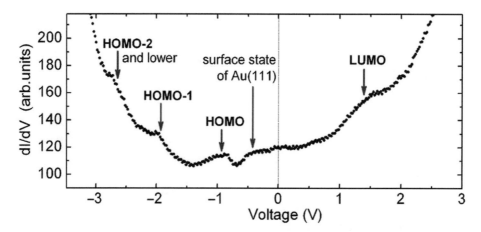

Fig. 8.15 STM can be used to observe molecular orbitals in conjugated molecules. Here data are taken for pentacene adsorbed onto a gold surface. A feedback system ensures a constant current, I, through the tip; by measuring the conductance $(\mathrm{d}I/\mathrm{d}V)$ at which this current is achieved for any given bias voltage, V, one can observe molecular orbitals. Because the molecular orbitals are associated with facile electron flow, they are visible as peaks in the $\mathrm{d}I/\mathrm{d}V - V$ plot. Although these experiments reveal the location of molecular orbitals, extracting the density of states from these data would be rather difficult. Similar experiments would be fiendishly difficult with conjugated polymers, which would not lie flat on the gold, and nor can they be deposited in such a fashion by vacuum sublimation, but in principle the same ideas would apply. Taken from Soe et al. *Phys. Rev. Lett.* **102** 176102 (2009). Copyright (2009) by the American Physical Society.

the optical properties of a surface can be mapped using light that is passed through an optical fibre (the SNOM tip) as it is scanned over that surface. In essence, SNOM is a form of optical or fluorescence microscopy whereby near-field imaging allows the diffraction limit to be circumvented.

The SNOM operates at a constant distance from the surface; this is achieved by keeping the amplitude of oscillation of the optical fibre constant: the fibre is oscillated laterally with an amplitude of a few nm, and this amplitude is damped as the surface is approached. The level of damping indicates the tip–surface distance. The fibre itself is a ~ 200 nm in diameter, but the internal diameter, through which light is directed, is typically ~ 50 nm, although some fibres can be considerably smaller. Light may be transmitted through, or reflected by, the sample, but, because the probe is only a few nm from the surface, the diffraction limit no longer applies.

Of particular interest is the possibility of investigating the optoelectronic properties of polymer films by the use of SNOM. In this case, a fluorescence microscopy mode is the tool of choice. With SFM it is difficult to determine the chemical constituents of a film, but if the film exhibits optical properties these can be determined by their absorption or fluorescence. In Fig. 8.16 we show data exemplifying those properties.

Scanning Kelvin probe microscopy

Whereas SNOM can use optical properties to map the location of polymers in a blend film, scanning Kelvin probe microscopy (SKPM) iden-

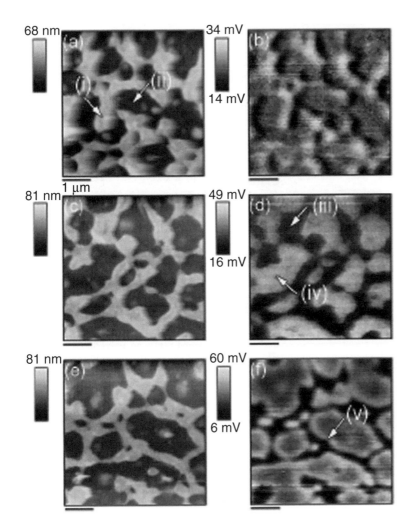

Fig. 8.16 Scanning near-field optical microscopy images of phase separated (1:1) blends of poly(9,9-dioctyl fluorene) (F8) and F8BT. The left-hand column (a), (c), and (e) shows topography information from different points on the film. These data were obtained simultaneously with the corresponding transmission (b) and fluorescence (d) and (f) images shown in the right-hand column. (SNOM is perfectly capable of working as a scanning force microscope.) The F8 absorbs the incident 362 nm radiation passed through the SNOM tip, and this absorption marks the location of the F8 in (b). Alternatively, F8 fluorescence at ∼ 425 nm may be measured (f), Where F8 and F8BT are mixed, Förster transfer occurs, and the F8BT fluoresces with a broad peak at ∼ 530 nm. Arrows (i) and (ii) show F8BT and F8 domains respectively, as determined from the absorption image (b). Note that there is more transmitted light through the F8BT-rich region than that of F8. Similarly arrows (iii) and (iv) also indicate respectively F8BT and F8-rich regions. The arrow (v) indicates F8 fluorescence at the boundary between phases. The scale bar in each image is 1 μm. Taken from Chappell et al. *Nature Mater.* **2** 616 (2003).

tifies the polymer by its work function (or surface potential), at least with respect to that of the tip. In the SKPM experiment, a current will flow between the tip and sample depending on the Fermi levels of the two materials. To stop that current flow, an offset potential difference is required, known as the contact potential difference, V_{CPD}. SKPM works by applying an oscillating potential at the tip, which responds to the interaction with the surface, and a feedback loop keeps the offset at V_{CPD}. SKPM is a non-contact scanning probe microscopy, in which the contact potential difference plays the same role as force in typical non-contact SFM. In Fig. 8.17 we show data illustrating the versatility of SKPM.

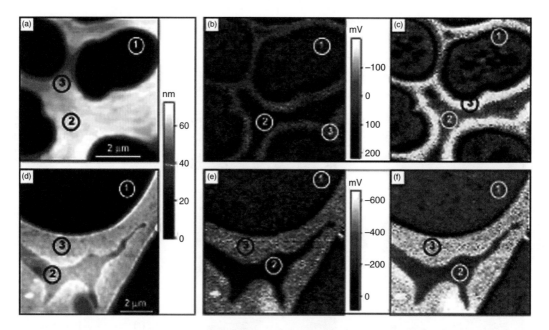

Fig. 8.17 Scanning Kelvin probe microscopy images of phase separated (1:1) blends of F8BT and poly(9,9-dioctylfluorene-*alt*-bis-N,N'-(4-butylphenyl)-bis-N,N'-phenyl-1,4-phenylenediamine) (PFB). Topography images are shown in (a) and (d), and (b), (c), (e), and (f) show information on the surface potential; (a), (b), and (c) were taken from the same point on the film, as were the images (d), (e), and (f). The differences between these images concern the sample preparation methodology. (SKPM, like SNOM, is capable of providing topographic information.) The images shown in (c) and (f) were obtained by illuminating the sample with 473 nm laser radiation. F8BT strongly absorbs this radiation, which causes its surface potential to shift due to the extra charges photogenerated by the light. The areas marked 1 are rich in PFB and those marked 2 are rich in F8BT, as are those marked 3, but in this case the F8BT is trapped at an interface between phases. Taken with permission from Chiesa et al. *Nano Lett.* **5** 559 (2005). Copyright (2005) American Chemical Society.

8.5   Further reading

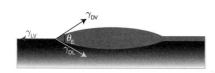

Fig. 8.18 A droplet on a liquid surface will deform the interface. The droplet has an equilibrium contact angle θ_E, and γ_{DV} and γ_{DL} are the surface (saturated vapour) and interfacial energies respectively. The surface energy of the liquid substrate is denoted γ_{LV}.

There are very few books on polymers at surfaces and interfaces at the right level for the reader of this book. The book of that name by Jones and Richards (1999) is perhaps an exception, and should be of interest to most readers. There are a number of review articles in this area, and modesty does not prevent the recommendation of that by Geoghegan and Krausch (2003). Further information is covered in the chapter by Geoghegan and Jones (2005). The latter two are given from a similar perspective to that in this chapter, so the reader might be interested in the review by Budkowski et al. (2002). From the point of view of the application of the science of thin films to conjugated polymers, one should consider the reviews by Moons (2002) and Leclère et al. (2006). Reviews on photoelectron spectroscopy can be found in the chapter by Ratner and Castner (1997) or the review by Salaneck et al. (2001). The latter review is particularly recommended for its concentration on conjugated polymers. Thin film characterisation techniques used in organic electronics are covered in the book by Petty (2007). The reader might find this book interesting because it covers other techniques to those

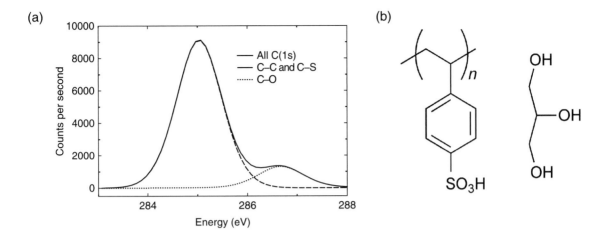

Fig. 8.19 (a) XPS data showing a high-resolution C(1s) scan of a film of a PEDOT:PSS mixture containing glycerol, which is known to improve the performance of such hole-injection layers. The data are separated into C–C and C–S corresponding to PSS and C–O (glycerol) peaks. (PEDOT, which contains C–S, C–O, and C–C is known not to be present at the surface.) (b) The chemical structures of PSS (left) and glycerol. These data are provided courtesy of Drs Ana Rodríguez and Tracie Whittle.

discussed in this chapter.

8.6   Exercises

8.1. A droplet is in contact with a liquid substrate as shown in Fig. 8.18. Under such circumstances, the equilibrium contact angle, θ_E is larger than the apparent contact angle with the liquid substrate, θ_A, because a component θ_B is hidden below the surface, so that

$$\theta_E = \theta_A + \theta_B. \tag{8.15}$$

Consider a droplet, the edge of which makes an angle with the horizontal of $\theta_A = 1°$, with relevant surface energies $\gamma_{LV} = 22.7 \text{ mJ/m}^2$, and $\gamma_{DV} = 5.3 \text{ mJ/m}^2$. If the work required to separate the droplet from the liquid substrate is given by $W_{DL} = 0.9 \text{ mJ/m}^2$, what is the equilibrium contact angle, θ_E?

We can define an effective interfacial energy,

$$\frac{1}{\gamma} = \frac{1}{\gamma_{DV}} + \frac{1}{\gamma_{DL}}. \tag{8.16}$$

Prove that the spreading coefficient can be written as $S = -0.5\gamma\theta_E^2$.

8.2. As well as calculating band structure, XPS can be routinely used for elemental analysis by which we mean a determination of how much of any one given element there is in a given sample. Typically, a scan over all photoelectron energies is performed, and this is followed by a high-resolution scan over a particular peak. One such peak that is commonly studied is the carbon C(1s) peak. In the example shown in

Fig. 8.19a we see such a C(1s) scan. Here the likely bonds are C–O, C–S, and C–C. Both C–S and C–C appear at 285 eV, whereas C–O appears at 286.5. Oxygen has a greater electronegativity than carbon or sulfur, so the outermost electrons in the carbon are drawn towards the oxygen, which reduces the shielding provided by these electrons and thus allows the photoelectrons to more easily escape, hence their greater energy than those for C–C and C–S, which overlap due to the similar electronegativities of carbon and sulfur.

Analysis software can be used to fit these peaks, and in the example in Fig. 8.19a the C–C peak has an area of 10584 eV counts s^{-1} and that of C–O of 1607 eV counts s^{-1}. It is known that the surface contains polystyrene sulfonic acid and glycerol and no other molecules. If the densities of PSS and glycerol are 1.30 and 1.25 g/cc respectively, what is the fraction by volume of glycerol at the surface? The chemical structures of PSS and glycerol are shown in Fig. 8.19b.

8.3. A suitably end-functionalized F8BT (Fig. 6.2) sample has been created *ex situ* in order to form a layer, chemically grafted to the required surface. Show that the maximum (dry) thickness of this layer is independent of the molecular weight of the polymer, and calculate this thickness, if the grafting were to be performed in a Θ solvent. The density of F8BT can be taken as 0.92 g/cc. Comment on your result.

Here, the end-to-end distance of the polymer is a less useful parameter than the radius of gyration, which is a more realistic estimate of the dimensions of a polymer, given that it is within this radius from the centre-of-mass that most of the molecular mass is located. The radius of gyration is given by

$$R_\text{g} = a\sqrt{\frac{N}{6}}, \tag{8.17}$$

where N is the number of monomers in the polymer and a is the monomer size.

Discuss how realistic your answer might be, considering both the polymer physics of the growth of polymer brush layers, and how you think the mechanical properties of conjugated polymers might affect the result.

Polymer transistors

9

Until now, we have not specifically considered practical polymer devices, although it is fair to say that they have never been far from our thoughts. Special attention in what has been presented has been allocated to optoelectronic behaviour, and we shall return to this in the next chapter. However, transistors have a massive role to play in polymer electronics because of the advantages that flexibility and processability can give to circuitry. Transistors are thought of very often in the context of Moore's law, and the number of transistors we can fit on silicon chips. This is all very well, but even now, the increased power of computers is not yielding commensurate benefits in terms of better software and more efficient performance from those using them. In fact, the current goal in terms of computing is price rather than performance. The arrival of netbooks, as well as the increasing market for smartphones and tablets suggests that all-out performance is not what is currently driving consumer needs. The important message from all of this is that *size isn't everything*. Massively powerful computers requiring smaller and smaller transistors are rapidly becoming a niche area dominated by large corporations and the military.

When one understands that transistor technology is not all about increased computing power, then one can appreciate the scope for a role played by polymer electronics. This will not necessarily be in computers but in other technologies such as RFID tags (Section 1.2.3) and display backplanes (Section 1.2.1). RFID tags are expected to soon replace the simple barcode in stores, and their usage will allow much more control over inventory than the barcode currently does. Transistors for display backplanes are likely to be used in flexible displays. In neither of these two examples is processing speed of the essence; the technology merely has to be good enough.

A transistor is simply a switch. Depending upon the signal, i.e. the input voltage, a current will or will not flow. The current therefore represents binary information in that, for example, a null output may be regarded as a binary 'zero'. The representation of transistors in this way is not the whole truth because the transistor is only responding to its input. Transistors may however be grouped together to form logic circuits, and this is the basis of computing.

For inorganic semiconductors, doping is necessary for optimal performance in many devices. This is not the case for semiconducting polymers, and is certainly not true of polymer transistors, which do not need to be doped in order to function. In fact, inorganic transistors that rely

Fig. 9.1 (a) Bottom gate field-effect transistor and (b) top gate FET. A voltage applied to the gate will create an electric field across the dielectric attracting charges from the active (semiconductor) layer. If the active layer has good electron mobility, then electrons will flow from source to drain, according to the arrangement shown here. If the layer has good hole mobility, then holes will flow in the other direction. If the voltage applied to the gate is negative or less than a certain threshold voltage then only a small current will flow, because charge carriers will not be attracted to the semiconductor–dielectric interface.

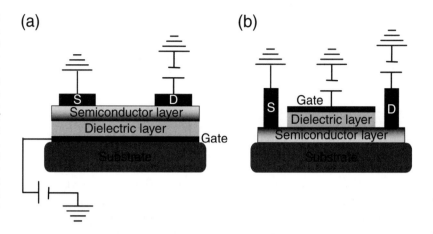

on doping cannot be mimicked in polymer electronics, so all polymer transistors are thin-film transistors (TFTs), which rely on the field effect (hence FET): the use of an electric field to overcome the effect of the band gap and to allow charge transport. The addition of dopants to polymer transistors is deleterious because dopants increase the current when the transistor is supposed to be 'off'.

9.1   The field effect

There are a variety of different transistor technologies based on inorganic semiconductors. Of these, only the field-effect transistor (FET) is popular with organic materials. These field-effect transistors are simple designs, with two electrodes sandwiching a dielectric layer and an active layer. Simple schematic diagrams of the basic components of field-effect transistors are shown in Fig. 9.1.

Charge transport can occur only when the electrons have been promoted to the LUMO, or when holes are present in the HOMO. Achieving either of these conditions requires effectively reducing band gap through doping. The gate electrode is an alternative to doping, because it changes the energy of the carriers relative to the band gap to allow electron injection into the LUMO or holes into the HOMO. If the semiconductor layer has good p-type mobility, then a negative voltage at the gate will attract charge carriers to that interface, and it is this excess of carriers at the interface through the electric field that means that doping is not necessary. This increase in charge carriers gives rise to a current when an appropriate voltage is applied between the source and the drain electrodes.

In order to behave like a switch, different phenomena need to occur, so that no current is passed under certain gate voltages, and a controlled current under others. Depending upon the gate voltage and the materials chosen, there are three distinct regions of behaviour in transistors: *accumulation, depletion,* and *inversion*—(, as schematized in

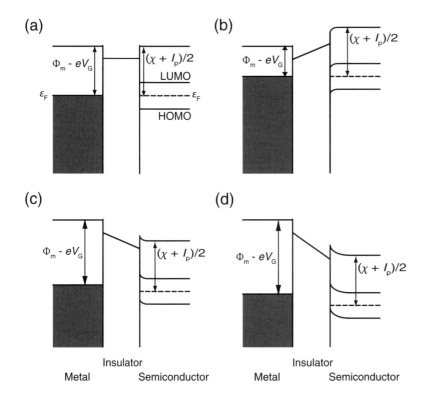

Fig. 9.2 Transistor behaviour is governed by the location of charges at the semiconductor–insulator interface. Here we show the behaviour of the energy bands for a polymer which has good electron mobility. When a positive bias is applied to the gate, the work function is essentially reduced to $\Phi_m - eV_G$, where we take the elemental charge, e, to be positive. Because current cannot flow across the insulator, $(\chi + I_p)/2$ remains unchanged. (a) The flat-band situation corresponds to a gate voltage for which there is no band bending. Under this condition $\Phi_m - eV_G = (\chi + I_p)/2$. (b) The accumulation regime occurs when a positive voltage attracts electrons from the polymer to the interface with the insulator. Formally, this is $\Phi_m - eV_G < (\chi + I_p)/2$. (c) The reverse situation (depletion), when a negative effective gate voltage repels electrons away from the interface, occurs when $\Phi_m - eV_G > (\chi + I_p)/2$. (d) Inversion occurs when $\Phi_m - eV_G \gg (\chi + I_p)/2$, and is generally not important in organic transistors.

Fig. 9.2. Different kinds of field-effect transistors rely on different regions for their behaviour. For example, traditional silicon-based FETs generally operate under inversion conditions, but the so-called thin-film transistors, in which the semiconducting layer is very thin, operate under accumulation conditions. Organic FETs are generally TFTs. A transistor that operates under inversion conditions does not use a thin film semiconductor layer, and so it can only have a top gate structure (Fig. 9.1b). TFTs can operate with both top and bottom gates. In organic transistors inversion is generally not observed except for a few exceptions. Polymeric materials rarely exhibit both hole and electron mobility of a good enough quality; if inversion were to be observed, oppositely charged carriers to those observed in the accumulation regime must populate the boundary between the gate dielectric and the semiconducting layer. Such a phenomenon is not often observed, but we discuss exceptions in Section 9.5.1.

9.2   The drain current

The requirements for a good transistor include a low threshold voltage (Section 9.3), a low subthreshold swing (Section 9.4), and a large drain current for small voltages at both the gate and between the source and drain. It is also desirable to have a small current between the source and the drain (known as the *off current*) when the transistor is supposed to be off. The drain current in most cases is dependent on the potential difference between the source and drain, V_D, and is given by

$$I_D = \frac{W}{L}\mu C_i \left(V_G - V_T\right) V_D, \text{ where } |V_G - V_T| \gg |V_D|, \quad (9.1)$$

where W is the width of the device, L, the channel length, i.e. the distance between source and drain electrodes and the key parameter in describing transistor size in the semiconductor industry,[1] C_i is the (areal) capacitance of the dielectric layer, V_G is the gate voltage, V_T is the threshold voltage, below which there is no transistor operation, i.e. $|V_G - V_T| > 0$, and μ is the charge carrier mobility. The requirement that $|V_G - V_T| > V_D$ reflects the increase in the number of charge carriers in the accumulation region with increasing V_G. This is known as the *linear regime*, and we turn to a derivation of eqn (9.1) here.

Polymeric field-effect transistor operate under accumulation conditions, in which case, the current is simply the areal accumulation charge, Q_A, divided by the time it takes to move from the source to the drain, t_D,

$$I_D = \frac{Q_A W L}{t_D}. \quad (9.2)$$

Q_A has units of C/m^2, so the accumulation charge needs to be multiplied by the area, WL, of the interface between the dielectric and the semiconductor in order to calculate the drain current. If this interface is uniform then the charge carriers will move at a constant speed (v_D) between the source to drain, so

$$v_D = \frac{L}{t_D} = \mu E = \frac{\mu V_D}{L}, \quad (9.3)$$

where E is the electric field applied between the source and the drain.

The accumulation charge is given by the simple relationship

$$Q_A = C \left(V_G - V_T\right), \quad (9.4)$$

which is valid for $V_G > V_T$ and where C is the capacitance of the dielectric layer. If we combine eqns (9.2), (9.3), and (9.4), we readily obtain eqn (9.1).

This linear relationship between V_G and I_D in eqn (9.1) means that this operational regime is sometimes known as *Ohmic*. However this is not always the case, and other models, most notably the *quadratic regime* show slightly different behaviours. The linear model requires constant areal accumulation charge, and if this is not the case a different model must be used.

If $|V_G - V_T| < |V_D|$, the accumulation layer starts to disappear (often known as *pinch-off*), at least between the gate and the drain. This is known as the *saturation regime* and commences when $V_G \approx V_D$. In this regime, the drain current is given by

$$I_D = \frac{W}{2L}\mu C_i \left(V_G - V_T\right)^2. \quad (9.5)$$

[1] For example, at the time of writing, modern computers are being introduced with processors based on transistors with 22 nm technology; the channel (or transistor) length is 22 nm.

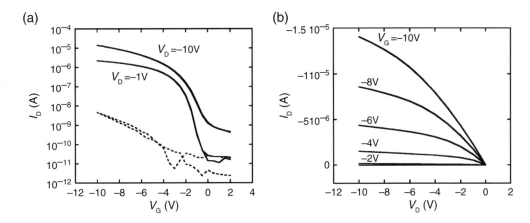

Fig. 9.3 The electrical characteristics are shown for an F8BT semiconducting layer on a 50 nm fluorinated polymer dielectric, known commercially as CYTOP®. (a) The transfer characteristics are the $I_D (V_G)$ curve, which is measured at constant V_D. The broken lines are the gate leakage currents at the two V_D values. (b) The output characteristics are the $I_d (V_D)$ curves, which are measured at constant V_G. Taken with permission from Cheng et al. *Chem. Mater.* **22** 1559 (2010). Copyright (2010) American Chemical Society.

The saturation regime results in a drain current independent of V_D, which is desirable from an engineering perspective, because it ensures that the transistor is predictable when operated in this mode. (Circuits are built up from multiple transistors, and if the use of the saturation regime means that differences in the drain voltage throughout the circuit, for example due to manufacturing imperfections, do not affect significantly the final current output.) The mobility obtained from eqn (9.5) is known as the *saturated* electron or hole mobility and is the usual *field-effect mobility* quoted for transistors. Because all the other parameters in eqn (9.5) are known, this is a relatively simple means of measuring mobility, as an alternative to the time-of-flight method discussed in Section 5.4.

The equations are identical to those for metal–oxide–semiconductor field-effect transistors (MOSFETs), which reflects their similar operation. The major difference between devices is that the inorganic semiconductors used with MOSFETs must be doped. More generically, all of these transistors are MISFETs (metal–insulator–semiconductor field-effect transistors); in the case of the MOSFET, the insulator is the oxide layer.

9.2.1   The linear regime

Transistor data are usually presented as either transfer or output characteristics. Transfer characteristics show the drain current against the gate voltage (at constant drain voltage) and output characteristics present the drain current as a function of the drain voltage (at constant gate voltage). For the transfer characteristic plot, the drain current is usually plotted on a logarithmic scale so that the subthreshold regime can

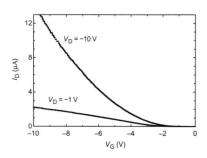

Fig. 9.4 The transfer data from Fig. 9.3a are plotted on linear axes. We see the linear regime for $V_D = -1$ V and saturation for $V_D = -10$ V.

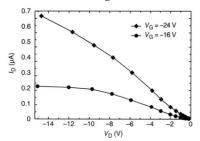

Fig. 9.5 The output characteristics of a transistor with a large (absolute) gate voltage should exhibit linear behaviour (Fig. 9.3b), but contact resistance causes a departure from this behaviour. Here, the output characteristics of a poly(9,9-dicotylfluorene-*alt*-bithiophene) (F8T2) transistor are shown; the linear behaviour is not observed at small V_D. These data are taken from Street and Salleo. *Appl. Phys. Lett.* **81** 2887 (2002).

be determined, but this is not necessary for the output characteristics. Examples of these transistor characteristics are shown in Fig. 9.3. We can examine these curves in detail to see how we can apply what we have learnt about the transistor to understand real data. The transistor in question is a top gate structure with a semiconducting layer (F8BT) with a large hole mobility. To attract the holes to the interface a negative gate voltage is required, as can be seen in the data shown. We also note that the drain current increases relatively rapidly as a negative gate voltage is applied, so this particular system has a rather small threshold voltage. From a brief observation of the transfer characteristics shown in Fig. 9.3a, it is difficult to observe the linear regime with the eye, because of the logarithmic scale. As mentioned, the logarithmic scale allows visualization of the subthreshold region and enabling a determination of how good the transistor is as a switch, i.e. the on/off ratio, which we discuss in Section 9.5 below, and the subthreshold swing. In Fig. 9.4 we show the data of Fig. 9.3a as a linear plot, and the difference between linear and saturation becomes very clear. For $|V_G| > |V_D|$, the linear regime is expected, and this is very clear for $V_D = -1$ V. Similarly, for $|V_G| < |V_D|$, we expect a parabolic curve, and the data do indeed show that the linear regime is no longer valid. In Fig. 9.3a we also see a steady growth in the gate leakage current for decreasing V_G (increasing $|V_G|$), which remains orders of magnitude smaller than I_D. The parabolic behaviour is clear in Fig. 9.6a, when $\sqrt{I_D}$ is plotted as a function of V_G.

If we consider the output characteristics (Fig. 9.3b), the linear regime (predicted by eqn 9.1) is visible for most devices, certainly for $V_G = -10$, -8, and -6 V. note how the departure from linearity moves to the right as smaller (absolute) gate voltages are used. The saturated regime is harder to see here, and the device when $V_G = -2$ V is off.

Other corrections to this linear relationship arise from contact resistance, with the result that the the drain current is reduced by a small amount, and the voltage drop across the layer is also decreased to account for the voltage drop across this contact resistance. Contact resistance is always present and is largely due to the electrode work function; if the semiconductor layer is hole transporting the electrode should have a high work function, and for an electron transporting layer it should have a low work function. The reasons behind this are the same as those discussed in Section 4.11 for a generic device. Contact resistance can be spotted in (linear) output characteristics, because rather than linear behaviour, curvature is visible at small drain currents (Fig. 9.5). Such matters are of course not relevant for the gate electrode, because that is in intimate contact with a dielectric. Nevertheless, the gate voltage does itself affect the contact resistance at the source and drain electrodes.

9.3   The threshold voltage

In comparison with other kinds of FET structures, the thin film transistor has a rather low threshold voltage. Nevertheless, it is necessary to apply a voltage to overcome differences in work function. The threshold voltage essentially (but not completely) corresponds to the voltage that needs to be applied to create the flat-band condition (Fig. 9.2a). Under these circumstances, the flat-band contribution to the threshold voltage is given by

$$V_{fb} = \frac{1}{2e}\left(2\Phi_m - I_p - \chi\right). \qquad (9.6)$$

It is usual, however, to include the effects of residual conductivity in the threshold voltage, which can be significant for doped polymer layers. This is not something that is easily formalized or calculated, and so is left in terms of n_0, the density of free charge carriers with no applied voltage. The threshold voltage then becomes

$$V_T = \frac{1}{2e}\left(2\Phi_m - I_p - \chi\right) \pm \frac{en_0 d_s W L}{C}, \qquad (9.7)$$

where d_s is the thickness of the semiconductor layer. The term $en_0 d_s / C$ is positive for layers with high hole mobility and negative for layers with high electron mobility. For high electron mobility semiconductor layers in which the flat-band contribution is zero, one notes therefore that V_T is negative; for all $V_G > 0$, a current can flow between the source and the drain. In fact, a current can also flow for $V_G > V_T$ (or $V_G < V_T$ for semiconductor layers with high hole mobility), but this mode of operation, the *depletion mode*, where the charge carriers are repelled from the interface, is rarely used in TFT devices although it is common in other transistors, of which the junction field-effect transistor is perhaps the best known. In organic FETs, the regime where $V_T > V_G$, but where V_G is such that inversion is not occurring, is known as the *subthreshold regime*, and we discuss this in Section 9.4 below.

The term 'threshold voltage' is often used interchangeably with *turn-on voltage*, although sometimes a different definition is used for the latter. They both refer to the point at which the transistor can be considered to function as a switch, i.e. at $V_G < V_T$, the transistor may be considered to be 'off'. We prefer to treat the threshold and turn-on voltages as identical, but the reader should be aware that they are not always treated the same.

The threshold voltage can be measured experimentally, although the result for a particular device would not necessarily be identical to that predicted by eqn 9.7. The threshold voltage can be determined in either the linear regime or the saturation regime, but if no stipulations is made as to which it is, then the saturation regime is assumed. The saturation threshold voltage is usually taken as the gate voltage at which the drain current is extrapolated to zero, and this is best achieved with a plot of $\sqrt{I_D}$ as a function of V_G (Fig. 9.6a). The threshold voltage in the linear regime is considered to be the point at which $I_D(V_G)$ (at fixed

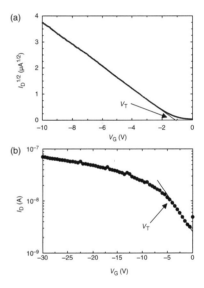

Fig. 9.6 (a) The threshold voltage can be calculated from the value of V_G at which the extrapolated drain current on a plot of $\sqrt{I_D}(V_G)$ reaches zero, which is here $V_T = -1.1$ V. (b) Alternatively, the threshold voltage can be taken from the point at which $I_D(V_G)$ ceases to be exponential, which occurs at $V_T \approx -4.5$ V. The data in (a) are taken from Cheng et al. *Chem. Mater.* **22** 1559 (2010) for a F8BT semiconducting layer on a 50 nm CYTOP for a saturated transistor ($V_D = -10$ V). The data in (b) are taken from von Hauff et al. *J. Appl. Phys.* **108** 063709 (2010) for a 240 nm P3HT layer for a linear transistor with a source-drain voltage, $V_D = -5$ V.

drain current) crosses over from exponential ($|V_G| < |V_T|$) to power law behaviour ($|V_G| > |V_T|$), as shown in Fig. 9.6b. The same determination of threshold voltage used in the linear regime may be used for saturated devices. The converse is, however, not true, because $\sqrt{I_D}$ is not proportional to V_G in the linear regime.

9.4   Subthreshold

The point at which the drain current reaches the level of system noise is referred to as the onset voltage, V_0. This is the voltage corresponding to the minimum in I_D as a function of V_G. For gate voltages between onset and threshold, the transistor is in the *subthreshold regime*. As $|V_G|$ is increased from V_0, the drain current increases exponentially (we recall that we were able to use departure from this exponential behaviour to define the threshold voltage in Fig. 9.6b) with V_G. Although this is not an important regime for device operation, it is useful for characterising device behaviour and quality.

The usual measure of device quality in the subthreshold regime is the subthreshold swing, S^{-1}, defined by

$$S^{-1} = \frac{\partial V_G}{\partial \log_{10} I_D}, \tag{9.8}$$

so that subthreshold swing is usually given in units of V/dec. A low subthreshold swing means that the subthreshold region is small and so the difference between V_T and V_0 is small. In such cases the distinction between 'off' and 'on' is very clear. Values of $S^{-1} < 100$ mV/dec are to be expected from a good transistor.

9.5   Optimizing transistors

We have been at pains to point out that transistors merely have to be good enough; they do not have to compete with silicon-based technologies. Nevertheless, transistor technology does mean that different materials will compete with each other for dominance. We have seen that transistor operation is best optimized for a low threshold voltage, which will usually lead to a low subthreshold swing, and a capability for effective operation at low gate and drain voltages, with as large as possible drain current. All of these are related to the quality of the semiconducting material either directly or indirectly. A material with lots of traps means that not all charges can travel from one electrode to the other. These traps contribute to the poor quality of the device and their presence can be measured in the subthreshold region (see Question 10.3 in the exercises at the end of this chapter). Finally, it is also useful to minimize contact resistance Section 9.2.1. Poor assembly will result in significant gate leakage and off currents.

Advantages of high mobility are related to switching speeds. The faster the charges move, the more operations that can be performed in

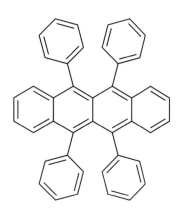

Fig. 9.7 Rubrene has an exceptionally high field-effect mobility for hole transport.

a second. This is particularly important in ring oscillator circuits,[2] but also, if one requires flexible screens of high resolution with an active display, then one needs to be able to address and switch the pixels quickly. The more pixels there are, the greater the processing power required.

The best organic transistor performance is with materials based upon acenes and their derivatives (pentacene and rubrene (Fig. 9.7) are striking examples), which have hole mobilities up to and beyond 10 cm^2 V^{-1} s^{-1}. Polymers are not as successful, with the best performers having hole mobilities of up to 1.0 cm^2 V^{-1} s^{-1}. To this end P3HT is generally considered as having excellent hole mobility, although other polymers are exhibiting better performance under certain conditions. Because of the likelihood of oxidation, there are fewer polymers that exhibit high electron mobility, and those that do are very often ambipolar. (Fortunately, ambipolar behaviour is a useful property.) The best electron mobilities can match those of holes in P3HT. One recent example is a polynaphthalenediimide-based polymer, which has an electron mobility of close to 1 cm^2 V^{-1} s^{-1} (Fig. 9.8).

Of course, high mobility is only one way to achieve rapid switching; miniaturization also plays an important role. Rearranging eqn (9.3) we see that

$$t_{\mathrm{D}} = \frac{L^2}{\mu V_{\mathrm{D}}}, \tag{9.9}$$

which we have already seen as eqn 5.23, and illustrates the importance of having as small a channel length as possible. The transit time of charges between the source and drain is a substantial lithographic challenge, because short-circuiting tends to be very common at shorter channel lengths. Given that a prerequisite is that the transistor be reliable, a compromise is invariably reached. Ink-jet printing and most lithographic techniques do not give particularly short channel lengths, so this is generally not a major problem when miniaturization is less important. Channel lengths of between 0.1 and 10 μm provide excellent performance, but if switching needs to be optimized, lengths better than 100 nm can be achieved by, for example, ink-jet printing or nanolithography. The best performance achieved with organic layers are graphene transistors with a metal oxide dielectric. Here $\mu > 20000$ cm^2 V^{-1} s^{-1} has been achieved. Graphene is impressive not just for its large mobility, but also because of the low channel lengths that can be achieved with this remarkable material.

Reliability also implies a high signal-to-noise ratio, which is best achieved by a large drain current. In the accumulation regime (eqn 9.1), as well as small L and large μ, we require, large C. (We should rather not have to increase too much V_{D} if at all possible.) Maximizing W limits the number of transistors that one can have in a given area, so the capacitance of the dielectric layer is a key component in transistor optimization. Here

$$C_{\mathrm{i}} = \epsilon_0 \epsilon_{\mathrm{r}} / d_{\mathrm{d}}, \tag{9.10}$$

where d_{d} is the thickness of the dielectric layer. (The reader is reminded that C_{i} is a capacitance per unit area.) It is these two parameters, ϵ_{r} and

[2] Ring oscillators are used, for example, in clock synchronization, electronic music (frequency generation), and radios. A ring oscillator is composed of NOT logic gates to provide an oscillating voltage output.

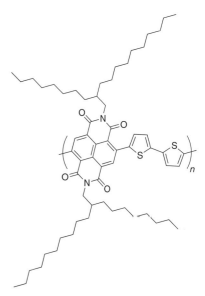

Fig. 9.8 Poly[(N, N'-bis(2-octyldodecyl)-naphthalene-1, 4, 5, 8-bis(dicarboximide)-2, 6-diyl)-*alt*-5, 5'-(2, 2'-bithiophene)] is an alternating copolymer of a naphthalenediimide unit with a bithiophene unit. The long chains not only confer solubility, but also allow ordering to occur, which helps the charge transport.

d_{d}, that one would wish to control. A large dielectric constant and small thickness are advantageous for any TFT dielectric layer. However, it is preferable that these two requirements are achieved in concert. Thick films of large dielectric constant materials are less advantageous because large dielectric constant materials are prone to traps (Section 5.3.2), with deleterious effects on the performance of the device.

A relatively straightforward metric for assessing the quality of a transistor is given by its *on/off ratio*, which is the ratio of I_{D} at its operating gate voltage with respect to I_{D} ($V_{\mathrm{G}} = 0$),

$$Q_{\mathrm{on/off}} = \frac{I_{\mathrm{D}}\left(V_{\mathrm{D}} = V_{\mathrm{Dop}}, V_{\mathrm{G}} = V_{\mathrm{Gop}}\right)}{I_{\mathrm{D}}\left(V_{\mathrm{D}} = V_{\mathrm{Dop}}, V_{\mathrm{G}} = 0\right)}, \tag{9.11}$$

where V_{Dop} and V_{Gop} are the respective operating (optimal) drain and gate voltages. Although we can measure output and transfer characteristics and obtain a large drain current, this is useful only if the drain current when no gate voltage is applied is low. Of course, one can change the gate voltage in the 'off' state to further minimize the drain current, but it is more practical to define the on/off ratio with respect to $V_{\mathrm{G}} = 0$, simply because this is how we should like to switch it off. One could imagine that we could increase V_{D} as much as possible to increase I_{D}, but if V_{D} is too large dielectric breakdown will occur, destroying the insulating layer. A good on/off ratio for a practical transistor would be $Q_{\mathrm{on/off}} = 10^6$ or better, and an excellent result would be $\sim 10^8$. The on/off ratio is not a property of the device alone, but depends on its stated operating conditions, which means that it is not enough to state $Q_{\mathrm{on/off}}$, but one needs also to state V_{Dop} and V_{Gop}. The on/off ratio for the device whose transfer characteristics are shown in Fig. 9.3a is $\sim 10^5$ for $V_{\mathrm{G}} = V_{\mathrm{D}} = -10$ V.

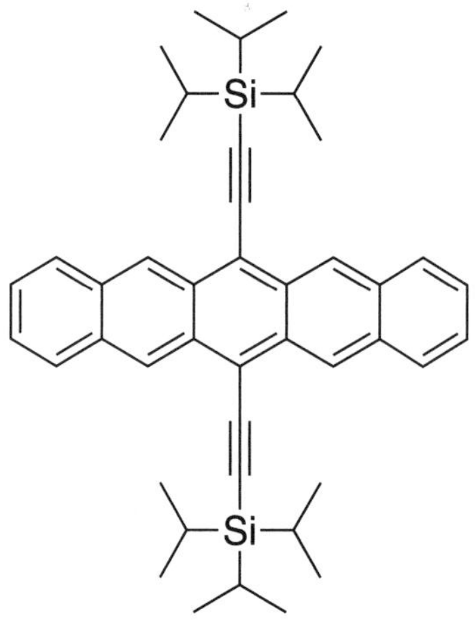

9.5.1 Blended semiconductor layers

Polymer blends have recently been introduced in transistor devices. The addition of another component essentially adds another processing step, which is not to be desired, and therefore a further layer of complication, which adds to the costs of production. However, under certain circumstances the addition of a non-conducting polymer to the active layer of a transistor can actually reduce costs (polymeric binders), and in other circumstances it is useful to mix two semiconducting layers to improve device performance.

Fig. 9.9 TIPS-pentacene.

Polymeric binders

It can be sometimes useful to add a (non-conducting) polymeric *binder* to the active layer in the hope that the electronic properties hold strong despite the addition of an insulating component. The usual rationale for this procedure is to improve processability, especially for technologies such as ink-jet printing, which require the high viscosity that polymer additives can bring. It is also helpful that the most expensive component remains the semiconductor, and so if adding a cheaper component does not impinge on device performance, then it is worth considering, especially if ease of processing has been improved by the addition of a polymer binder. Polystyrene has been shown to not deteriorate performance when added to polytriarylamine, for example. More remarkable results have been shown when isotactic polystyrene (Fig. 7.6b) is blended with a pentacene derivative, 6,13-bis(triisopropylsilylethinyl)pentacene, or TIPS-pentacene (Fig. 9.9). The reason the hole mobility remains so large in the active layer is because the TIPS-pentacene segregates to the interface with the gate dielectric. Data showing binder performance for different polymers with TIPS-pentacene is included in Fig. 9.10.

The key reason why isotactic polystyrene works so well with TIPS-pentacene is that it crystallizes more slowly than the pentacene, allowing the active component to assemble at the interface before the polymer. The crystallinity of the polymer is nevertheless important, because it inhibits an interaction with the pentacene, which can cause dislocations. There is a real need for the physics discussed in Chapter 7 to be applied to blends applied for transistors, so that we can optimize device performance.

Ambipolar behaviour

In Section 9.1 we noted that inversion is not common in polymeric transistors, because both good hole and electron mobility are needed for this phenomenon. The need for ambipolar behaviour is actually more practical. An interesting application of ambipolar behaviour is in the creation of light-emitting transistors, which we turn to in Section 10.2. Good electron mobility is usually the main challenge in such devices, because electron carriers are susceptible to oxidation from water, atmospheric oxygen, or any other contaminant that acts as an oxidizing agent. There are of course some exceptions, such as BBL (Section 3.7), but these are dwarfed by the number of available polymers with good hole mobility.

The blending of two carriers is one route to ambipolar behaviour, and has been achieved for a few different materials. Most commonly, the n-type material in ambipolar blend devices is the small-molecule PCBM, or, more formally, [6,6]-phenyl-C_{61}-butyric acid methyl ester (Fig. 9.11). This can be mixed with polymers such as P3HT and PPV to create a blend of hole and electron transporters. These will form two separate phases, and thus create distinct pathways for charge transport. Example output characteristics data for a blend of PCBM and a PPV derivative are shown in Fig. 9.12. In this particular case the hole characteristics are better than the electron characteristics, but this is not a general rule.

9.5.2 The dielectric layer

Gate insulating layers are generally either silicon dioxide or polymeric. Silicon dioxide is the classic gate dielectric used in most MOSFETs, and has a role to play in polymeric transistors. Its dielectric constant is 3.9,

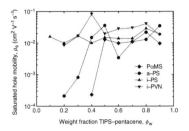

Fig. 9.10 Hole mobility data for TIPS-pentacene, solution cast with a binder polymer from tetralin. Slowly crystallizing isotactic polystyrene (i-PS) retains good hole mobility, even when the TIPS-pentacene is only 10% of the film. Isotactic poly(vinyl naphthalene) (i-PVN) also shows good performance, but the amorphous polymers poly(α-methyl styrene) (PαMS) and atactic polystyrene (a-PS) are less impressive. These data are taken from Madec et al. *J. Mater. Chem.* **18** 3230 (2008).

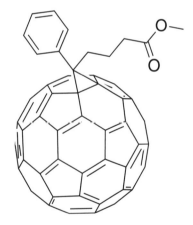

Fig. 9.11 PCBM ([6,6]-phenyl-C_{61}-butyric acid methyl ester) has good electron transporting properties.

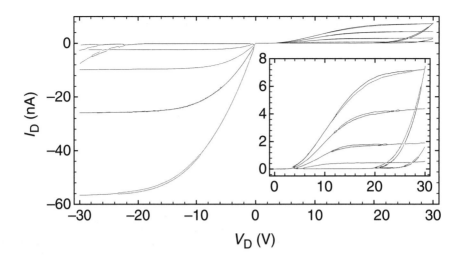

Fig. 9.12 Output characteristics characteristics for a FET consisting of PCBM as the electron carrier and a PPV derivative, poly[(2-methoxy-5-(3′,7′-dimethyloctyloxy))-*para*-phenylene vinylene] as the hole carrier. The inset gate voltages used in these experiments were 40, 35, 30, 25, 20, and 15 V for the n-channel (i.e. for the electron carrier) and −30, −25, −20, −15, and $V_G = 0$ V for the p-channel. The inset shows a magnification of the n-channel data. A decreasing magnitude of V_G corresponds to a decreasing $|I_D|$ plateau value, although the data for $V_G = 15$ V increases much more than that for $V_G = 20$ V. Although the blend exhibits ambipolar behaviour, the hole characteristics are better than those for the electron carrier. In these particular devices the hysteresis is extremely low, which is a useful quality of an FET. These data were taken from Meijer et al. *Nature Mater.* **2** 678 (2003).

which is quite high in comparison to many polymeric materials, and SiO$_2$ layers can be readily grown using commonly available technology. (Briefly, steam or oxygen are reacted with silicon at $\sim 1000°$C to produce SiO$_2$.) Even silicon oxide does not escape the need for treatment with organic materials, because pendant hydroxyl groups at its surface are a major source of traps, as discussed at the beginning of Chapter 5. Coating the surface of the silicon oxide with a self-assembled (organic) monolayer eliminates contact with these traps, and substantial effort is now being engaged in the general area of self-assembled monolayer field-effect transistors (SAMFETs).

Thin dielectric layers are an important part of miniaturization. The channel length must typically be two to five times longer than the thickness of the dielectric layer. If the dielectric layer is too thick then it becomes difficult to control the 'pinch-off' of carriers in the device. Thinner dielectric layers also allow smaller gate voltages to control the areal charge density at the interface with the semiconducting layer.

Polymeric insulating layers can take many different forms, of which PMMA is probably the most common because it has a relatively large dielectric constant, $\epsilon_r = 3.0$ and is easily processed. Other polymers can be used, and protecting them from pinholes by crosslinking (i.e. chemically attaching one polymer to another) has also been tried with considerable success; crosslinked PVA and poly(4-vinyl phenol) are routinely used as gate dielectric layers, and the dielectric layer in Fig. 9.3 was crosslinked for this purpose. Given that PMMA has one of the larger polymer dielectric constants commonly in use (although $\epsilon_r > 4$ is not rare), there is generally little scope for improvement, although hybrid polymer–inorganic systems, which involve nanoparticles dispersed in a polymeric matrix have shown some promise.

A different means of creating a gate dielectric with high dielectric constant is the use of polyelectrolytes as the gate layer. These can be crosslinked too, if necessary. The electrolytes have mobile ions within them, which results in a large polarizability and consequently high capacitance. The polyelectrolyte consists of the ions on the polymer, which are essentially fixed to the polymer, even if they have limited mobility on that molecule, and the counter-ions, which are chosen to be positively or negatively charged so as to be attracted to the gate electrode during operation. Other related developments include the use of solid polymer electrolyte layers, which consist of a neutral polymer (such as poly(ethylene oxide) and an appropriate salt). These electrolytic capacitors are very well adapted for fast switching operations.

There are numerous difficulties to be overcome in achieving the perfect insulating layer. A very thin layer must not be so thin that a current starts to flow between the gate and the semiconductor. Short circuiting (source-to-gate leakage) is often a problem with various film casting techniques (such as those discussed in Section 8.3) because impurities can often create pinholes in the layer, or provide conductive routes between the semiconducting layer and the gate electrode. The thinnest layers for which reasonable transistor behaviour has been observed are as small as 2.5 nm, so extremely thin layers can be possible. In these very thin layers the molecules have been designed so that they pack tightly on a silicon substrate, allowing very little leakage from the semiconducting layer to the gate. Although a breakthrough, costs and ease of preparation will dictate whether such methods will be important in the future.

9.6 Logic circuits

Transistors can be used in many things because of their ability to switch behaviour dependent on the signal. A small input signal can be amplified because the drain current when the transistor is switched on does not bear any causal relation to the input signal. This amplification can be used in sensors, for example, or indeed in any circuit where confirmation of the input signal is required. Although this is very useful, it is nothing compared to the phenomenal success of logic circuits. Logic circuits allow for output on the basis of one or more inputs, and this is, of course, the basis of computing. These are achieved with logic gates. An AND gate, for example, will give a *true* output if two input signals are both true. We can define what we please to be a 'true' input; it is given the binary value 1, and we might for example say that a positive input is true, and a negative input is false. If the input terminals of an AND gate are both positive, the output will also be positive, i.e. true, otherwise a

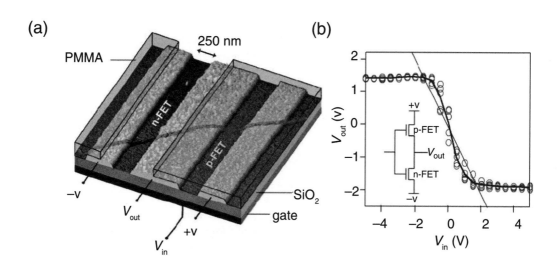

Fig. 9.13 (a) Hybrid SFM/schematic image of the device. Three gold electrodes (the two terminals and the output) are deposited on an SiO$_2$ gate. The gate electrode (input) is situated underneath the gate. A bundle of single-walled nanotubes is deposited across this, and one half covered by PMMA. The other half is treated with potassium vapour to give it n-type characteristics. (b) NOT gate switching characteristics. The straight line indicates what would be achieved for a gain of unity. Note that the gain of 1.6 is suitable for driving further circuits. Taken with permission from Derycke et al. *Nano Lett.* **1** 453 (2001). Copyright (2001) American Chemical Society.

false (negative) output (0) will be returned.

Logic gates must be able to amplify the original signal in order to have use in driving further logic gates; if an AND gate receives two 1 V signals and its true output is 0.1 V, then, after not many logic operations, noise will dominate the final output signal.

The simplest form of logic gate is the NOT gate, which is the only logic gate requiring only one input. We recall from Section 9.5 that ring oscillators are built up from NOT gates; indeed they are probably the most important application of organic inverters. A NOT gate changes a true input (binary 1) into a false output (binary 0), or a false input into a true output, and is consequently known as a *complementary inverter*.[3] In order to illustrate how organic transistors can be used to create logic circuits, we first describe an early experiment with a NOT gate based upon a single carbon nanotube. The single molecule experiment in many ways can be considered state of the art, but logic gates based on polymeric materials are routine. The nanotube NOT gate allows us to see more clearly the need for both n- and p-type behaviour in the TFTs.

We show in Fig. 9.13a the structure of the nanotube NOT gate and the switching characteristics in Fig. 9.13b. Here the input is at the gate, which is connected to an SiO$_2$ dielectric. Three gold electrodes are evaporated onto the gate; two of these are the positive and negative terminals, and the other is the output electrode. A bundle of single wall-ed nanotubes is deposited across the electrodes. (A bundle is a collection of nanotubes intertwined, rather like the fibres on a rope.) Single-walled nanotubes are p-type, but they are amenable to chemical treatment. In

[3]'Complementary' refers to the use of n-type and p-type materials making up the logic gate.

this case, exposure to potassium vapour renders the nanotubes n-type (potassium is a good electron donor). If this exposure is made on only one transistor, then the requisite hole and electron mobile transistors can be achieved. (A layer of PMMA is allowed to cover both devices and this is then removed over one transistor using a high-resolution lithographic technique, such as electron-beam ablation.)

The output characteristics in Fig. 9.13b show that for a negative input voltage, a positive output voltage will be achieved, and for a positive input voltage, a negative output voltage results. The negative input voltage cannot bring electrons to the interface in the n-type FET; it repels them, but it does attract holes in the p-type FET. As a result, only one of the two transistors is able to operate. Because of the good conductivity of the p-type FET under a negative input, the voltage drop over this FET is commensurately small. Therefore V_{out} is mostly positive; hence a negative input gives rise to a positive output. To see this, one can think of the transistors as resistors in series with resistances R_{p} and R_{n} for the p- and n-type FETs respectively. Since the total voltage drop across the device must be $2V = +V - (-V)$, we have

$$V_{\text{out}} = -V + 2V \frac{R_{\text{n}}}{R_{\text{n}} + R_{\text{p}}}. \tag{9.12}$$

For large R_{n}, which is the result of this arrangement with a negative input voltage, it is clear that V_{out} must be close to $+V$. The reader can examine the diagram and see that the opposite result must be obtained for a positive input, namely that a negative output will result.

In this particular example we see that the maximum gain is greater than one. The straight line in Fig. 9.13b shows a gain of unity. This NOT gate can be connected to other logic devices and noise would not constitute a problem, because each input would lead to a larger output. The gain is the derivative of the switching characteristics. Another example, this time for an all-polymer inverter, is shown in Fig. 9.14. Here the gain is quite impressive, and indeed, better results have been obtained with this system.

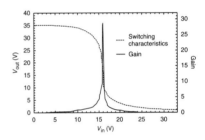

Fig. 9.14 Example switching characteristics of an all-polymer inverter fabricated from a p-channel P3HT transistor and an n-channel poly[(*N,N*'-bis(2-octyldodecyl)-naphthalene-1, 4, 5, 8-bis(dicarboximide)-2,6-diyl)-*alt*-5,5'-(2,2'-bithiophene)] (often referred to as PNDI2OD-T2) using the same commercial polymer dielectric layer. The gain is given by $dV_{\text{out}}/dV_{\text{in}}$, and is also shown. The data are taken from Yan et al. *Nature* **457** 679 (2009).

9.7 Further reading

The largest source of information on organic transistors is probably the book edited by Bao and Locklin (2007); this book mostly covers small-molecule transistors, however. Polymeric transistors are covered in the final chapter of the text by Heeger, Sariciftci, and Namdas (2010) or the chapter by Horowitz (2007). The reader interested in going beyond the material contained in this chapter may well find the highly readable review by Schwierz (2010) on graphene transistors rewarding.

Table 9.1 Transfer characteristics (V, μA) of a device of which the active layer is F8T2 and the insulator is poly(4-vinyl phenol).poly(4-vinyl phenol) The data have been extracted from Sirringhaus et al. *Appl. Phys. Lett.* **77** 406 (2000).

−60.0, 1.22	−56.1, 1.03	−52.1, 0.873	−48.1, 0.738
−44.4, 0.599	−40.1, 0.486	−36.2, 0.378	−32.2, 0.282
−28.1, 0.210	−24.2, 0.147	−20.1, 0.0930	−16.2, 0.0529
−12.1, 0.0260	−8.2, 9.71×10^{-3}	−4.1, 1.82×10^{-3}	0, 6.27×10^{-5}
3.9, 8.59×10^{-6}	7.9, 1.36×10^{-5}	—	—

9.8   Exercises

9.1. A simple transistor is fabricated with a BBL semiconducting layer and a platinum electrode. The ionization potential of BBL may be taken to be 6.0 eV, and its electron affinity to be 4.2 eV. The device is used with platinum electrodes, which have a work function of 6.4 eV. Without seeking additional information, estimate the gate voltage at which the transition from the linear to the saturation regime occurs. State any approximations that you make.

9.2. A transistor to measure the properties of the hole transporting polymer poly(9,9-dioctylfluorene-*alt*-bithiophene) (F8T2) is used in which the ratio of channel width to length is given by $W/L = 10$ and in which gold electrodes are used, so the work function is 5.1 eV. A 280 nm thick layer of silicon oxide (with dielectric constant $\epsilon_r = 3.9$) is used. A voltage across the source and drain of 6 V is applied, and saturation behaviour occurs at a gate voltage given by $V_G > 1.05$ eV. When $V_G = 2.0$ eV, a drain current, $I_D = 3.9 \times 10^{-10}$ A is measured. You are told the ionization potential of F8T2 is 5.5 eV. What is the electronic band gap of F8T2 and what is its field-effect (hole) mobility?

9.3. The subthreshold swing can be related to the electrical properties of the transistor by

$$S^{-1} = 57 \left(1 + \frac{C_s}{C_i} \right), \tag{9.13}$$

where S^{-1} here has units of mV/dec, and C_s is the total (areal) capacitance of traps both in the semiconductor, and at its interface with the insulator. Transfer characteristics for an example device are listed in Table 9.1. Assuming the insulating layer is 1.0 μm thick, calculate the trap capacitance for this device. (The dielectric constant of the insulator layer may be taken as $\epsilon_r = 4.2$.)

Calculate the threshold voltage for the same device. Do the data correspond to a linear or saturated device?

Optoelectronic devices

10

The presence of an energy gap is intimately linked with optoelectronics because light can be absorbed and emitted at wavelengths commensurate with this gap. The emission of light is not necessarily as straightforward as its absorption, because there are non-radiative processes competing with the emission of photons, as discussed in Chapter 4. This band gap leads to enormous market opportunities for new technologies. The main two areas of research are in the area of photovoltaic cells and light-emitting diodes. Because of their processability and relatively low expense, polymeric devices are attracting a great deal of attention.

Organic light-emitting diodes have many attractive qualities beyond solution processing. They can be seen in more and more laptop screens and televisions. Although most of these are created by using smaller organic molecules, full-size television sets based upon polymer light-emitting diode technologies have been demonstrated. Polymer LEDs remain very efficient, consuming little power, although their limited lifetime does remain a challenge to scientists and engineers despite recent progress. Ultimately, their low power requirements mean that they should have much to offer as we try to become much more energy efficient in future years. A 2006 report from the International Energy Agency pointed out that light accounted for $\sim 17.5\%$ *of the world's energy use*; some 2200 TWh/year (7.9×10^{18} J/year). Breakthroughs in LED technology, especially in those involving white light, will make a huge difference to our energy demands.

Of course, reducing our demand on fossil fuels is an even greater challenge, and photovoltaic technology is an important element of this. Solar cell technology is presently dominated by inorganic semiconductors, but these technologies can be expensive, not just due to sample fabrication, but also due to the availability of some materials such as selenium. Compared to these technologies, conjugated polymers are attractive candidates for the development of solar cells. However, the (power conversion) efficiency of polymer photovoltaic devices is currently quite unimpressive and significantly less than 10%.

10.1 Light-emitting diodes

LEDs are simple devices. As schematized in Fig. 8.3b, holes are injected at an anode and electrons at a cathode. They meet to form an exciton, which decays radiatively. In an ideal LED there would be as many holes as electrons, and they would all meet to form excitons. Again, in an ideal LED, these excitons would all decay radiatively, rather than decay through collisions. Certainly, their efficiency would not be restricted to 25% due to the formation of triplets. The photons emitted would also be emitted in the direction we should require; i.e. the out-coupling would be optimized. Finally, they would work at low driving voltages and would also last for as long as they were needed.

10.1.1   Units and definitions

The basic requirements for LED technologies are that they be bright with good colour saturation (colour purity), efficient, have a long lifetime, and work with low operating voltages. *Brightness* is a concept that needs to be considered in a little more detail, because it can be confusing. The sense in which we might perceive an object to be 'bright' is perhaps best thought of in terms of luminance, which is the emitted luminous intensity per unit area and has SI units of cd/m^2. The candela (cd) is one of the seven base units in physics (alongside K, m, kg, A, mol, and s), and 1 cd has the rather convoluted definition of being the luminous intensity of a source that delivers 1.464 mW over an (solid) angular spread of 1 steradian at 5.4 THz. (Some readers might want to be reminded that there are 4π steradians over all directions.) The 1.464 mW ($1.464 = 1/683$) is an artefact of an older definition, and 5.4 THz corresponds to green light, and is where the human eye is particularly sensitive. A cheap candle emits about 1 cd of luminous intensity, hence its name; *candela* is the latin word for candle.

It is perhaps interesting to note that the candela can be replaced with W/sr (sr being steradians), which might make one query its role as a base unit. However, the two should not be confused. The candela is the measure of luminous flux, with units of lm/sr, and 1 W is not the same as 1 lm.[1] The Watt is a satisfying unit to a physicist, because it is indisputable. The lumen (lm) is a satisfying unit for an engineer because it is practical. One needs more power than 1.464 mW of red light to equate to 1 lm of green light, simply because the eye is less sensitive to red light. Ultraviolet and infrared do not contribute at all, because they cannot be seen. The weighting factors for each colour to equate to 1.464 mW at 5.4 THz are predefined by a standard governed by the *Commission Internationale de l'Éclairage* (CIE) to represent the sensitivity of the human eye. This dimensionless function is shown in Fig. 10.1. To sum up the importance of this paragraph, a computer screen has many pixels; for example, a small XGA screen has 1024×768 pixels. If a pixel emitting red were adjacent to one emitting green, we should not see it if their emissive power were of the same *radiant intensity*. We therefore need a system of units so that the sensitivity of the human eye can be accounted for.

For an LED there are more practical issues too, such as their current efficiency and the current density associated with the device. The *current efficiency* has units of cd/A, which is clear enough. (Sometimes the current efficiency is referred to as the luminance efficiency.) A large

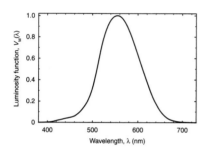

Fig. 10.1 The CIE standard (photopic) luminosity function, $V_M(\lambda)$, illustrates the sensitivity of the eye to different wavelengths in good light. The function is defined so as to have a value of unity for $\lambda = 555$ nm.

[1] The candela and lumen are not unique in physics for having apparently identical units to another property; one should remember from mechanics that torque has units of Nm, but 1 J = 1 Nm. Torque is a vector and energy a scalar so the two cannot be identical.

current driving a device is clearly inefficient. Another efficiency that is associated with LEDs is the *luminous efficiency*, sometimes known as the *power efficiency*, which has units of lm/W. A candle has a luminous efficiency of 0.3 lm/W compared to sunlight with 93 lm/W, which would be about the same as a very good LED. (The reader should spend a few moments to convince him or herself that the luminous efficiency of an ideal 5.4 THz radiator is 683 lm/W.) A point worth noting here is that the current and luminous efficiencies have dimensions, whereas many other efficiencies such as the electroluminescence efficiency defined in eqn (4.10) are dimensionless.

The current density (with units A/m^2) is used to describe the operation of the device. The current density is generally proportional to the luminance, with the current efficiency being the constant of proportionality. It is often used in the $I(V)$ behaviour of devices. The power density (with units W/m^2) is a measure of the energy output of a device, and this is perhaps a more important parameter in photovoltaic devices, where it represents the light input into the device and is required for the efficiency (Section 10.3.4). The power density of the sun when directly overhead at the tropics (but without atmospheric absorption) is taken as 137 mW/cm^2, although with an atmosphere and away from the tropics this will decrease to perhaps a little less than 100 mW/cm^2. Above the atmosphere, we refer to an air mass of zero (AM0), and we return to this in Section 10.3.1.

10.1.2 Single-layer devices

Polymer LEDs are certainly currently good enough for virtually all applications in terms of their luminous qualities and lifetimes. For example, a high-quality computer display or television might have a brightness of 400 cd/m^2, whereas polymer LEDs can be created with brightness of better than ten times this amount. (100 times is even possible but would generally be at undesirably large drive voltages.) Polymer LEDs are perhaps not yet as efficient as their inorganic competitors, and 100 lm/W is something of a goal for the community, but is certainly not unobtainable and OLEDs are approaching this value through improvement in out-coupling methods.

The first paper on conjugated polymer light-emitting diodes was published in 1990 and demonstrated the emission of a green-yellow light from a single layer of PPV sandwiched between indium oxide and aluminium electrodes. The electroluminescence quantum yields of these devices were at best 0.05%—a very poor performance by today's standards. Much has changed to improve the quality of device, including improved polymer chemistry, better hole-injection layers based on PE-DOT:PSS on ITO electrodes, and more air-stable cathodes. However, the fact that the device was a single layer of one component is a significant weakness. Firstly, PPV is primarily a hole-transporting material, so many holes make their way to the cathode without meeting an electron with which to form an exciton. This means that the efficiency of

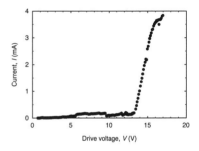

Fig. 10.2 The current as a function of drive voltage for the original PPV-based LED. Taken and adapted from Burroughes et al. *Nature* **347** 539 (1990).

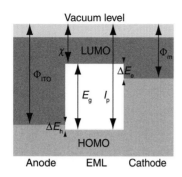

Fig. 10.3 The band structure of a single-layer device. For a material with good hole mobility, the band structure for an emitting layer (EML) between a metal cathode and an ITO anode would be reasonably well schematized by the above diagram. ΔE_h and ΔE_e are the injection barriers for holes and electrons respectively. Otherwise, the symbols have the same meanings as those shown in Fig. 4.11.

the device is determined by the electron mobility of the polymer, rather than its hole mobility. Secondly, the band gap of most hole transporters is such that the electron injection barrier is rather large, which means that significant driving voltages must be applied across the polymer film in order to obtain a current. The $I(V)$ data for the original PPV device are shown in Fig.10.2, which reveal that 13 V needs to be applied across the device in order for it function. Such a large voltage would not be acceptable for commercial use.

The band structure (ignoring the effects of band bending, discussed in Section 5.5.1) of an LED is shown in Fig. 10.3 (essentially the same as Fig. 4.11), highlighting the different injection levels, ΔE_h for holes and ΔE_e for electrons. The reader may wish to refer to Section 5.5.3 for a brief discussion of the pertinent matters concerning the choice of cathode. At the other side of the device, ΔE_h may be as low as zero, in which case one has an Ohmic contact. (ITO, which is the most common anode used in LEDs, can contaminate the device, so it is also advisable to ensure that the exciton formation takes place away from the anode.)

Because electrons are often less mobile than the holes, excitons formed in these devices are usually formed near the cathode. These excitons are often and easily quenched in a radiationless process. It is understood that excitons within about 10 nm of the cathode are very capable of decaying by energy transfer to the cathode, which causes a depletion in excitons close to the cathode, and acts as a chemical potential gradient, causing more excitons to drift within range of an interaction with the cathode. The energy transfer is therefore much more complicated than that discussed in Section 4.5 because the cathode functions like a mirror and the excitons quench themselves.

A working (biased) single-layer LED is essentially that schematized in Fig. 5.10. We require efficient injection of the holes and electrons into the device. Once injected, we require them to meet, to form a singlet state, and to decay via only a radiative process, all of which are characterized by the internal electroluminescence efficiency defined by eqn (4.10).

10.1.3 Double-layer devices

The problem with single-layer devices is that it is hard to find true ambipolar behaviour, and so the charges moving in each direction are not balanced. One solution to this problem is to create a bilayer, with one layer optimized for electron transport, and the other layer for hole transport, as illustrated in Fig. 10.4. This way excitons will be formed away from the cathode, and mainly at the interface between the two components. Only one of the two layers needs to be emissive, although putting this layer at the cathode better protects the device against ITO contamination (Section 5.5.3). This ITO contamination problem is largely solved by using an additional hole-injection layer such as one of PEDOT:PSS.

With bilayer devices, electron injection is less of a problem for device performance, because polymers with good electron mobility can be used as the cathode layer and these tend to have a larger electron

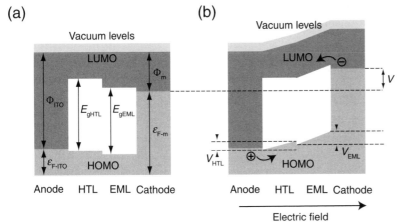

(a)   (b)

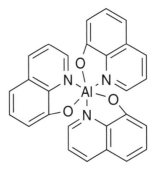

Fig. 10.4 A double-layer device, consisting of a hole-transporting layer (HTL) and an electron-transporting layer, which also acts as the emissive layer (EML). The energy levels of the device are shown in (a) and how they respond to an applied electric field are illustrated in (b). A voltage V is applied across the device which results in voltage drops V_{HTL} and V_{EML} across the HTL and EML respectively. The Fermi energies of the cathode and (ITO) anode are ϵ_{F-m} and ϵ_{F-ITO} respectively.

affinity, which brings the LUMO closer to the work function of the cathode. A common example of an emissive layer in organic LEDs is tris(8-hydroxyquinoline)aluminum(III), which is usually denoted as Alq3 and is shown in Fig. 10.5.

We note from Fig. 10.4 the effect of the applied voltage across the device. In the event of a positive voltage at the cathode and the anode remaining at an earth voltage, an electric field in the direction of the cathode will be formed. A positive charge will travel in the direction of the electric field, so this corresponds to a forward bias. The applied voltage, V, shifts the Fermi level of the cathode up by an amount eV. Note that the work function does not change; it still requires the same energy to eject an electron as it did before the electric field is applied. If ejected, that electron will have greater potential energy than one ejected with no applied voltage. There is no change to the anode under the bias shown in Fig. 10.4b. The voltage falls across both the hole-transporting (HTL) and emitting layers (EML), so

$$V = V_{HTL} + V_{EML}, \tag{10.1}$$

where V_{HTL} and V_{EML} are the voltage drops across the two respective layers. Here the two layers are shown to have the same thickness, which is not necessary, but is helpful if one has balanced charge injection. In this situation, we see from Fig. 10.4b that $V_{HTL} < V_{EML}$. Again, $V_{HTL} < V_{EML}$ need not be the case but it is not uncommon because it is a result of a greater hole mobility in the HTL and a greater injection barrier to electrons at the cathode than the hole-injection barrier at the anode. If, for example, the hole mobility was greater than the electron mobility, there would be more holes being transmitted than electrons through the device, which would have a detrimental effect on its efficiency. One could increase the EML thickness to increase the number of electrons in the device, but since they are largely expected to combine at the interface between the HTL and EML, this would not improve performance significantly, because the electron current would remain low. The

Fig. 10.5 Tris(8-hydroxyquinolinato)-aluminium (Alq3) is an organometallic compound with good electron mobility that emits green light, which makes it a popular choice in conjunction with a polymeric or small-moleculehole carrying layers in OLEDs. The reader may note the six bonds to the central aluminium atom. This apparently contradicts the valency of aluminium; Alq3 is a bidentate coordination complex (Section 6.3).

larger voltage drop across the EML does have the benefit of improving electron injection because it decreases the potential barrier compared to a smaller V_{EML}.

Band-structure diagrams such as the one shown in Fig. 10.4 are forms of energy-level diagrams and show the direction electrons will travel in order to reduce their energy. Of course, holes have a different charge, and the diagram is designed with electrons in mind. Holes reduce their energy by moving *up* the diagram. If we consider Fig. 10.4 as an example, under operating conditions, there is a (small) barrier to hole injection, but once injected it can flow unimpeded to the cathode. The hole has no problem in accessing the emissive layer, because there is an energetic benefit in crossing the boundary. Electrons on the other hand have to overcome an, again small, energy barrier to be injected. Once they have tunnelled through this injection barrier, they face another barrier to the hole transporting layer. This barrier should be significantly bigger than the injection barrier because it is desirable that both holes and electrons be contained within the emissive layer. The injection barriers contribute to the turn-on voltage of the device. Below this voltage the electric field across the device is too small for both electrons and holes to be injected into the device. Once injection barriers are overcome, electroluminescence can occur.

It is possible to use PEDOT:PSS as the hole transport layer in an OLED without another transport layer, but this is generally not done, because an additional HTL has the benefit of confining the excitons at the required interface. Sometimes one refers to these hole-transporting layers as *electron-blocking layers* because their poor electron transport properties keep the electrons in the region whereby light emission is desired. The formation of an exciton anywhere other than close to or within the emissive layer is undesirable in an LED because light of the wrong wavelength (or no light at all) is emitted when the exciton decays. A simple criterion for a good electron blocking layer is a large band gap, because this provides an energy barrier to electrons from easily moving towards the anode. It is also helpful if the voltage drop is largely across the EML. A small drop across the HTL means that the tunnelling barrier for electrons to travel into the HTL is also large. The example shown in Fig. 10.4 might indicate that the HTL does not have a large enough band gap because it is only slightly larger than that of the EML, but, nevertheless, the tunnelling barrier at the HTL/EML heterojunction might still be good enough for this layer to work effectively as an electron-blocking layer.

Finally, we remind the reader of the caveats noted above. The emissive layer does not have to be the electron-transporting layer; in the case of the use of a PEDOT:PSS hole-injection layer, there is no practical reason why we should not have an emitting layer that favours hole transport. There is also no reason why the hole-transporting layer should have better performance than the electron-transporting layer, although we accept that this is the case in the majority of situations. Nevertheless, the schematic diagram shown in Fig. 10.4 illustrates the basic physics of charge injection and transport across a two-layer LED. One point that we have so far neglected is that, as for transistors, contact potentials can play a detrimental role in these devices. We note, for example, that there is no potential drop across any of the interfaces; the drop falls across the entire layer equally. However, there may well be traps at these interfaces which can leave residual charge present, and a potential may form. We have also ignored band bending here. These effects can either improve or deteriorate device performance, but their control is a rather complicated matter.

A requirement in the fabrication of a solution-processed double-layer device is a good choice of solvents. The normal method of preparation is to coat the anode with the first film (HTL), by one of the techniques discussed in Section 8.3, with doctor-blading (Section 8.3.1) and spin-coating (Section 8.3.3) popular because of their ease of use and the relatively good quality films that result from being formed in these ways. The same methods can be used to deposit the EML, but clearly it is not desirable for the solvent for the EML to dissolve the HTL. It is therefore important to use an orthogonal solvent (Section 8.3.3) for the second layer. This solvent does not have to be wholly insoluble in the first layer; a small amount of solubility will cause interfacial mixing between the two layers, which should increase the size of the exciton recombination region, which increases the device efficiency. Such miscible structures are discussed below.

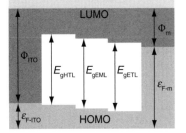

Fig. 10.6 Energy-level diagram of a triple-layer device, which consists of a hole-transporting layer (HML), an electron-transporting layer (ETL), sandwiching an emissive layer (EML).

10.1.4 Other structures

It is possible to consider even more complicated structures such as three-layer structures. In this case, one would sandwich the emitting layer between the hole-transporting layer and an electron-transporting layer (Fig. 10.6). Under such a geometry, the region where excitons would be formed is within the emitting layer. This has benefits for device efficiency, but is rather difficult to fabricate because of the solvent requirements (two orthogonal solvents would be required for the deposition of the EML and ETL), and also simply because it adds an extra step to the device preparation.

Blend films represent an attractive alternative for efficient device performance. We showed in Fig. 8.3b how this might work. The morphology of hole-injection layers is an area in which some research has been performed, but the current discussion is still somewhat speculative. These smooth gradients from anode to cathode mean that traps can be avoided. They also spread out the region whereby electrons and holes can recombine. (This is the same reason why it can be advantageous to use a not completely orthogonal solvent in Section 10.1.3 above.) The reason why this discussion remains speculative is that the requirements for the structure shown in Fig. 8.3b are two miscible polymers, one of which has an affinity for the ITO cathode and the other for an air interface. Should these requirements be realizable, it is not unrealistic to imagine such a scenario. Unfortunately,

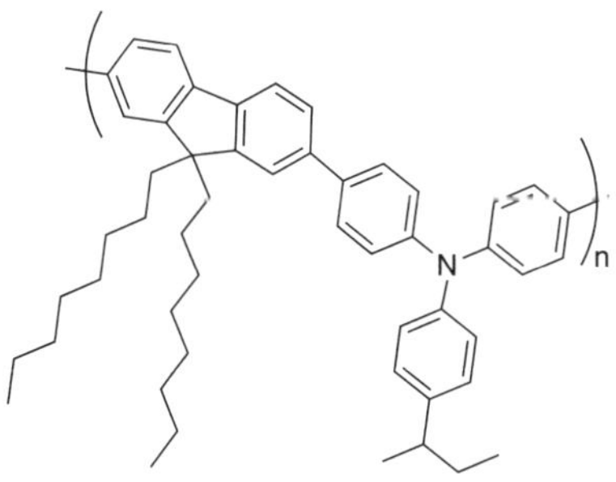

Fig. 10.7 The chemical structure of poly(9,9-dioctylfluorene-*alt*-*N*-(4-butylphenyl)diphenylamine) (TFB), a common hole-transporting polymer in LEDs. This polymer is not fully conjugated and so its charge transport properties are not expected to be due to band transport.

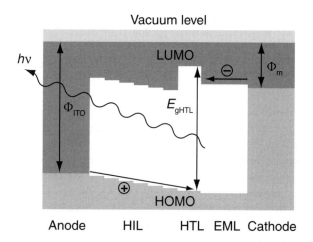

Fig. 10.8 LED energy levels for a device with good performance due to graded hole injection. The gradient in HOMO in the injection layer (HIL) allows facile transport for holes. The hole-transporting layer (HTL) has a large band gap, which means that it functions as an electron blocking layer, confining electrons to the interface where recombination should occur. If the HIL is based on a synthetic metal such as PEDOT, as is often the case, a LUMO and band gap (as shown here) is inappropriate and should be replaced by a work function.

however, most polymeric mixtures are immiscible. An important blend that works for an all-polymer LED is of F8BT and F8, for which we have already shown images in Fig. 8.16. This system is nevertheless immiscible. For LED experiments the best performance is obtained with 95% by weight of F8. The large asymmetry in this 95% blend reduces the degree of immiscibility, although that would depend on the relative chain lengths of the two polymers. Another immiscible polyfluorene-based blend is that of F8BT with poly(9,9-dioctylfluorene-*alt*-*N*-(4-butylphenyl)diphenylamine) (TFB, shown in Fig. 10.7). In both of these blends, F8BT is the material with the better electron mobility.

A demonstration how good device performance can be achieved with a controlled gradient hole-injection layer has been achieved with the layer-by-layer technique that we discussed in Section 8.3.4. An ideal LED based upon a graded hole-injection layer would have energy bands that look like those shown in Fig. 10.8. The hole-transporting layer is able to inhibit electron transport by its large band gap. Electrons do not have the energy to tunnel into the HTL and so are confined to the region near the interface between the EML and HTL. A layer-by-layer LED based on a graded hole-injection layer would have the form as shown in Fig. 10.9. Note that the injection layer exhibits a decrease in the depth of the HOMO, meaning that it is energetically unfavourable for a hole to travel through into the HTL and subsequently the EML. A forward bias voltage will of course decrease the thickness of this barrier, allowing tunnelling.

Example data illustrating the performance of devices using a graded HIL are shown in Fig. 10.9b. This performance is impressive, but it is clearly impractical to use the layer-by-layer method for the large-scale production of devices. Nevertheless, self-assembly routes involving wetting (Section 8.2) can in principle be used to create a smoothly graded layer. Nevertheless, good as this result is, one can easily question whether or not it is worth investing much effort into morphological

(a)

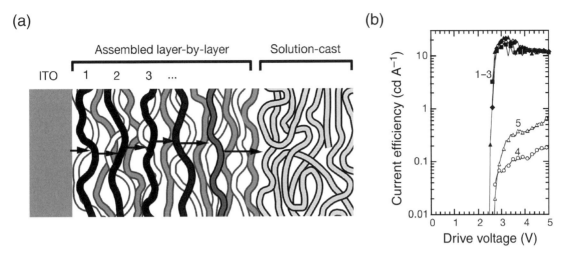

(b)

Fig. 10.9 The layer-by-layer method has been used to create a hole-injection layer that is particularly efficient. In this case, the PEDOT (black) and poly(p-xylylene-α-tetrahydrothiophenium) (grey) are cationic, whilst poly(styrene sulfonic) acid is anionic. (a) The layer-by-layer technique is used to deposit the polymers, but each time, the PEDOT is partially de-doped, so that it is less charged than during the previous deposition, signified in the schematic illustration as a lightening of the shading. The solution-cast layer is a PPV derivative, designed to emit orange or green light. (b) The current efficiency of such devices (1-3), with each having different doping, is measured in comparison to a device with no hole-injection layer (4) or one with an equivalent thickness PEDOT:PSS layer (5). Although the PEDOT:PSS layer improves efficiency, there is an enormous improvement in the performance with the graded hole-injection layers. Taken and adapted from Ho et al. *Nature* **404** 481 (2000). Adapted by permission from Macmillan Publishers Ltd, copyright (2000).

control or whether it is better to use research effort into choosing the best materials. The right choice of electrodes can provide the best hole and electron injection so that a single-layer device can equal this level of performance; for example, a single F8BT layer has been shown to have a current efficiency of 23 cd/A.

10.1.5 White light

White-light emission is a crucial area of current research because of the importance that white light plays in general lighting, and of course the large amount of energy used in lighting our homes, streets, and places of work. White organic light-emitting diodes (WOLEDs) are dominated by small molecules rather than their polymeric counterparts (WPLEDs), but nevertheless there is still a place for polymeric white LEDs because of the intrinsic processability advantages.

Blends

No molecule emits white light, although broad emission is possible. To this end a mixture of emitters is required, which raises some problems. If one emitter is particularly intense, one can reduce its concentration, but if one mixes a blue emitter with a red emitter, red emission might dominate through the blue de-exciting via non-radiative energy (Förster) transfer. The design of such devices is crucial. Block copolymers offer

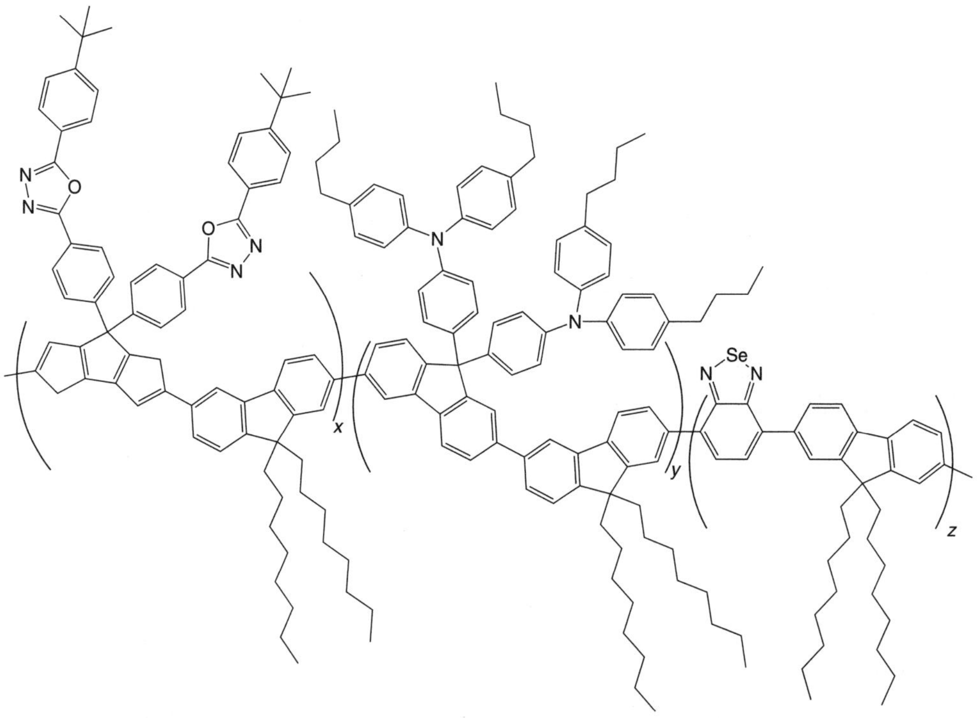

Fig. 10.10 It is possible to obtain white-light emission with a mixture of statistical copolymers of different polyfluorenes. This particular copolymer, without the benzoselenadiazole-containing component ($z = 0$), when $x = y = 0.5$, emits blue light, but when $x = y = 0.45$ and $z = 0.1$ it emits orange light. The two copolymers (i.e. the blue and orange emitters), when mixed in the right combination, emit white light. The triphenylamine pendant groups (on the second monomer denoted y) offer very good hole transport properties, and balance between electrons and holes is achieved by the electron-transporting oxadiazole groups on the first monomer, denoted x. (The reader might note the oxygen present in these moieties, whose purpose was described in Section 3.7.) The full chemical title for this molecule is not for the faint-hearted or for those easily short of breath: poly([9,9-bis(4-(5-(4-*tert*-butylphenyl)-[1,3,4]-oxadiazol-2-yl)phenyl)-9',9'-di-*n*-octyl-[2,2']-bifluoren-7,7'-diyl]-*co*-[9,9-bis(4-(*N*,*N*-di(4-*n*-butylphenyl)amino)phenyl)-9',9'-di-*n*-octyl-[2,2']-bifluoren-7,7'-diyl]). The benzoselenadiazole-9,9-dioctylfluorene unit (z) terminates the polymer at both ends. (The reader will note that 9',9'-di-*n*-octyl-[2,2']-bifluoren-7,7'-diyl is the same as 9,9-dioctylfluorene.) This polymer is prepared by a standard Suzuki coupling procedure, as described in Section 6.3.2.

one possible solution because they will generally microphase separate rather than mix. However, the length scales of block copolymers (a few nm) do not preclude Förster transfer, merely limit it. Of course, the sharp interfaces inherent in block copolymers can also give rise to unwanted traps, reducing the efficiency of the diode. Similarly, the alternative possibility of having mesoscopic discrete layers is also imperfect because it is not optimal for current flow within the diode. Graded structures, such as the one shown in Fig. 8.3b, remain relevant; clearly the intermixing would encourage Förster transfer, although a two-component blend layer is not a particularly efficient route to creating white light. Nevertheless, such limitations have not stopped research in the area; over the past few years there have been reports of blended polymers as a route to white-light emission.

Single-layer WPLED devices can be fabricated with conjugated polymers functionalized with appropriate units; for example polyfluorene is a blue emitter, but it can be mixed with dye-functionalized polyfluorenes that emit at longer wavelengths, for example, orange dyes are good for two-component white light emission (Fig. 10.10). (Mixtures of red, green, and blue emitters are also possible for white light.) The orange-emitter acts as a dopant for the blue-emitting polyfluorene. The amount of dopant present will affect the level of energy transfer taking place, and thus the spectrum. The internal quantum efficiency of these devices is generally less than 5%, with the main limitation being the (thermal) loss of triplet excitons in collisional processes. (Data for a good blend device are shown in Fig. 10.11.) Another important limitation would be unbalanced charge transport. Good quality white light depends on control over the energy transfer between the components, which in turn depends on their respective concentrations. Ambipolar polymers can be used to improve the balance of hole and electron transport, which are usually achieved by adding appropriate electron-transporting moieties to the side chains. Of course, one should always consider the out-coupling as a factor for improvement in all such devices.

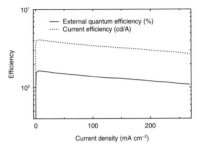

Triplet harvesting

Despite the WPLED blend shown in Fig. 10.10 being no more efficient than standard polymer LEDs, WOLEDs (which by definition includes WPLEDs) generally have much better efficiency than the one-colour alternatives discussed above because of their use of triplet harvesting, which is employed in WOLEDs to circumvent the problem of the spin-forbidden radiative decay of excitons that we discussed in Sections 4.1 and 4.2. We know that the decay of a triplet to a singlet state is spin-forbidden. That does not mean that the decay can never take place, in which case radiative decay is through phosphorescence. Phosphorescence occurs more rapidly when heavier elements than those usually found in organic molecules are involved. To this end it is worthwhile thinking a little about the physics behind why such transitions might take place, albeit slowly.

Degeneracy is often broken in atomic or molecular physics through coupling interactions with something else. The triplet and singlet states are not degenerate because triplet states have lower Coulombic energy than singlet states since the electrons are further apart (Section 4.1). Similarly, these triplet states are not degenerate because the electrons interact with nuclei through spin–orbit coupling. In the rest frame of the electron, the nucleus is a big charge orbiting it, which means that the electron is in the middle of a current, from which a magnetic field must exist. We recall that any moving charge gives rise to a magnetic field, and the relation between the two is given by Ampère's law,

$$\mathbf{J} = \frac{1}{\mu_0} \nabla \times \mathbf{B}, \qquad (10.2)$$

where $\mathbf{J}$ is the current density, μ_0 is the magnetic constant (permeability of free space), and $\mathbf{B}$ is the magnetic field resulting from the current.

Fig. 10.11 Luminescence and external quantum efficiencies of the blend of the polymer shown and described in Fig. 10.10. The dopant is the orange emitting polymer ($x = y = 0.45$ and $z = 0.1$) and is present at 9% by weight. The maximum EQE obtained by this device is quite low, only 1.64%, indicating the limitations of the technology. The reader is reminded not to confuse the external and internal quantum efficiencies, with the latter given by eqn (4.10). The current efficiency is typical for an LED, with values intermediate between the optimized and unoptimized PPV devices shown in Fig. 10.9b. Taken and adapted from Shih et al. *Adv. Funct. Mater.* **16** 1582 (2006).

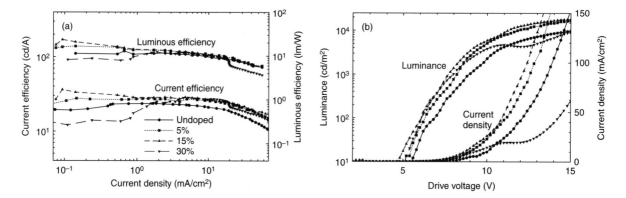

Fig. 10.12 A good-quality white polymer LED made up of an emissive layer, a hole-transporting layer (PEDOT:PSS), and an electron-transporting layer (a polyfluorene derivative doped with lithium carbonate). The emissive layer contains poly(vinyl carbazole) mixed with a small-molecule electron carrier for charge balance. The dyes used are osmium (orange) and iridium (blue) phosphorescent complexes. (a) Luminous and current efficiencies. The best performance is the material in which the electron-transporting layer is doped with 15% Li_2CO_3 by mass. (b) Luminances and current densities as a function of the drive voltage. These data were kindly provided by Professor Fei Huang and were originally published in Huang et al. *Adv. Mater.* **21** 361 (2009).

The spin of the electron (its intrinsic angular momentum) can interact with the magnetic field through its magnetic dipole moment, $\boldsymbol{\mu}$ to obtain a splitting energy

$$\Delta E_L = -\boldsymbol{\mu} \cdot \mathbf{B}. \tag{10.3}$$

A correction is added to account for the curved motion of the electron. The size of the splitting is dependent upon the size of the magnetic field, and a larger nucleus (i.e. with more protons) creates a larger magnetic field, and so a greater splitting occurs. The size of the splitting scales with atomic number as Z^4. Also, the greater the splitting, the faster the (spin forbidden) radiative transition becomes, meaning that thermal de-excitation is less likely.

The addition of inorganic components of large mass increases this spin–orbit coupling, and therefore the amount of phosphorescence occurring is also increased. Iridium is a common component in WOLEDs, but others such as platinum can also be used. These metals are generally decorated with an organic component, which improves compatibility, impedes aggregation, and reduces susceptibility to energy transfer, as we discuss below. In principle the use of organometallic compounds can increase the efficiency of a device by a factor of four if the triplet states decay radiatively with a similar probability to the singlet states. These are different transitions, so the colour may not be as pure as that coming from an OLED without triplet harvesting. For this reason triplet harvesting molecules are mainly used in WOLEDs. These can be mixed with other emitters to obtain the desired white spectrum; we generally prefer white light to be as close as possible to daylight. The distribution of wavelengths that make up white light is defined by the CIE as coordinates on a colour plot. We shall not discuss this definition any further.

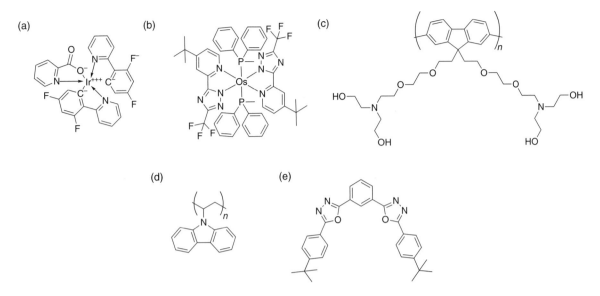

Fig. 10.13 The molecules used for the device data shown in Fig. 10.12.   (a) Bis(4,6-difluorophenylpyridinato-N,C^2)picolinatoiridium (FIrpic) is a blue-emitting phosphorescent iridium complex. The arrows can be taken as bonds; the molecule is not planar. (b) Os-O is an orange emitter, which combines with FIrpac to give white light. Its full chemical name is osmium(II) bis[3-(trifluoromethyl)-5-(4-*tert*-butylpyridyl)-1,2,4-triazolate]diphenylmethylphosphine. (c) The polyfluorene host, poly[9,9-bis(2-(2-diethanolamino ethoxy)ethoxy)ethyl) fluorene], makes up the electron-transporting layer. This polymer is doped with Li$_2$CO$_3$. The host material for the iridium and osmium white-emitting complexes consists of (d) PVK, repeated from Fig. 5.2c, and (e) 1,3-bis[(4-*tert*-butylphenyl)-1,3,4-oxidiazolyl]phenylene (OXD-7). The OXD-7, which has a good electron mobility, is used to balance charge transport in the emitting layer.

Triplet harvesting compounds are typically organometallic, decorated with a macromolecular hydrocarbon which allows efficient charge transport to the metal, as well as (entropically) inhibiting aggregation, which we discussed in the case of brushes in Section 8.3.7. Sometimes the organic component is useful in keeping these broader emitters away from other emitters so that energy transfer is not a major concern.

The typical structure of a white-light-emitting diode with triplet harvesting is no different than for ordinary LEDs such as those schematically illustrated in Fig. 10.4 whereby a hole transport layer and emissive layer are used in conjunction with a transparent anode and an appropriate metal cathode. Often an electron-transporting layer is used (Fig. 10.6) for improved performance. Under these circumstances very good performance can be obtained, as shown for example in the data of Fig. 10.12.

The devices from which the data shown in Fig. 10.12 were obtained are well characterized, so this system is a particularly useful example of a white polymer LED. We shall discuss these data in a little more detail as a case study. The molecules used in this device are shown in Fig. 10.13. For these devices the electrodes are standard, with an ITO anode and a metal cathode, here consisting of a very thin layer of barium and aluminium making the bulk of the cathode. The hole transport layer is PEDOT:PSS, which we have already discussed. The electron-transport layer is poly[9,9-bis(2-(2-(2-diethanolamino ethoxy)ethoxy)ethyl) fluorene] (PF-OH) doped with Li_2CO_3. The use of salt as a dopant in the ETL is a common procedure. In this case, the Li_2CO_3 effectively adds electrons to the polymer by increasing the Fermi level, whilst the HOMO and LUMO remain unaltered. This leaves the Fermi level no longer in the middle of the band gap, effectively improving charge transport. Note the large amount of oxygen in the solubilizing chains in the PF-OH; this is in itself not likely to improve the electron mobility since it is not part of the conjugation, but it is likely to cause the formation of dipoles, which aid electron injection by reducing the barrier height between the electrode and the emitting layer. The emitting layer should provide balanced charge transport with impressive hole and electron mobilities, so that excitons may form around the phosphorescent molecules. PVK mixed with 1,3-bis[(4-*tert*-butylphenyl)-1,3,4-oxidiazolyl]phenylene (Fig. 10.13e) satisfies this requirement.

These WPLEDs have a maximum external quantum efficiency of 14.2% for forward viewing, or 21.4% in all directions. Given that the internal quantum efficiency of eqn (4.10) is the product of three different efficiencies and does not include parameters external to the emitting layer, and given that without the phosphorescent materials $\eta_{EL} < 0.25$, ignoring the effect of possible intersystem crossings to the single state, this is a very impressive result. The forward-viewing EQE reflects the expectation that the device be used in an environments other than hanging from the ceiling or in a lamp, like incandescent bulbs.

The maximum luminous efficiency from these devices is 31.5 lm/W in all directions, which is not as good as daylight (93 lm/W) but is nonetheless a very acceptable performance. From Fig. 10.12b we note a turn-on voltage of 5 V, which is rather large, but not excessively so, being between those for the two PPV values shown in Figures 10.2 and 10.9. The brightness (luminance) of these devices can be huge, but large drive voltages are required to achieve values in excess of 10^4 cd/m^2. (A good laptop computer display will have a brightness of more than 300 cd/m^2.)

With device structures such as the one described above, the concentration of triplet harvesting complexes must be kept low so as to restrict Förster transfer. Efforts are thus being made to isolate these complexes by encapsulation techniques so that better and more reliable control over device performance can be achieved, as well as brighter devices that use less power. How this may best be achieved is currently speculation, and so we shall not pursue it here.

Field-induced polymer electroluminescence

Field-induced polymer electroluminescent devices are capable of using an alternating current input. They consist of dielectric layers at the two electrodes, one of which is transparent, with one or more active layers in the middle of the device. Some designs require only one dielectric layer. The active layer consists of a large energy gap polymer (e.g. PVK), a triplet harvester, and carbon nanotubes (other nanostructured materi-

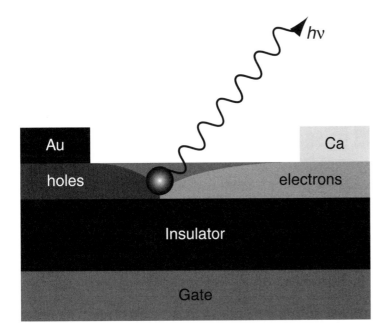

Fig. 10.14 A light-emitting field-effect transistor requires ambipolar transport in the active layer. The relative voltages between the two electrodes and the magnitude of the gate voltage control the amount of holes and electrons in the active layer. At the point where the electrons and holes meet, an exciton is formed and light is emitted.

als suitable for charge generation can be used instead of the nanotubes). The choice of dielectric layers is important because they must be able to enhance the electric field in the device, otherwise the applied voltage will be dropped across the dielectric. The class of dielectrics used can polarize charge at the interface with the active layer, creating an internal field which can be used to aid charge generation. The geometry of the nanoparticles or nanotubes is furthermore likely to facilitate charge generation, although the exact mechanism for this is unclear. The charges thus created then can undergo electroluminescence. These devices can produce good-quality light, and may also have a long lifetime. This technology is not restricted to white light, although that is where its likely applications will be found.

10.2   Polymer light-emitting transistors

The FET consists of a current moving from the source to the drain, depending on whether the applied field between these two electrodes allows current flow. If we have electrons flowing in one direction and holes in the other they may recombine when they meet, allowing the possibility of radiative decay. Such light-emitting field-effect transistors (LFETs) are an important area of research, and two transistor designs can be used to achieve light emission. The first requires an ambipolar active layer; the second requires separate layers of electron and hole transporting materials, either (or both) of which can be used as an emissive layer, or a separate emissive layer can be placed between the transporting layers. The latter multilayer structure, although more complicated, does

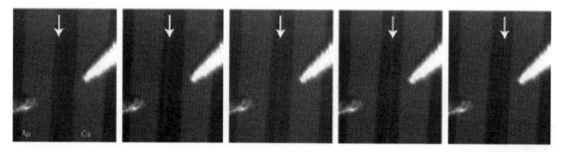

Fig. 10.15 Images showing the emission from a light-emitting field-effect transistor at a constant source-drain current (-30 nA, as the gate voltage is scanned in -5 V increments from -65 V (left) to -85 V (right image). The emission is in a line, indicated by the arrow, and moves from the hole-injection (gold) electrode) on the left to the electron-injecting calcium electrode on the right-hand side of each image. As the gate voltage becomes more negative it attracts more holes into the active layer, forcing the electron–hole recombination zone closer to the electron-injecting electrode, and hence the emission moves to the right. The channel length (the distance over which the light moves) is $\sim 80\mu$m. Taken from Zaumseil et al. *Nature Mater.* **5** 69 (2005). Reprinted by permission from Macmillan Publishers Ltd, copyright (2005).

allow more possibilities for device design, and consequently better performance. If such devices can be manufactured competitively and to the required quality, there should be important implications for the display industry. Currently, active matrix displays require a transistor to store the image state of each pixel, which is a separate light-emitting device. If it were possible to use the transistor as the light-emitter then devices could be made even thinner, with commensurate implications for flexible electronics.

10.2.1   Ambipolar light-emitting transistors

Ambipolar charge transport through the active layer allows excitons to be formed when electrons and holes meet, somewhere between the source and drain. (Formally, for ambipolar behaviour both electrodes act as source electrodes, because charge is injected at both.) For this to occur they must travel at the same side of the active layer, in contact with the gate insulator layer. One can imagine that if the gate is negatively charged, hole transport would occur at the gate-active layer interface, leaving the electrons to move at the other side of the active layer. In such circumstances, the carriers would not be able to recombine because they would not come into contact with each other. To avoid such problems, a thin active layer (~ 50 nm) is required. Thin active layers permit recombination because all of the charge carriers will be near enough to each other for this to occur. A hole accumulation layer will exist close to the hole-injection electrode, and an electron accumulation layer near the electron injection electrode (Fig. 10.14). This means that the source and drain electrodes cannot be the same metal. Gold is usually appropriate for hole injection, and calcium is typical for electron injection. In Fig. 10.15 we can see how changing the gate voltage changes the location of the emission by altering the balance of holes and electrons in the active layer, which shows that both electrons and holes are present in the active layer and ambipolar behaviour is indeed responsible for the optical emission in these transistors.

Of course, for optical emission band gaps of greater than ~ 2 eV are required, which makes efficient injection more difficult and limits the range of materials that can be used in the active layer. With many hole transporting materials, electron injection is only achieved through tunnelling, and devices are not very efficient. Nevertheless, good quality ambipolar behaviour has been observed in some materials, particularly the phenylene vinylenes, which can be modified by the addition of oxygen-bearing moieties to improve electron conduction (Section 3.7).

10.2.2   Multilayer structures

Optimization of LFET devices is limited with single-layer structures, because that active layer is a balance between good hole and good electron transport, but not optimized for either. For example, the PPV derivative used for the device shown in Fig. 10.15 had respective electron and hole mobilities of 3×10^{-3} and 6×10^{-4} cm^2 V^{-1} s^{-1}, a factor of 5 difference between the two. One could consider blending different electron- and hole- transporting material, but for now, the most successful method in optimizing LFET performance is to create multilayer structures. These can be a bilayer of hole- and electron-transporting layers, in which case, emission would be expected to take place in the layer with the better optical properties, although there is no reason why emission cannot take place from both layers. (This is generally not desired if monochromatic emission is required.) Alternatively, a specialized emissive layer can be created between the two transporting layers, for a trilayer structure (Fig. 10.16). In such devices inversion of the hole- and electron-transporting layers can be controlled by the gate. In the device shown in Fig. 10.16, a negative gate voltage would be expected to draw both holes and electrons towards the emissive layer interface, whereas a positive gate voltage will keep the two apart. As long as the emissive layer is thin enough (a few nm) for some electron and hole transport, the device can function. Trilayer structures have been demonstrated to be very efficient (outperforming equivalent LEDs) with small molecules, but currently little development has been performed with polymers. As an aside, we note that in these devices, the layer in contact with the electrodes is the same, and thus we do not need different electrodes. Since the hole transporting-layer is in contact with the electrodes in the device schematized in Fig. 10.16, gold electrodes are appropriate. If the electron-transporting layer were on top (which is unlikely, given the fact that a buried electron-transporting layer is offered some protection from oxidation), then calcium electrodes would be more appropriate.

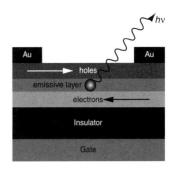

Fig. 10.16 In a trilayer LFET structure, the active layer is separated into optimized hole-transporting and electron-transporting layers, with an emissive layer between the two.

10.3   Photovoltaics

The population of the Earth is using an average of 15 TW of power, which is a lot. The sun generates some 4×10^{26} W, which is a bit more.

Of course, not all of the sun's power ends up as radiation incident on the Earth. Our planet, at 1.5×10^{11} m from the sun, with a radius of 6.4×10^6 m, absorbs (and re-emits) some $5 \times 10^{-8}\%$ of this. (The calculation of the fraction of the sun's energy absorbed by the Earth is easy; the reader is invited to verify it.) The total power from the sun incident on the Earth is some 1.5×10^{17} W, or 10,000 times more than we need. The surface area of the Earth is 5×10^{14} m^2, so a 20% efficient solar energy farm would need to cover an area of 2.5×10^{11} m^2 to supply the world with all its energy needs. (Lake Superior has a surface area of 8×10^{10} m^2, so its sacrifice might do wonders for the long-term viability of the planet.)

With a little imagination one can achieve great things with numbers. We have to bear in mind the limitations of the calculations performed above on a 15-year old photovoltaic cell-powered calculator. The sun's energy is not uniformly distributed on the Earth's surface, and so Lake Superior is probably far enough north that it would be a poor location for a giant floating solar farm. Not all power requirements are uniformly distributed across the globe; for example, there are more people in the northern hemisphere than south of the equator, so the requirements during the north's winter would presumably be greater than that of the southern part of the globe. Similarly, not all useful power can be generated by solar energy. Some forms of energy are going to remain specific to certain technologies. A nuclear-powered submarine will not be replaced by one covered in solar cells any time soon, or indeed ever.

The real challenge in energy technology is perhaps less its generation rather than its storage. If we could store and easily move energy we should be much better equipped to deal with problems concerning the sustainability of the planet. In any case, the contribution of polymers to large-scale solar cell technology is not necessarily going to be huge. Inorganic solar cells currently have a much greater efficiency (up to 25%) than those based on organic materials ($< 10\%$). They last longer too; oxidation is a problem for polymer electronics, and photovoltaics are no different. Unfortunately inorganic cells cost rather a lot to manufacture and install, but they will remain preeminent for some time to come. Polymer photovoltaic materials will retain their use in small-scale technologies, where price is paramount, and where flexibility will play a role. This is the same reasoning that we had at the start of Chapter 9 for RFID tags. Despite these caveats, it is not impossible for polymer or other organic photovoltaics to make a significant contribution to large-area solar energy generation, but the technological lead enjoyed by companies using inorganic materials is unlikely to be overcome in the near future.

The difficulties involved with polymer photovoltaic cells does not mean that the subject is not important; many researchers would like to see organic devices catch up with their inorganic counterparts. If this were to happen, then the cost of polymeric photovoltaics would mean that they would dominate the market. Even without large-scale energy production being part of the polymer photovoltaic portfolio, this is a huge area. Small electronic devices such as calculators and mobile phones could be powered by these materials. Flexible technology could even allow the recharging of mobile telephones by incorporating photovoltaic technology directly in clothing; phones could be left to charge whilst in a pocket, even when the clothes are being worn. Flexible newspapers based on passive materials (such as some electronic book display platforms) require an external source of light to be read (passive devices cause less eye strain), and this source of light can be used for its power. The market for such devices is well worth the investment; even where organic devices do not currently dominate the market, cost savings in manufacturing might cause a revision of that situation. The proponents of polymer photovoltaics in large-area devices note that the processability of polymers does allow effective ease of coating by roll-to-roll printing methods (similar to that used for the mass production of newspapers), or spray deposition. Such advantages mean that inexpensive devices are not inconceivable. If polymer (or other organic) solar panels can be manufactured easily and cheaply, there is no great barrier to their replacement at the end of their useful life. Time will tell whether or not polymer devices will dominate large-area markets, but that they will be important in many areas of energy creation is generally accepted.

One should also note that although solar cells drive much of the research, cells used to generate current from radiation have other uses. Photodiodes can be used to detect radiation, and their purpose is not to generate power. Under such circumstances they can be operated under a bias voltage to increase their sensitivity. In fact, a photovoltaic device is simply a photodiode with a small positive bias.

10.3.1 Efficiency of solar cells

Before discussing the science behind polymer photovoltaics, it is important to appreciate that different devices have different optical properties. There is therefore a problem of comparison between photovoltaic cells. For example, some cells have an internal efficiency of up to 100% at certain wavelengths, for specific absorption frequencies. Of course, devices requiring monochromatic radiation in order to perform well are not particularly useful.

Each claim of improved efficiency has to be externally verified, and this requires a standard set of test conditions. Here the device must be measured at a temperature of 298 K under an irradiance of 1 kW/m^2 with a light spectrum corresponding to that of the sun at an air mass of 1.5 (AM1.5). AM1.5 is an optical path length for solar radiation that is 50% greater than that at sea level at the tropics, which is AM1. (AM0, or air mass zero, is taken as the absence of air, and would be considered to be the radiation above the Earth's atmosphere.)

Fig. 10.17 The p-n junction is formed when two doped (inorganic) semiconductors of different kinds (a) are brought together. When they are connected with no voltage applied across them, the Fermi levels align (b). ϵ_F is the Fermi level of the relevant semiconductor and E_a and E_d are the respective donor and acceptor energy levels. (The dopants affect the location of the Fermi level in the semiconductor, as well as reducing the band gap.) When these are brought together, band bending occurs and a depletion zone results.

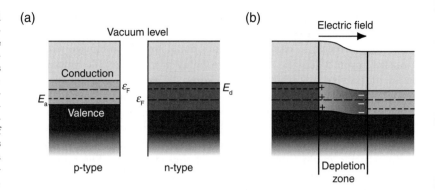

10.3.2   Photovoltaic effect

The photovoltaic effect is the creation of electrical current from photons, but there are significant differences between inorganic and organic semiconductors. In inorganic semiconductors the flow of charge from a p-n junction plays the crucial role, whereas in organic materials the dissociation of excitons to electrons and holes at a heterojunction is critical. Inorganic semiconductors must be doped, organic materials do not have this requirement. It is first worth recalling the photovoltaic effect in inorganic materials so that there is no confusion about the difference between the two processes.

The p-n junction

We have shown in Fig. 5.9 the interface between a metal and semiconductor. In the case of doped inorganic semiconductors, this interface is known as a *Schottky barrier*. The photovoltaic effect in inorganic materials is a special case of the junction between two doped semiconductors: one p-type and the other n-type. The physics of what happens at such an interface is an extension of that discussed in Section 5.5.1. The energy levels associated with the p-n junction are illustrated in Fig. 10.17. When the two semiconductors are brought together, the Fermi levels must equalize because there are mobile charges that can allow this to happen. In this situation the voltage drop across the interface between the two semiconductors is known as the *open circuit voltage*.

Doped semiconductors contain free (mobile) charges, and these respond to chemical as well as electric potentials. When the two semiconductors are placed in contact, there is a chemical potential gradient across the boundary and so holes will diffuse from the p-type material into the n-type, and electrons will diffuse the other way. When this happens, there is a net positive charge on the n-type (doped semiconductors have no net charge) and a negative charge on the p-type. This happens regardless of the Fermi levels of the two components and the height of their respective valence and conduction bands. The mobile charges that have diffused across the boundary do not contribute to this charge

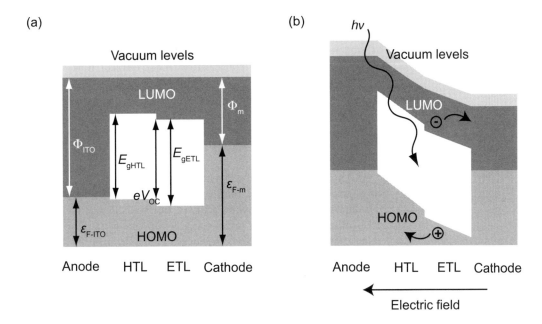

Fig. 10.18 Energy-level diagram of a two-layer photovoltaic device. We use the same nomenclature as previously, with the relevant parameters shown in (a), which is essentially the same structure as that shown in Fig. 10.4a. For photovoltaic operation, the device may be short-circuited (b) causing the LUMO and HOMO to bend in a different direction resulting in an electric field directed towards the anode. When a photon of energy $h\nu$ is incident on the material, an exciton is formed which can dissociate (generally at the interface between the two layers) to send an electron and hole to the cathode and anode respectively, under the electric field caused by the equalization of the Fermi levels of the two electrodes. V_{OC} is known as the open-circuit voltage, which is the maximum voltage that lies across the cell when no current flows through the device, although this is rarely achieved in practice.

build-up because they are eliminated by recombination with the other dopants. The electric field thus formed is due solely to the net charge caused by the migration of free carriers. Naturally this migration cannot go on forever, or even until the dopants are all exhausted, and it will be arrested when the electric field formed during the creation of the p-n junction is large enough to oppose further charge movement across the boundary. The boundary is known as the *depletion zone* or the region of *space charge*.

This p-n junction is critical to the (real-world) application of the photovoltaic effect in inorganic semiconductors. Incoming light will excite electrons into the conduction band on both sides of the depletion zone. The corresponding holes will also be created. These holes and electrons will be mobile; there is no applied electric field, so they can diffuse in all directions. If they meet a defect, which will act as a trap, they are likely to recombine and be lost, so high purity of photovoltaic materials is a prerequisite for acceptable performance. If these electrons and holes arrive at the depletion zone, then electrons generated within the p-type material and holes within the n-type will drift across it to the other side. This flow of charge represents the current within the device. In fact, it is even better than that, because the holes remaining in the p-type component, and the electrons in the n-type can flow, with this drift current, towards the respective electrodes. Therefore, for each photon that creates an electron–hole pair (not an exciton, because they are not bound), a charge of $2e$ contributes to the current flow.

Of course, as is the case for organic photovoltaic devices, most photons do not result in such useful current. Many will pass through the material. This is especially likely if their energy is less than that of the band gap. The band gaps, and the donor and acceptor levels, are only important in this regard in the photons that are required to excite an electron to the conduction band.

Although we are not concentrating on inorganic materials, a few general remarks should be made. First of all, a p-n junction is not a prerequisite for the photovoltaic effect nor for solar cells in general. Very thin film solar cells do not require a p-n junction. The optimization of these devices is a massive research challenge, and the best approaches involve multilayered epitaxially grown thin films, which can achieve external quantum efficiencies in excess of 40%—a figure well in excess of what the most optimistic polymer photovoltaic researcher would see as being commercially likely.

(a)

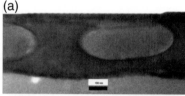

(b)

Fig. 10.19 Transmission electron micrograph of a cross-sections of a (a) blend of polyvinyltriphenylamine (donor) poly(perylene bisimide acrylate) (acceptor) and (b) a block copolymer of the same components. The size of domains is significantly smaller in the block copolymer film, which also exhibits better photovoltaic performance, although neither forms a good quality device. The dark regions in these images are perylene bisimide, which is stained for contrast in the electron microscope. These images are taken from Lindner et al. *Angew. Chem. Int. Ed.* **45** 3364 (2006).

Organic heterojunctions

The photovoltaic effect in polymers does not require doped materials, and takes places through a different mechanism. The basic principles have already been stated in Section 4.2; light excites an electron to the LUMO, which is bound to its corresponding hole as an exciton. This exciton can decay by reradiation, or by being separated into its corresponding hole and electron, which can diffuse to their respective electrodes. This exciton dissociation requires a heterojunction (i.e. an interface), rather than an impurity which may well act as a trap for electrons or holes, impinging upon the collection of charge at the electrodes. The energy barrier between the donor and acceptor levels is ideal for this role. Consequently, the right choice of organic heterojunctions is important for the optimization of device performance.

Assuming the exciton dissociates to form its constituent electron and hole, there is no p-n junction with associated electric field to direct them. However, if the two electrodes are short circuited, the equalization of their Fermi levels means that the cathode essentially donates electrons to the anode by increasing the Fermi level of the anode relative to the cathode; this creates an electric field, which will help guide the charges to their relevant electrodes. For this reason photovoltaic devices can work in thin film structures (it is the same for inorganic solar cells which work without p-n junctions), but these suffer from poor photon absorption. More complex structures with electron and hole transport layers are therefore desirable.

If we consider a two-layer structure (Fig. 10.18), the holes require a higher HOMO on their route to the anode than would exist in an electron-transporting layer, and the electrons require a lower LUMO on their route to the cathode than exists on the hole-transporting layer. This means that if the LUMO on the ETL were higher in energy than that of the HTL, an electron would not be able to travel from the LUMO to the HOMO, unless it were able to tunnel through the barrier. Of course, for excitons formed on the electron-transporting layer this would not be a problem, although the requirement that its HOMO be lower than that of the hole-transporting layer should be taken into consideration. In principle, with a two-layer device, one would want both layers to contribute to the performance, because that would maximize the EQE, because fewer photons would be lost through not being absorbed by the device. This is particularly important because double-layer devices must be rather thin, of perhaps 100 nm or less in thickness. Thicker devices would not help greatly because more excitons would be formed further away from the interface between the two layers. This interface is the location in the device where it is most efficient for excitons to dissociate into the component holes and electrons.

Heterojunctions exist in blends too, and the discussion in Section 8.2 is relevant here because Figure 8.3 reveals how the two-layer structure can be retained but with thicker films because of the increase in interfacial area. In such cases it is desirable that the two different components of the blend wet the two interfaces, with a strong phase separation in the middle of the film. Because isolated phase-separated domains are common in thicker films, it is unlikely that a useful blend structure can be made that is much thicker than a two-layer structure, although, because of the extended interfaces inherent in blends, the efficiency might well be better than that of a bilayer.

One alternative to a blend is to use a donor–acceptor block copolymer. Block copolymers microphase separate over length scales commensurate with the exciton diffusion length. The microphase separation allows good opportunities for exciton dissociation, although it is probably less effective at allowing effective charge transport towards electrodes. An example of a blend and block copolymer morphology is shown in Fig. 10.19.

Although a truly effective donor–acceptor block copolymer device has not yet been demonstrated, the concept certainly works in principle and provides an appealing route for further development. Nevertheless, control of domain size and morphology may be tailored separately by use of what is known as supramolecular chemistry, which involves the coordinated self-assembly of molecules to form structures with the required properties. Supramolecular chemistry is, depending on perspective, both simpler and more difficult than traditional routes. Its simplicity arises from the need to mix components that have the required functionalities, rather than create complex molecules that may have their own synthetic challenges. The final structure is therefore not built up of covalent bonds, but rather of weak interactions involving van der Waals, hydrogen, or hydrophobic bonds. The difficulty with supramolecular chemistry arises from the need to ensure that no unwanted assembly occurs. In photovoltaic cells, supramolecular chemistry can be used to effectively control

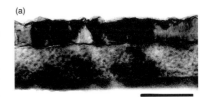

(a)

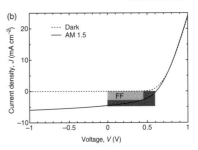

(b)

Fig. 10.20 Supramolecular chemistry allows the assembly of donor and acceptor species into an ordered structure. (a) Cross-sectional image of a supramolecular assembly of PCBM with a block copolymer of P3HT and poly(4-vinylpyridine), P3HT-*block*-P4VP. The aluminium cathode is shown in the upper portion of the image and the scale bar is 200 nm. PCBM and P4VP mix to form a blend that acts as an electron acceptor, with the P3HT acting as an electron donor. dark regions show an iodine stain, which is taken up by the P4VP block. (b) Data under both dark and AM1.5 conditions for a device in which the active layer was annealed to allow the morphology to develop. The overall efficiency of the device is 1.20%, but this is without any further optimization. The fill factor (FF) is explained in Section 10.3.4. The images and data are taken from Sary et al. *Adv. Mater.* **22** 763 (2010).

domain size. An example is shown in Fig. 10.20.

10.3.3 Open-circuit voltage

If the device terminals are connected, we have a short-circuit; if they are not connected, we have an *open-circuit*. For an open circuit (Fig. 10.18) there is no equalization of the Fermi levels of the electrodes across the device, and so a potential must exist within it. In the absence of any current, this is known as the open-circuit voltage, and is taken as the maximum voltage across the device. The above does not offer a unique definition of the open-circuit voltage. One could consider the difference between the Fermi levels of the two electrodes, or the differences between the HOMO and LUMO levels of the two layers. (Note that the difference between the HOMO and LUMO of an individual layer is of no interest here, because this cannot be considered as across the device.) The difference between the Fermi levels of the two electrodes is less than that between HOMO and LUMO levels, so cannot correspond to the open-circuit voltage. We then need to consider which of the voltages between the HOMO of the ETL and LUMO of the HTL, or the LUMO of the ETL and the HOMO of the HTL better corresponds to the open-circuit voltage. The former of these is not interesting, because electron transport is not associated with the HOMO and hole transport is not associated with the LUMO. The most practical definition, as shown in Fig. 10.18a, is for the difference in energy between the LUMO of the ETL and the HOMO of the HTL to be given by eV_{OC}, where V_{OC} is the open-circuit voltage.

When the exciton is formed, it is in a metastable state and will easily dissociate. Since the interface between the two layers is the most likely point for dissociation to happen, the electron and hole are formed at an energy eV_{OC} apart. This is not necessarily the case for all of the excitons formed in the process. An exciton that decays within the HTL will have a greater energy than eV_{OC}, which in turn would be greater than the energy of any exciton dissociating within the ETL. Nevertheless, the open-circuit voltage is something that can be measured, and experiments have shown that the (maximum) open-circuit voltage corresponds to the difference between ETL LUMO and HTL HOMO, as shown in Fig. 10.18a. In most practical situations the open circuit voltage is ~ 0.3 V less than V_{OC}, due to a compensating effect of dark current and also because the electric field associated with the open-circuit voltage does not completely fall across the heterojunction.

Some earlier photovoltaic devices were made with a single active layer, and the open-circuit voltage thus depends upon its nature, i.e. whether it is a hole or electron-transporting layer. If the layer is primarily hole transporting, then the exciton will be less effective if it dissociates at an anode, because the electron will find it difficult to make its way to the cathode. In that case, eV_{OC} is the difference between the Fermi level of the cathode and the HOMO of the active layer. Similarly, in the less likely case of a primarily electron-transporting active layer, the open-circuit voltage corresponds to the difference between the LUMO of the active layer and the Fermi level of the anode. The key point is that the open-circuit voltage corresponds to the level at which one would expect the exciton formed within the device to dissociate. The reader is invited to deliberate over where the open-circuit voltage would be in the case of an ambipolar active layer.

10.3.4 Current characteristics

The performance of photovoltaic devices is measured in terms of a current–voltage curve, shown in Fig. 10.21. Of course, if one doubles the size of the device, one doubles the current that can flow through it, so the areal current density, J, is the relevant parameter, rather than simply current. The open-circuit voltage was considered in Section 10.3.3, and is marked in Fig. 10.21. Equally important is the short-circuit current, J_{SC}, which is, from Fig. 10.21, the current at which the voltage across the (short-circuited) device is zero. In other words, the incident photons create enough current to counter the voltage created by the two electrodes being short-circuited.

The *fill factor* (FF) is a measure of device quality, or specifically, how efficiently charge is extracted from the device. The fill factor is the power generated when JV is a maximum divided by $J_{SC}V_{OC}$. Ideally one would like a device in which this was unity (100%), but this cannot be achieved, and the best devices have values of FF of not much more than 50%.

The power conversion efficiency of a photovoltaic is given by

$$\eta_P = \frac{J_{SC}V_{OC}FF}{P_A}, \tag{10.4}$$

where P_A is the areal light power (W/m^2) incident on the device. If the fill factor is given as a percentage, then η_P will also be a percentage. If we take the data from Fig. 10.21 as an example, then $J_{SC}V_{OC} = 7.3$ mW/cm^2, and so FF = 51%. In this particular case, $P_A = 85$ mW/cm^2 (not too dissimilar to the direct sunlight value mentioned in Section 10.1.1), so $\eta_P = 4.4\%$, which is characteristic of a good-quality device. Indeed, the blend of P3HT with PCBM used for the device, the data of which are shown in Fig. 10.21, is currently very popular for use in photovoltaic development because of the ease with which it can be processed, the good-quality data that can be reliably obtained, and its overall stability.

In operation, a forward bias is applied to photovoltaic cells so that the device is run under conditions by which the maximum power can be extracted. This bias means that the cell is connected to an external power supply, and thus must use power to operate. Assuming no heat loss, the power drawn by the device is exclusively through its dark current, and this affects the performance of a device under illumination; the dark current is not a current that exists only in the absence of light; when the device is illuminated, the dark current exists to cancel out some of

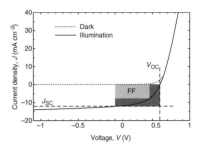

Fig. 10.21 Current–voltage plots are used to demonstrate device performance and obtain important parameters describing the device, such as its open-circuit voltage (V_{OC}) and its short-circuit current (J_{SC}) The maximum power obtained from such a device is given by the fill factor, FF, which is the ratio of the area under the $J(V)$ curve whereby the product of J and V are maximized compared to the product $J_{SC}V_{OC}$. In the absence of photons, the device can be treated like a normal diode, and its dark performance is also shown. These data are for a device using a blend of P3HT and PCBM and are taken from Kim et al. *Nature Mater.* **5** 197 (2006).

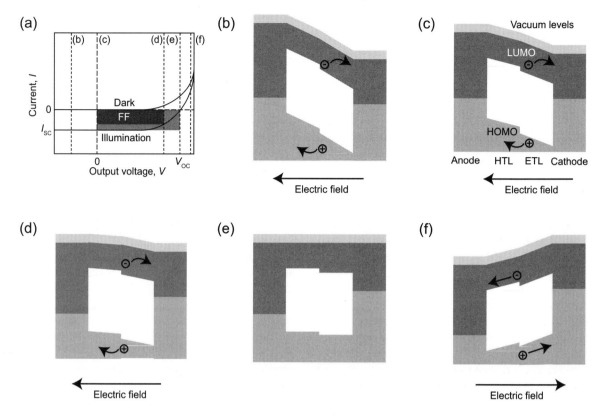

Fig. 10.22 (a) The current–voltage properties of a photovoltaic device are schematized with the dark and illuminated (photocurrent) shown with the upper and lower solid lines respectively. The dark shaded region corresponds to the fill factor (FF), whereas the lighter shaded region corresponds to the product of the short-circuit current and open-circuit voltage, $I_{SC}V_{OC}$. The vertical lines indicated (b)–(f) indicate the energy-level diagrams of a double-layer device under the associated current–voltage conditions for a device under illumination. (b) A reverse biased diode, which may be used as a photodetector. (c) When no voltage is applied, photocurrent may be generated. Here the current generated is the short circuit current, I_{SC}. The anode, hole transport layer (HTL), electron-transport layer, and cathode are indicated in this plot, along with the vacuum levels. (d) Under forward bias, a point is reached at which IV is a maximum. This is the best operating condition for the device, because at this point it is at its most efficient. (e) When no current is measured, $V = V_{OC}$, the energy bands are horizontal. V_{OC} counters the internal electric field within the device, and no current flows. (f) Under a strong forward bias, electrons will travel to the anode and holes to the cathode. Under these circumstances the device is operating as a simple diode.

the current that would be usefully drawn by the device. In principle the device should be designed so that the only contribution to dark current is black-body radiation, as discussed below.

The behaviour of a double-layer device under different operating conditions is schematized in Fig. 10.22. The ideal operating condition for a device is with a small forward bias so that $J_{SC}V_{OC}$ is maximized in order for the most efficient conversion of photons to electron–hole pairs. If no voltage is applied, the device may still operate as a photodetector, although photodiodes are generally operated under reverse bias in order to improve their efficiency. A device operating under zero bias could still be considered a photovoltaic cell in principle, but it is not efficient because there is no potential difference across the two electrodes and so no means of drawing the current generated apart from its own internal field. Clearly, an efficiency of unity would mean that the maximum power drawn, $(JV)_{\mathrm{max}} = J_{\mathrm{SC}}V_{\mathrm{OC}}$. The reverse bias shown in Fig. 10.22b is appropriate for photodiodes, and when a strong reverse bias is exhibited, the device could perform as an *avalanche photodiode*, which is capable of detecting individual photons. Polymeric photocells are not currently used for these purposes, because there is no market need for the advantages that polymers convey (processability), but their disadvantages of relatively short lifetime are possibly prohibitive, especially under the high-voltage operating conditions of such devices. In Fig. 10.22e the device is forward biased to such a point that the field across the device cancels out the internal field, and so $J = 0$. Here, no photocurrent can be generated by the device. For all applied voltages, $V < V_{\mathrm{OC}}$, the device will provide no current, excluding leakages, under dark conditions. In Fig. 10.22f, a strong forward bias is applied, and the device is behaving as a normal diode. The photoelectric effect still exists, but there is little difference between the dark and illuminated currents; the current contains only the dark component, and the current due to exciton dissociation is negligible. Here, the use of electroluminescent polymers for either of the active layers would permit the device to act as an LED. Such devices are generally not efficient, because the injection barriers for electrons and holes are not optimized for LED operation.

A final point worth noting is that the internal electric field of the device does not determine whether or not it is forward or reverse biased. The relative position of the Fermi levels determines the bias of the device. So, in Fig. 10.22b, where the Fermi level of the anode is raised with respect to the cathode, the device operates under reverse bias, but in Fig. 10.22d the device is operating under forward bias, despite the internal electric fields being in the same direction.

Dark current

The dark qualities of the device reflect the black-body properties of the device environment. The device operating in forward bias in dark conditions, at least for $V < V_{\mathrm{OC}}$, should not ideally let any current pass, but thermal generation of holes and electrons is possible. The Nobel laureate William Shockley took the approach that a solar cell could be considered as being contained within a black body, and subjected to radiation of the relevant (Planck) distribution of a black-body radiator at 300 K, or any other appropriate temperature. 300 K black-body radiation corresponds to the infrared, and would not lead to the creation of excitons for materials with optical band gaps. However, a black-body distribution will have a high energy tail that could create such excitons. Planck's law gives the emitted power of a black body at absolute temperature T per unit wavelength λ per unit solid angle as

$$I\left(\lambda, T\right) = \frac{2hc^2}{\lambda^5} \frac{1}{\exp\left(\frac{hc}{\lambda k_{\mathrm{B}} T}\right) - 1}. \tag{10.5}$$

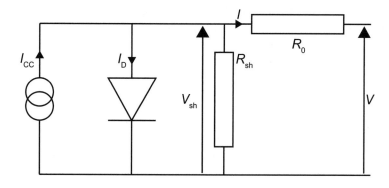

Fig. 10.23 The equivalent circuit for a photovoltaic cell can be broke up into a shunt and series resistance, R_{sh} and R_0 respectively; a diode, through which a current, I_D passes; and a source of electromotive force, V_{sh}, from which I_{CC} flows. A current, I and voltage, V are output from the device, delivering a useful power, IV.

This can be compared at operating temperature and with a colour temperature of a white incandescent light bulb, which may be taken as 2850 K. Clearly there is a large difference between the exponential term at 300 K and at 2850 K—a factor of 10^{-31} for 2 eV photons. This points to a negligible leakage current, but this is mitigated by the black-body radiation subtending a solid angle of 4π, although this will increase the factor by less than an order of magnitude. Any dark current in excess of these values can be attributed to energy losses due to any energy requirement in holding the voltage, V, at a finite value. These might be due to leakage currents and contact resistances, to which we turn next.

10.3.5   Equivalent circuit

The photovoltaic cell in operation can be viewed as an equivalent circuit, as shown in Fig. 10.23. A source of electromotive force provides the bias to the device. The photovoltaic device itself is symbolized by the diode, whereby the direction of its symbol indicates forward bias. The voltage that comes out of the device is not necessarily the same as that that is across the device, because of losses, which are represented by a series resistance, R_0. This series resistance is due largely to contact resistances between the individual components, as well as the resistance of the different layers that make up the device. The series resistance can be obtained from the $I(V)$ curve in Figures 10.21 and 10.22 by

$$R_0 = \left(\frac{\mathrm{d}V}{\mathrm{d}I}\right)_{I \to 0}. \tag{10.6}$$

The series resistance drains otherwise useful energy from the device, and ideally would be zero. Perhaps more interesting from a scientific point of view is the shunt resistance, R_{sh}, which is given by

$$R_{\text{sh}} = \left(\frac{\mathrm{d}V}{\mathrm{d}I}\right)_{V \to 0}. \tag{10.7}$$

This resistance is due to leakage in the device, from which imperfections in the films (such as pinholes), carrier traps, and exciton recombination are important contributions. These leakage phenomena decrease the

Vacuum levels

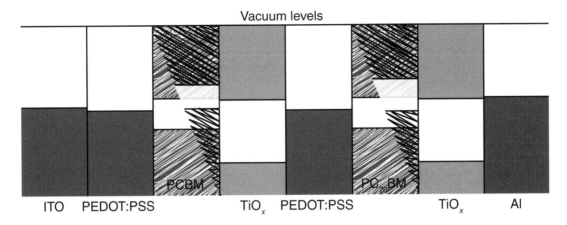

ITO   PEDOT:PSS   TiO$_x$ PEDOT:PSS   TiO$_x$ Al

Fig. 10.24 Energy-level diagram for a tandem solar cell containing a blend of PCPDTBT and PCBM in one device, and P3HT and PC$_{70}$BM in the other. The PEDOT:PSS and titanium oxide form hole-transport and electron-transport layers respectively. The diagram is drawn to scale with the Fermi level of PEDOT:PSS at 5.0 eV, shown at the midpoint in the diagram. The energy levels of the hole carriers in the blends (PCPDTBT and P3HT) are shown in the darker hatching. P3HT has a slightly wider band gap (1.9 eV) than PCPDTBT (1.6 eV), with P3HT also having the smaller electron affinity (3.2 eV compared to 3.5 eV).

resistance, by giving alternative routes for charges to flow. Such phenomena need to be prevented, and the shunt resistance should, ideally, be infinite.

If we first consider the device under dark conditions, and apply a reverse bias to it, then no current will flow through the diode, and in an ideal device ($R_{\mathrm{sh}} = \infty$) the closed circuit current would flow through R_0. Under a positive bias, for example whereby $V_{\mathrm{sh}} > V_{\mathrm{OC}}$, current will flow both through the device and through the series resistance. If $V_{\mathrm{sh}} < V_{\mathrm{OC}}$, then the device would behave as if under reverse bias, and no current (or just some leakage current) would flow. Practically, leakage currents do flow, so under reverse bias the resistance of the device is going to be large, but not infinite.

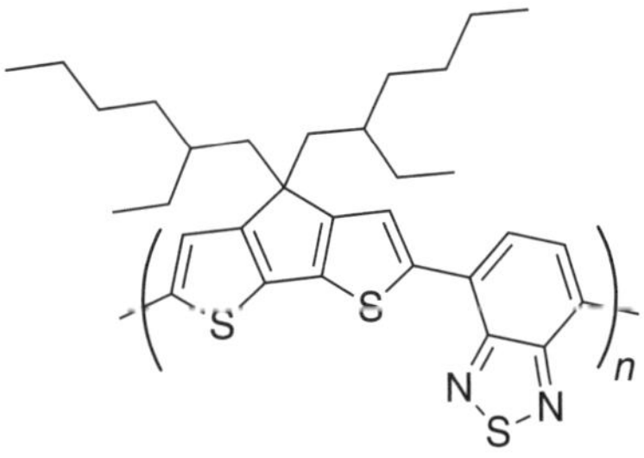

Under illumination, and with a strong forward bias for $V_{\mathrm{sh}} > V_{\mathrm{OC}}$, any photocurrent is in the same direction as that from the voltage source. However, under operating conditions, i.e. when $V_{\mathrm{sh}} < V_{\mathrm{OC}}$ the photocurrent is in the opposite direction to that which the diode would normally allow to pass. This current is created within the device, and will run counter to that from the source. If we consider the limit of a forward bias, at $V = 0$, then

$$V_{\mathrm{sh}} + IR_0 = 0. \tag{10.8}$$

Because $R_0 \ll R_{\mathrm{sh}}$, the shunt resistance dominates, which is why we can define it using eqn (10.7). Under forward bias, the diode resistance is not the limiting effect, and the resistance R_0 dominates the output characteristics of the device, and so we define R_0 as in eqn (10.6). These shunts and series resistances are not properties of the device that are valid under all operating environments. Different shunt and series resistances are valid under dark conditions, for example. Under illumination

Fig. 10.25 The chemical structure of poly[2,6-(4,4-bis-(2-ethylhexyl)-4H-cyclopenta[2, 1-b; 3, 4-b']dithiophene)-alt-4,7-(2,1,3-benzothiadiazole)], which is usually abbreviated to simply PCPDTBT.

they are unlikely to vary much, however, although one should in general state the nature of the illumination when quoting these resistances.

10.3.6 Improving devices

The data shown in Fig. 10.21 are for a device of good performance characteristics but with efficiency $\eta_P < 5\%$, so clearly there is some way to go before these cells can compete with inorganic devices. As has been stated, different efficiencies exist for different spectra so a band gap that can respond to a wide range of wavelengths is required. In particular, most of the polymers used have relatively high band gaps, which means that the red part of the solar spectrum is not so well catered for. Recent developments have addressed this, with commensurately improved device performance.

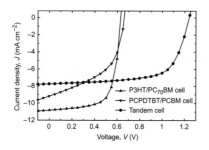

Fig. 10.26 Current density-voltage, $J(V)$ plots for P3HT/PC$_{70}$BM and PCPDTBT/PCBM blend solar cells and a tandem cell made of the the two cells. The power conversion efficiencies for these cells are 4.7%, 3.0%, and 6.5% respectively. These data are taken from Kim et al. *Science* **317** 222 (2007).

Tandem cells

One means of addressing limited band gap problems is to use more than one donor or acceptor. One might think that this could be achieved by creating blends of three or four components, widening the range of wavelengths over which efficient absorption can occur. In fact, this impedes charge transport in the device, for example, by creating traps.

Multiple donors and acceptors can be achieved by linking solar cells together to form *tandem* solar cells. Tandem solar cells require two different donor–acceptor pairs linked in series. The energy-level diagram of one example is shown in Fig. 10.24. One of the cells is a blend of PCBM with poly[2,6-(4,4-bis-(2-ethylhexyl)-4*H*-cyclopenta[2,1-*b*;3,4-*b'*]dithiophene)-*alt*-4,7-(2,1,3-benzothiadiazole)] (PCPDTBT), the chemical structure of which is shown in Fig. 10.25, and the other cell is

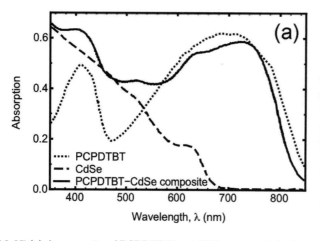
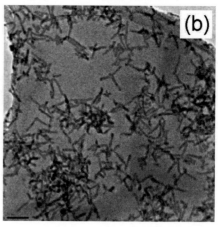

Fig. 10.27 (a) A composite of PCPDTBT and CdSe nanoparticles improves the absorption of the device; the polymer absorbs stronger in the red giving an extended absorption range. The CdSe has much better absorption properties in the region around 500 nm, and the spectrum of the composite device reflects this improvement. (b) The CdSe nanoparticles aggregate, giving rise to an interconnected structure. The absorption data and electron microscope image are both taken with permission from Dayal et al. *Nano Lett.* **10** 239 (2010). Copyright (2010) American Chemical Society.

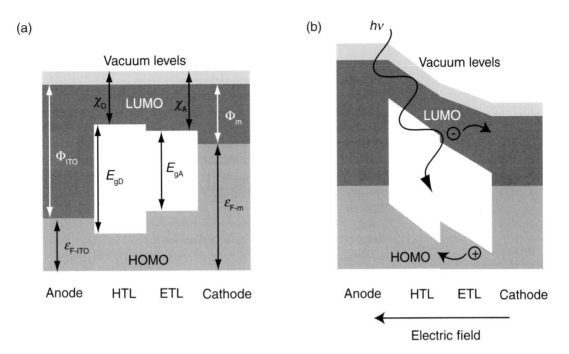

Fig. 10.28 Energy-level diagram of a double-layer diode operated as a photovoltaic device. As for the LED and photovoltaic cell shown in Fig. 10.4 and Fig. 10.18 respectively, the electron affinity of the acceptor layer is less than that of the donor, i.e. $\chi_A < \chi_D$. Here, however, the ionization potentials of donor ($E_{gD} + \chi_D$) and acceptor ($E_{gA} + \chi_A$) are such that $E_{gA} + \chi_A > E_{gD} + \chi_D$.

a blend of P3HT with PC$_{70}$BM. (PC$_{70}$BM or [6,6]-phenyl-C$_{71}$ butyric acid methyl ester is identical to PCBM except that there are more carbons in the fullerene, giving it a slightly prolate shape.) Both devices are capable of working as photovoltaics, but the tandem cell is substantially more efficient than either of the individual cells, as demonstrated in Fig. 10.26. Indeed tandem solar cells have been certified as having exceeded 10% power conversion efficiency.

Hybrid cells

Alternative methods of improving device problem have focused on hybrid (polymer-inorganic) systems such as the addition of nanoparticles to the device. Nanoparticle–polymer composites have the advantage of being solution-processable, and some, such as cadmium selenide, have excellent optical properties, which means that they can effectively harness the light. An example of a composite device is a blend of the alternating copolymer PCPDTBT with CdSe. Both of these will absorb light, increasing the wavelength range over which the device can function (Fig. 10.27a). The nanoparticle shape affects their aggregation, resulting in the structure shown in Fig. 10.27b. We know from Section 10.19 that interconnected structures act less as traps, and more effectively as pathways to the relevant electrodes than other structures, so the choice of shape is important. In this case the particles are branched, which gives rise to their interconnected string-like structure.

10.4   Further reading

LEDs and photovoltaics are well-reviewed in the literature, but the book by Heeger, Sariciftci, and Namdas (2010) is an obvious place to start. Similarly, one could refer to chapters in Hadziioannou and Malliaras (2007). The review by Friend et al. (1999), while old, is still relevant. More than anything else in this book, photovoltaics is possibly the fastest-developing area in terms of state-of-the-art. The reader should find the review on the subject by Coakley and McGehee (2004) interesting, but it is not so recent. A more recent review with a perspective dominated perhaps by characterization is that by Nicholson and Castro (2010); although general to organic materials, it does contain substantial information on polymer-based photovoltaics. A review discussing both LEDs and photovoltaics (transistors too) is given by Facchetti (2011). For slightly more specific reviews, white-light emission is covered in the review by Gather et al. (2011), and light-emitting polymer transistors are covered in the review by Muccini (2006). Finally, the application of block copolymers to photovoltaics is covered by Segalman et al. (2009).

10.5   Exercises

Table 10.1 $J(V)$ data for a photovoltaic device, formatted as $V(\mathrm{V}), J(\mathrm{mA/cm^2})$.

−1, −6.06	−0.96, −6.02	−0.92, −5.95	−0.88, −5.89
−0.84, −5.83	−0.8, −5.76	−0.76, −5.71	−0.72, −5.64
−0.68, −5.57	−0.64, −5.49	−0.6, −5.42	−0.56, −5.34
−0.52, −5.26	−0.48, −5.18	−0.44, −5.08	−0.4, −4.98
−0.36, −4.9	−0.32, −4.8	−0.28, −4.7	−0.24, −4.59
−0.2, −4.48	−0.16, −4.37	−0.12, −4.25	−0.08, −4.12
−0.04, −4	0, −3.89	0.04, −3.76	0.08, −3.6
0.12, −3.48	0.16, −3.33	0.2, −3.19	0.24, −3.04
0.28, −2.88	0.32, −2.71	0.36, −2.52	0.4, −2.3
0.44, −2.04	0.48, −1.7	0.52, −1.24	0.56, −0.61
0.6, 0.26	0.64, 1.39	0.68, 2.83	0.72, 4.53
0.76, 6.51	0.8, 8.72	0.84, 11.14	0.88, 13.73
0.92, 16.46	0.96, 19.31	1, 22.21	—

10.1. How well would the bilayer system shown in Fig. 10.28 function as a photovoltaic device?

10.2. A device made using the same supramolecular system as that shown in Fig. 10.20 (but before being annealed) has $J(V)$ data given in Table 10.1. Calculate the fill factor for this device. Given that the device is subjected to an irradiance of 1000 W/m^2 (AM1.5), what is its power conversion efficiency, η_P?

10.3. You are given that P3HT has an ionization potential of 5.1 eV and an electron affinity of 3.2 eV, and that PCBM has an ionization potential of 5.9 eV and an electron affinity of 4.2 eV. If an exciton dissociates into a hole and electron at the P3HT:PCBM interface in a bilayer device, what is the maximum wavelength of light that would correspond to this exciton? Is such a photon likely to have been absorbed by such a device, and if so, in which layer?

10.4. A bilayer is used in a light-emitting diode where one layer is F8BT, and the other TFB. The electron mobility of F8BT is 1×10^{-7} m^2 V^{-1} s^{-1} and the hole mobility of TFB is 1×10^{-6} m^2 V^{-1} s^{-1}. The dielectric constants for these polymers can be taken as 3.2 (F8BT) and 3.0 (TFB).

(a) If the layers are of thickness 90 nm (F8BT) and 120 nm (TFB), and 10 V is used to drive the device, what is the voltage drop across the individual layers, assuming that 100% recombination occurs at the interface?

(b) Although TFB is an efficient electron-blocking material, F8BT has similar hole and electron mobilities. How might this affect the results you obtained in part (a) above?

Charge injection into a device requires a potential barrier to be overcome. Moving a charge away from a metal, such as an electrode needs, the charge to have an energy greater than that of its work function. This is not the only requirement, but others, such as the necessity for it to have a minimum velocity in the direction of the surface, are related and do not need to be considered here. If that charge is taken away from the surface without the requisite energy, a *mirror* charge will attract it back. This mirror charge is the origin of the barrier. However, devices generally work under an applied bias and the electric field associated with that bias, will act as to reduce the barrier height. This is known as the Schottky barrier, and in this Appendix we shall derive the result shown in eqn (5.29).

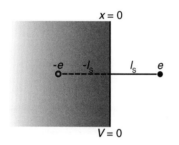

Fig. A.1 The interaction between a test charge e and a flat planar conducting surface can be reduced to one of a charge and an equal and opposite charge, located the same distance behind the plane as the first (test) charge is in front of it.

A.1   The method of images

The first step to calculate the decrease in the height of the potential barrier due to an applied electric field across the device is to consider the potential energy of a charge distance $x = l$ from a flat planar surface located at $x = 0$. This surface, which in our case will be that of an electrode, will be at a potential V, which we can set to zero. The potential is also zero at infinity. We now want the potential at which our charge sits. This is not a problem that can be solved very easily, but we can reduce it to a different problem whereby the solution is relatively trivial. Let us replace the surface by a point charge of opposite sign located a distance x behind the plane, as shown in Fig. A.1. This might seem a little arbitrary, but all we are interested in is the potential of our charge. Given that the boundary conditions in our system are well-defined (i.e. the potential at infinity, and that of the surface being zero), there can only be one potential associated with the location of our charge.[1] So, in three dimensions, we have our charge, e located at (l, y, z), and the mirror charge, $-e$ located at $(-l, y, z)$. The potential, $\varphi(0, y, z) = 0$, as is that at infinity. The potential of our system of two charges becomes

$$\varphi(x, y, z) = \frac{e}{4\pi\epsilon_0\epsilon_r} \left(\frac{1}{\sqrt{(x-l)^2 + y^2 + z^2}} - \frac{1}{\sqrt{(x+l)^2 + y^2 + z^2}} \right),$$
(A.1)

which vanishes at $x = 0$ and when $x^2 + y^2 + z^2 \to \infty$.

[1] Formally, this is a result of the *uniqueness theorem*, which states that if a potential distribution (or electric field) satisfies the boundary conditions that we must impose upon it, then that potential distribution is the (unique) correct one. The uniqueness theorem is a consequence of the Poisson equation in electrostatics, and the reader is directed to relevant texts for a complete discussion.

The potential energy of our charge and its image is simply the Coulombic potential energy due to two point charges separated by a distance $2l$, and is given by

$$U_{\mathrm{m}} = \frac{-e^2}{2\left(4\pi\epsilon_0\epsilon_{\mathrm{r}}\right)l}. \tag{A.2}$$

This image charge system contains two charges, whereas the real system in which we are interested, contains only one. If we now consider the general case, the potential energy of a charge distance x from a flat planar (conducting) surface is half that of the image system, and is thus

$$U_{\mathrm{fps}} = \frac{-e^2}{4\left(4\pi\epsilon_0\epsilon_{\mathrm{r}}\right)x}. \tag{A.3}$$

A.2   The Schottky barrier

Now we know the potential energy of a charge removed from a planar surface, we can consider the situation when an electric field, E is applied,

$$U_{\mathrm{Schottky}} = \frac{-e^2}{4\left(4\pi\epsilon_0\epsilon_{\mathrm{r}}\right)x} - Eex. \tag{A.4}$$

This energy (eqn A.4) and that of its components are shown in Fig. A.2. The reduction in the energy required for charge injection caused by the applied electric field will occur at a distance, $x = l_{\mathrm{S}}$, which is when U_{Schottky} is at its maximum.

$$\frac{\mathrm{d}U_{\mathrm{Schottky}}}{\mathrm{d}x} = \frac{e^2}{4\left(4\pi\epsilon_0\epsilon_{\mathrm{r}}\right)x^2} - Ee = 0, \tag{A.5}$$

so that

$$l_{\mathrm{S}} = \frac{1}{2}\sqrt{\frac{e}{4\pi\epsilon_0\epsilon_{\mathrm{r}}}}. \tag{A.6}$$

If we then substitute l_{S} into eqn (A.4), we have

$$U_{\mathrm{Schottky}}\left(x = l_{\mathrm{S}}\right) = W_{\mathrm{S}} = e\sqrt{\frac{eE}{4\pi\epsilon_{\mathrm{r}}\epsilon_0}}. \tag{A.7}$$

which completes the proof of eqn (5.29).

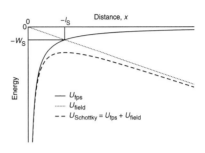

Fig. A.2 The potential energy of a charge located a distance x from a flat planar conducting surface, U_{Schottky} in an electric field is made up of a component due to the electric field, U_{field}, and that due to the induced charge on the surface, U_{fps}. The potential barrier required for the charge to escape is no longer located at an energy commensurate with the function of the metal (here denoted zero energy) but a distance W_{S} below it.

B

Dispersity in step-growth polymerization

Calculating the dispersity of a polymer mixture is possible to obtain analytically for polymers formed from condensation reactions (or more generally for polymers formed in step-growth reactions), as is often the case for semiconducting polymers. The dispersity is defined as the ratio of the weight average and number average molecular mass of the polymer. The number average molecular mass is easily defined, because it is essentially the Carothers equation (eqn 6.3),

$$\overline{M}_n = M_{\mathrm{w,m}} X_n = \frac{M_{\mathrm{w,m}}}{1-p}, \tag{B.1}$$

where $M_{\mathrm{w,m}}$ is the molecular weight of a monomer, and reference to the time dependence has been omitted.

The total number of chains with x monomers is given by

$$N_x = n_0 \left(1-p\right)^2 p^{x-1}, \tag{B.2}$$

where $n_0 = n\left(0\right)$. This equation (B.2) is the product of the number of molecules remaining, given by $n_0 \left(1-p\right)$ and the probability $P\left(x\right)$ that a molecule containing x monomers exists, given by

$$P\left(x\right) = \left(1-p\right) p^{x-1}. \tag{B.3}$$

Here, p^{x-1} is the probability of finding x connected monomers (i.e. $x-1$ bonds) and $1-p$ is the probability that the next reactive group remains unreacted. The pre-factor $1-p$ is important because polymers with more than x monomers will also satisfy the requirement for $x-1$ bonds.

The mass fraction is given by

$$w_x = \frac{x N_x}{n_0} = x \left(1-p\right)^2 p^{x-1}, \tag{B.4}$$

and, using eqn 6.6, the weight average molecular weight by

$$\overline{M}_w = M_{\mathrm{w,m}} \left(1-p\right)^2 \sum_{x=1}^{\infty} x^2 p^{x-1}, \tag{B.5}$$

since $M_x = x M_{\mathrm{w,m}}$. We can simplify the sum with the known mathematical relation

$$\sum_{x=1}^{\infty} x^2 p^{x-1} = \frac{1+p}{\left(1-p\right)^3}, \tag{B.6}$$

which means that by combining eqns (B.1), (B.5), and (B.6) we obtain eqn (6.8), $D_{\mathrm{M}} = 1 + p$.

B.1 Further reading

A more detailed derivation of this proof is given in several books for example, the text by Young and Lovell (1991). This book presents a generalized introduction to polymers at a level that all readers of this book should find comfortable.

C | Regular solution theory

We consider a mixture of two molecules on a cubic lattice. Our goal is to calculate the entropy of mixing of this binary mixture. The approach here is in contrast to that of Section 7.5.1, where we calculated the entropy per lattice site. Here we shall calculate the entropy of a lattice of n sites explicitly. For a lattice containing n sites, there are $n!$ configurations for n molecules on this lattice. We have two components, with n_A and n_B molecules of each species, where $n = n_A + n_B$, so $\phi = n_A/n$ and $1 - \phi = n_B/n$. The entropy of mixing is simply the (natural logarithm of the) number of arrangements of the two molecules compared to the number of possible configurations on the lattice. The number of configurations is simply

$$C\left(n_A, n_B\right) = \frac{n!}{n_A!n_B!} = \frac{n!}{(\phi n)!\left((1 - \phi)\,n\right)!}, \tag{C.1}$$

and is illustrated for a 12×12 lattice in Fig. C.1. We recall Stirling's approximation

$$\ln n! \approx n \ln n - n, \tag{C.2}$$

which is valid for large n, and use it to calculate the entropy of mixing, ΔS_l for the whole lattice,

$$\begin{aligned}\Delta S_\mathrm{l} &= k_\mathrm{B} \ln\left(\frac{n!}{n_A!n_B!}\right) \\ &= k_\mathrm{B} n\left(\ln n - \phi\ln\left(\phi n\right) - (1 - \phi)\ln\left((1 - \phi)\,n\right)\right).\end{aligned} \tag{C.3}$$

(The reader is reminded that Boltzmann's relation, $S = k_\mathrm{B} \ln \Omega$, links entropy, S, to the number of configurations, Ω.) We can thus simplify eqn (C.3) to obtain

$$-\frac{\Delta S}{k_\mathrm{B}} = \phi\ln\phi + (1 - \phi)\ln(1 - \phi), \tag{C.4}$$

where $\Delta S = \Delta S_\mathrm{l}/n$. By considering the whole lattice we nevertheless recover eqn (7.19) from eqn (C.4) when $N = 1$. In Section 7.5.1 we explicitly included the entropy of the pure states, whereas in the calculation outlined in this Appendix, the entropy of the pure states is implicitly included as $n!$.

We cannot easily extend this theory to a polymer in a solvent. Firstly, it is not possible to replace n_A by $\phi n/N$ because $C\left(n_A, n_B\right)$ requires $n_A + n_B = n$, but $\phi n/N + (1 - \phi)\,n \neq n$. Secondly and related, a polymer molecule can have different configurations on the lattice, whereas a

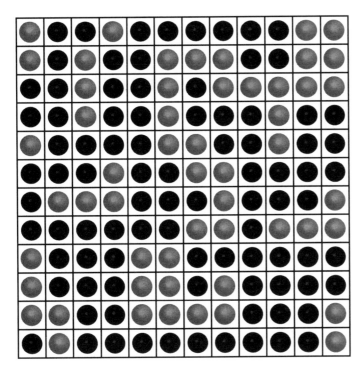

Fig. C.1 There are $144!/(54! \times 90!) \approx 1.62 \times 10^{40}$ different combinations of spheres on this lattice. If we consider the lighter-shaded spheres, then $\phi = 0.375$. The calculated entropy of mixing using eqn (C.3) is within 5% of the actual entropy of mixing. For bigger lattices this (already small) error will diminish.

solvent just takes up one lattice position with only one configuration, and the additional effect of this polymer configurational (conformational) entropy needs to be accounted for. For these reasons, it is necessary to follow the method presented in Section 7.5.1 when polymers are considered.

Answers to selected problems

D.1   Chapter 2

2.1. 12 electrons.
2.2. The length of the C–C single bond is 0.145 nm and the energy at which the density of states is a minimum is $E_k = 1.22$ eV.
2.3. (a) $E = 1.00$ eV. (b) $k_{2E}a = 1.359$ rad. (c) $k_E a = 1.050$ rad.

D.2   Chapter 3

3.2. (a) 278, 720, 90, and 90.5 g/mol. (b) 176 and 3.4×10^{21}. The volume of one mole of P3HT monomer is 151 cm^3.
3.3. $n = 6$ and $m = -3$.

D.3   Chapter 4

4.1. $\Delta E = \frac{\hbar^2 \pi^2}{4m_e a^2 N} + \frac{3\hbar^2 \pi^2}{8m_e a^2 N^2}$.
4.3. (a) 11. (b) 8.

D.4   Chapter 5

5.2. $\mu_2 = 5 \times 10^{-10}$ m^2 V^{-1} s^{-1}.
5.4. $j_1 = 164$ kA/m^2 and $j_2 = 0.161$ A/m^2.

D.5   Chapter 6

6.5. 0.40 mol.

D.6   Chapter 7

7.1. (a) 10.0 nm. (b) 147.8°. (c) 20.1 nm. (d) 5.9 nm.
7.2. $\Delta H_{\mathrm{m}} = 151$ kJ/kg.

D.7 Chapter 8

8.1. $\theta_{\mathrm{E}} = 4.4°$.
8.2. 13%.
8.3. 1.9 nm.

D.8 Chapter 9

9.1. $V_{\mathrm{G}} = -2.5$ eV.
9.2. $E_{\mathrm{g}} = 2.9$ eV and $\mu = 7.0 \times 10^{-3}$ cm^{2}V^{-1}s^{-1}.
9.3. $V_{\mathrm{G}} = -4.2$ V and $C_{\mathrm{s}} = 0.18$ nF/cm^2.

D.9 Chapter 10

10.2. FF = 40% and $\eta_{\mathrm{P}} = 0.9\%$.
10.3. 1.4 μm.
10.4. (a) $V_{\mathrm{F8BT}} = 6.7$ V and $V_{\mathrm{TFB}} = 3.3$ V.

Bibliography

Akcelrud, L. (2003) *Prog. Polym. Sci.*, **28**, 875-962.

Babudri, F., Farinola, G.M., and Naso, F. (2004) *J. Mater. Chem.*, **14**, 11-34.

Barford, W. (2013) *Electronic and Optical Properties of Conjugated Polymers.* (2nd edn). Oxford, Oxford.

Bao, Z. and Locklin, J.J. (editors) (2007) *Organic Field-Effect Transistors.* CRS, Boca Raton.

Bässler, H. (1997) in *Primary Photoexcitations in Conjugated Polymers: Molecular Exciton versus Semiconductor Band Model*, (editor Sariciftci, N.S.) World Scientific, Singapore

Budkowski, A., Bernasik, A., Cyganik, P., Rysz, J., and Brenn, R. (2002) *e-Polymers*, no. 006.

Blom P.W.M., Tanase, C., and van Woudenbergh, T. (2007) in *Semiconducting Polymers.* (editors Hadziioannou, G. and Malliaras, G.G.) (2nd edn). Wiley-VCH, Weinheim.

Blythe, A.R. and Bloor, D. (2005) *Electrical Properties of Polymers.* (2nd edn). Cambridge, Cambridge.

Bolognesi, A. and Pasini, M.C. (2007) in *Semiconducting Polymers.* (editors Hadziioannou, G. and Malliaras, G.G.) (2nd edn). Wiley-VCH, Weinheim.

Cowie, J.M.G. (1991) *Polymers: Chemistry and Physics of Modern Materials.* (2nd edn). Chapman and Hall, London.

de Gennes, P.G. (1979) *Scaling Concepts in Polymer Physics.* Cornell University Press, Ithaca.

Donald, A.M., Windle, A.H., and Hanna, S. (2006) *Liquid Crystalline Polymers.* (2nd edn). Cambridge, Cambridge.

Flory, P.J. (1955) *Principles of Polymer Chemistry.* Cornell University Press, Ithaca.

Friend, R.H., Gymer, R.W., Holmes, A.B., Burroughes, J.H., Marks, R.N., Taliani, C., Bradley, D.D.C., Dos Santos, D.A., Brédas, J.L., Lögdlund, M., and Salaneck, W.R. (1999) *Nature*, **397**, 121-128.

Facchetti, A. (2011) *Chem. Mater.*, **23**, 733-758.

Gather, M.C., Köhnen, A., and Meerholz, K. (2011) *Adv. Mater.*, **23**, 233-248.

Geoghegan, M. and Jones, R.A.L. (2005) in *Nanoscale Science and Technology*, (editors Kelsall, R.W., Hamley, I.W., and Geoghegan, M.) Wiley, Chichester.

Geoghegan, M. and Krausch, G. (2003) *Prog. Polym. Sci.*, **28**, 261-302.

Grell, M. (2005) in *Nanoscale Science and Technology*, (editors Kelsall, R.W., Hamley, I.W., and Geoghegan, M.) Wiley, Chichester.

Hadziioannou, G. and Malliaras, G.G. (editors) (2007) *Semiconducting Polymers.* (2nd edn). Wiley-VCH, Weinheim.

Heeger, A.J., Sariciftci, N.S., and Namdas, E.B. (2010) *Semiconducting and Metallic Polymers*. Oxford, Oxford.

Hiemenz, P.C. and Lodge, T.P. (2007) *Polymer Chemistry*. (2nd edn). CRC, Boca Raton.

Hoffmann, R., Janiak, C., and Kollmas, C. (1991) *Macromolecules*, **24**, 3725-3746.

Horowitz, G. (2007) in *Semiconducting Polymers*. (editors Hadziioannou, G. and Malliaras, G.G.) (2nd edn). Wiley-VCH, Weinheim.

Jones, R.A.L. and Richards, R.W. (1999) *Polymers at Surfaces and Interfaces*. Cambridge, Cambridge.

Jones, R.A.L. (2002) *Soft Condensed Matter*. Oxford, Oxford.

Kaiser, AB (2001) *Rep. Prog. Phys.*, **64**, 1-49.

Kittel, C. (2004) *Introduction to Solid State Physics*. (8th edn). Wiley, New York.

Lakowicz, J.R. (1999) *Principles of Fluorescence Spectroscopy*. (2nd edn). Springer, New York.

Leclère P., Surin, M., Brocorens, P., Cavallini, M., Biscarini, F., and Lazzaroni, R. (2006) *Mater. Sci. Eng. R*, **55**, 1-56.

Moons, E. (2002) *J. Phys.: Condens. Matter*, **14**, 12235-12260.

Muccini, M. (2006) *Nature Mater.*, **5**, 605-613.

Nicholson, P.G. and Castro, F.A. (2010) *Nanotechnology*, **21**, no. 492001.

Odian, G.G. (2004) *Principles of Polymerization*. (4th edn). Wiley, Hoboken.

Petty, M.C. (2007) *Molecular Electronics*. Wiley, Chichester.

Pron, A. and Rannou, P. (2002) *Prog. Polym. Sci.*, **27**, 135-190.

Ratner, B. and Castner, D. (1997) in *Surface Analysis – The Principle Techniques*, (editor Vickerman, J.C.) Wiley, Chichester.

Rothburg L. (2007) in *Semiconducting Polymers*. (editors Hadziioannou, G. and Malliaras, G.G.) (2nd edn). Wiley-VCH, Weinheim.

Rubinstein, M. and Colby, R.H. (2007) *Polymer Physics*. Oxford, New York.

Salaneck W.R., Lögdlund, M., Fahlman, M., Greczynski, G., and Kugler, T. (2001) *Mater. Sci. Eng. R*, **34**, 121-146.

Scholes, G.D. and Rumbles, G. (2006) *Nature Mater.*, **5**, 683-696.

Schwierz, F. (2010) *Nature Nanotechnol.*, **5**, 487-496.

Segalman, R.A., McCulloch, B., Kirmayer, S., and Urban, J.J. (2009) *Macromolecules*, **42**, 9205-9216.

Søndergaard, R.R., Hösel, M., and Krebs, F.C. (2013) *J. Polym. Sci. B: Polym. Phys.*, **51**, 16-34.

Strobl, G. (2004) *The Physics of Polymers*. (3rd edn). Springer, Heidelberg.

Su, W.P., Schrieffer, J.R., and Heeger, A.J. (1980) *Phys. Rev. B*, **22**, 2099-2111.

Yamamoto, T. (2010) *Bull. Chem. Soc. Jpn*, **83**, 431-455.

Young, R.J. and Lovell, P.A. (2011) *Introduction to Polymers*. (3rd edn). CRC Press, London.

Index